Science on Screen and Paper

Visual and Media Cultures of the Cold War and Beyond

Published by Berghahn Books and the DEFA Film Library at the University of Massachusetts Amherst

This interdisciplinary series focuses on a range of visual and media cultures in and beyond the Cold War period (1945–1991) in both social and transnational contexts. It explores ways in which film and other media, their creators and audiences, and industries and states participated in, were shaped by and, in turn, shaped cultural relations during the Cold War. Beyond 1991, this series also welcomes interdisciplinary explorations of the legacies of the Cold War and its ongoing cultural impact in a global context.

Volume 2
Science on Screen and Paper: Media Cultures and Knowledge Production in Cold War Europe
Edited by Mariana Ivanova and Juliane Scholz

Volume 1
Screened Encounters: The Leipzig Documentary Film Festival, 1955–1990
Caroline Moine

SCIENCE ON SCREEN AND PAPER

Media Cultures and Knowledge Production in Cold War Europe

Edited by
Mariana Ivanova and Juliane Scholz

berghahn
NEW YORK • OXFORD
www.berghahnbooks.com

Published in 2024 by
Berghahn Books
www.berghahnbooks.com

Library of Congress Cataloging-in-Publication Data

Names: Ivanova, Mariana, editor. | Scholz, Juliane, 1983- editor.
Title: Science on screen and paper : media cultures and knowledge production in Cold War Europe / edited by Mariana Ivanova and Juliane Scholz.
Description: New York : Berghahn Books, 2024. | Series: Visual and media cultures of the Cold War and beyond; vol 2 | Includes bibliographical references and index.
Identifiers: LCCN 2024014172 (print) | LCCN 2024014173 (ebook) | ISBN 9781805396352 (hardback) | ISBN 9781805396369 (epub) | ISBN 9781805396376 (adobe pdf)
Subjects: LCSH: Science in popular culture. | Science and state. | Cold War—Social aspects—Europe. | Cold War—Social aspects—Europe, Western. | Cold War—Social aspects—Europe, Eastern.
Classification: LCC Q172.5.P65 S46 2024 (print) | LCC Q172.5.P65 (ebook) | DDC 306.4/5—dc23/eng/20240521
LC record available at https://lccn.loc.gov/2024014172
LC ebook record available at https://lccn.loc.gov/2024014173

British Library Cataloguing in Publication Data

A catalogue record for this book is available from the British Library

ISBN 978-1-80539-635-2 hardback
ISBN 978-1-80539-636-9 epub
ISBN 978-1-80539-637-6 web pdf

https://doi.org/10.3167/9781805396352

Contents

Figures

Acronyms

AEF	Afrique Equatoriale Française [French Equatorial Africa]
AOF	Afrique Occidentale Française [French West Africa]
APW	Akademie der Pädagogischen Wissenschaften der DDR [Pedagogical Sciences Academy of the GDR]
ASTP	Apollo Soyuz Test Project
Barch	Bundesarchiv [German National Archives]
Barch-Farch	Bundesarchiv-Filmarchiv [Film Archive of the German National Archives]
BIPOC	Black, Indigenous, and People of Color
CAI	Consortium Audiovisuel International [International Audiovisual Consortium]
CIA	Central Intelligence Agency [US]
CNC	Centre National de la Cinématographie [French National Center of Cinematography]
COMACICO	Compagnie africaine de Cinématographie industrielle et commerciale [African Industrial and Commercial Cinematography Company]
DEFA	Deutsche Film Aktiengesellschaft [(East) German Film Company]
DEGEWI	Deutsche Gesellschaft für wissenschaftliche Filme [German Society for Scientific Film]
DFF	Deutscher Fernsehfunk [(East) German Television Broadcasting]

DPZI	Deutschen Pädagogischen Zentralinstitut [(East) German Central Pedagogy Institute, GDR, 1962–70]
DR	Deutsche Reportagefilm [West Germany]
DRA	Deutsches Rundfunkarchiv [German Broadcasting Archive]
DW	Deutsche Wochenschau [West German newsreel producer]
DZL	Deutsches Zentralinstitut für Lehrmittel [German Central Institute for Teaching Aids]
ESI	Educational Services Incorporated
FAC	Fond d'Aide et de Coopération [French Aid and Cooperation Fund]
FCK	VEB Fotochemisches Kombinat Wolfen [synthetic fiber division of the Wolfen Filmfabrik]
FPO	Federal Press Office [FRG]
FRG	Federal Republic of Germany (West Germany)
FWU	Institut für Film und Bild in Wissenschaft und Unterricht [Institute for Film and Image in Science and Education]
GDR	German Democratic Republic (East Germany)
HTW	Hochschule für Technik und Wirtschaft [University of Applied Sciences, FRG]
IFBT	Institut für Film, Bild und Ton [Institute für Film, Image, and Sound, GDR, 1970–90]
ISFA	International Scientific Film Association
IWF	Institut für Wissenschaftlichen Film [Institute for Scientific Film, FRG, 1956–2010]
KWI	Kaiser-Wilhelm-Institut für Strömungsforschung [Kaiser Wilhelm Institute for Fluid Mechanics Research]
KWS	Kaiser-Wilhelm-Gesellschaft [Kaiser Wilhelm Society]
MPAA	Motion Picture Association of America
MPS	Max-Planck-Gesellschaft [Max Planck Society]
NCFMF	National Committee for Fluid Mechanics Film

NSDAP	Nationalsozialistische Deutsche Arbeiterpartei [National Socialist German Workers' Party]
NSF	National Science Foundation [US]
NVWF	Nationale Vereinigung für den Wissenschaftlichen Film [National Association for Scientific Film, GDR, 1962–91)
ORWO	"Original Wolfen" [raw film brand made at the Wolfen Filmfabrik, as of 1964]
PCI	Partito Comunista Italiano [Italian Communist Party]
PSSC	Physical Science Studies Committee
REM	Reichsministerium für Wissenschaft, Erziehung und Volksbildung [Reich Ministry for Science, Education, and National Culture]
RfdU	Reichsstelle für den Unterrichtsfilm [Reich Office for Educational Films], later the RWU
RM	Reichsmark [German currency of the Third Reich]
RTF	Radiodiffusion-télévision française [French Radio and Television Broadcasting]
RWU	Reichsanstalt für Film und Bild in Wissenschaft und Unterricht [Reich Office for Film and Image in Science and Education. Formerly the RfdU]
SA	Sturmabteilung [Stormtroopers, or paramilitary wing of NSDAP]
SATEC	Société d'aide technique et de cooperation [French Technical Assistance and Cooperation Society]
SECMA	Société d'exploitation cinématographique africaine [African Film Company]
SED	Sozialistische Einheitspartei Deutschland [Socialist Unity Party of Germany]
SFA	Staatliches Filmarchiv der DDR [National Film Archive of the GDR]
TF Unit	Abteilung Technischer Forschungsfilm [Technical Research Film Unit at RWU]

TIB	Leibniz-Informationszentrum Technik und Naturwissenschaften Universitätsbibliothek [Leibniz University Library's Information Centre for Science and Technology]
UFA	Universum Film AG
USSR	United Soviet Socialist Republics
VEB	Volkseigener Betrieb [East Germany: people/state-owned company]
ZFB	Zentralinstitut für Film und Bild in Unterricht, Erziehung und Wissenschaft [Central Institute for Film and Image in Teaching, Education, and Science, GDR, 1950–54]
ZBS	Zentralbildstelle [Central Film Office, GDR, 1946–50]

Introduction

Science and Media in Cold War Europe

Mariana Ivanova and Juliane Scholz

Representations of scientific knowledge and scientists on screen and paper provoked both awe and anxiety among audiences during the Cold War—and continue to do so today.[1] After the end of World War II, scientists, governments, and the public alike sought to reevaluate the impact of war and genocide as well as of the most destructive force in the history of mankind—the atomic bomb. With this reevaluation arose the questions of how to preserve the peace and how new advances in science and technology would contribute to this goal.[2] For the West, the very notion of a Cold War—famously coined by British author George Orwell in a much-cited 1945 essay "You and the Atomic Bomb"—carried the fear that it would be displaced from its control over geopolitical economies and, ultimately, world peace.[3] From the very inception of the term, Orwell—following James Burnham's political philosophy—linked knowledge production and circulation via public media to science and power. Indeed, then as now, print and audiovisual media, as well as feature and documentary films, responded to societal aspirations and anxieties and made scientific knowledge more accessible to the wider public. After World War II, theoretical and popular science journals, radio broadcasts, public relations (PR) films, and educational and other media reflected not only on the threat of nuclear war but also on the need to harness science for the advancement of humanity.

The "dual use" of technology and science, which could both harm and aid humanity, was a dilemma that soon became central to two new disciplines: science studies and the history of science.[4] Nevertheless, the military sponsorship of scientific research—and the resulting entanglement of science with media and governmental bodies as well as armament efforts—remained insufficiently explored. Concurrently, scientific organizations not only continued to interact with expert audiences, including researchers in industry and the academy, but also began making some of their research available to the public via mass media. Theoretical nuclear physicists who had been involved in wartime atomic programs in Nazi Germany, for example, were now able to reestablish their careers and advertise their expert knowledge in relation to new "democratization" projects, be they in the service of either a communist or democratic society.

A prominent example for the new awareness of the duality of science and technology is the signing, in April 1957, of the so-called Göttinger Erklärung (Göttingen Manifesto) by eighteen leading West German scientists (the Göttingen Eighteen). Among them were nuclear experts who both rejected the production of nuclear armaments and advocated for the peaceful use of science to reduce global conflicts.[5] Also in 1957, the international Pugwash Conferences on Science and World Affairs, still taking place today, were founded to "bring scientific insight and reason to bear on the catastrophic threat to humanity posed by nuclear and other weapons of mass destruction" and to appeal for the reduction of armed conflicts by scientific and diplomatic means to secure international peace.[6] Thus, scientists on both sides of the Iron Curtain contributed, on one hand, to advancing nuclear research and, on the other hand, to achieving a "nuclear balance".[7]

By the 1960s, with the arms and space races in full swing, public interest in scientific knowledge and scientists at the service of the state escalated. The decade saw the release of Mikhail Romm's drama *Nine Days in One Year* (1962, USSR), about a physicist in the socialist bloc, and of Stanley Kubrick's classic *Dr. Strangelove or: How I Learned to Stop Worrying and Love the Bomb* (1964, US/UK). Though quite different in formal and narrative terms, both films visualized the widespread fear of experiments with fusion power and the detrimental effects of radiation on the human body. Focusing on the moral dilemmas faced by scientists, these pictures invited viewers to question the boundless potential of science as well as its limitless capacity for destruction. These topics continued to shape the popular imagination of science and scientists throughout the Cold War and beyond. That scientists' dilemmas remained in focus until the very end of Eastern European communisms is indicated by the 1990 premiere of the East German documentary *Väter der Tausend Sonnen* (*Fathers of a Thousand Suns*, dir. Joachim Hellwig, 1989, GDR). Indeed, the film's US premiere, at the World Fellowship Center in New Hampshire in 2023, inserted it into current discussions of pacificism and political action as well as renewed fears of nuclear warfare fueled by ongoing global crises like the Russian invasion of Ukraine.

Fathers of a Thousand Suns tells the story of Klaus Fuchs, a participant in the Manhattan Project and infamous Cold War spy. Before joining the highly secretive Los Alamos Laboratory in 1944, Fuchs had escaped from Nazi Germany and pursued a research career as a nuclear physicist in Bristol, in the UK. A committed communist since his youth, he became a spy for the Soviet Union and, for several years, transmitted classified information that aided the Soviets in designing their own atomic bomb. Tried and convicted of being a spy in 1950, Fuchs was nevertheless released after spending only nine of fourteen years in prison. In 1959, he started a new life in East Germany (GDR), where he led the Central Institute for Nuclear Physics near Dresden until 1979. Here, he achieved considerable prominence and recovered his reputation.

Upon closer inspection, the film is one example, among many others, of how Cold War media construed the image of the scientist. Hellwig's documentary represents more than twenty physicists from the United States, the UK, and the USSR, as well as from East and West Germany, who not only worked with Fuchs but also were intimately involved in creating the atomic bomb before and after the end of World War II. The film not only sheds light on the moral dilemmas of this international community of physicists but also documents, on camera, paths of transmitting scientific knowledge from the capitalist to the communist camp during the Cold War. A microhistory of the Cold War, *Fathers of a Thousand Suns* went beyond laboratory epiphanies to thematize the impact of nuclear research on individual lives; in addition to recording interviews with witnesses, the director used unpublished footage and photographs from international archives and private collections on both sides of the Iron Curtain. *Fathers of a Thousand Suns* thus confirms that both confrontation and cooperation shaped the exchange and exploitation of scientific knowledge between the Cold War's ideological blocs.[8]

As the present volume shows, various media during the Cold War fulfilled a similar function. Their role in inserting, representing, reshaping, and construing scientific knowledge in relation to people's everyday lives is emphasized in the subtitle: "media cultures of knowledge production." More importantly, the present volume takes its cue from perennial discussions about competing discourses and definitions media representation of science and the role of the scientist as a public figure during the Cold War.[9] The chapters in this volume both interrogate the entanglements of science institutions and scientific discovery with politics and media and also explore the role and contributions of a multiplicity of media products to the Cold War.

Science and Expert Knowledge in the Cold War

Science on Screen and Paper: Media Cultures of Knowledge Production in Cold War Europe explores in what ways science and media were central to the making of the Cold War as well as to the lived experiences of persons in divided Europe. The chapters illuminate the impact of the ideological and scientific competition between the United States and the USSR on individual lives, as well as geopolitical and cultural differences between societies on both sides of the Iron Curtain. The production, circulation, and reception of knowledge, as well as control over these processes, shaped events and experience on both sides. As historian Michael Gordin has argued, the Cold War was as much about "knowledge about knowledge"—what kind of knowledge was circulated, who possessed it, and who did not—as it was about the knowledge itself, as understood in a political, scientific, and media production context.[10]

Historian Odd Arne Westad has emphasized that the Cold War was as much about the "control and improvement" of the Global South and other areas as it was about politics and ideology. He notes: "It is to me less meaningful to talk about patterns of US or Soviet domination as 'empires,' than to describe them in a specific temporal sense."[11] Building on Westad's insight that we must see the Cold War as part of a longer trajectory of human history—shaped by colonialism, imperialism, discrimination, racism, and the geopolitical divisions between the Global North and South, as well as East and West—this volume also contextualizes the production and exchange of knowledge within preexisting patterns of dominance and control, oppression and restricted access. The studies of transnationally produced media and films by Evan Torner, Sarah Stein, and Jan Uelzmann importantly shed new light on the control of knowledge and information as well as on marginalizing and oppressive representations of non-European peoples and Persons of Color in Cold War–period media.

In short, hegemony over knowledge and information was not only contested between the two ideological camps of state socialism and capitalism. In the last decade, the authors of *The Cambridge History of the Cold War* have accordingly examined the origins, crises, and ending of the Cold War and determined that it was not a monolithic era, frozen in time, but rather one shaped by evolving, dynamic political and cultural processes as well as transnational protagonists and institutional relations.[12] Historians of science Naomi Oreskes and John Krige have advocated for looking at science more broadly and at knowledge production more specifically, taking into account cultural origins in dialogue with given societies.[13] Adopting media and media representations as our lens is how we propose to contribute to such (re)contextualizations.

An important intervention made by Oreskes and Krige has been to raise the question of the extent to which the space and arms races, ideological animosities, and instances of collaboration changed the content and/or the character of the scientific knowledge that was produced during the Cold War.[14] Sheila Jasanoff, a leading voice in science and technology studies, concurs, arguing that scientific knowledge and society are coproduced and coexist.[15] In other words, knowledge production and its material carriers are products of social life and the media that shapes it. This perspective must become central to our understanding of media knowledge production in the Cold War. Jasanoff further suggests that we may only gain explanatory power and insight into scientific processes by thinking of natural and social orders as produced together.[16] In agreement with Jasanoff, Lorraine Daston argues that disciplinary boundaries between the history of science, science and technology studies, and other fields must be challenged and overcome.[17] Our volume responds to the appeals of these scholars to establish vital connections that link science and knowledge production to media and politics in the second half of the twentieth century.

Media Cultures of Knowledge Production

Within the framework of the Cold War, "media cultures of knowledge production" emerge at the intersection of media studies and science and technology studies. The multidimensional relationship between science and media—or, as we are saying, media cultures of knowledge—is not a new phenomenon, and it impacted politics and the public throughout the Cold War period. Focusing on this approach allows us to reflect on the ways in which scientific discourse was adapted or critiqued in film, audiovisual, and print media in divided Europe.

Scientific knowledge and its distribution have been significantly reconceptualized since the 1990s, as Peter Weingart and other scholars identified a "new arrangement between science and media."[18] Weingart has characterized the new status quo by the "pre-publication of results in the media, the role of media prominence in relation to scientific reputation, and the initiation of catastrophe discourses in order to catch public attention."[19] In contrast, the ideal form of knowledge production in the twentieth century, according to historian Paul Forman, was a combination of cultural core values, such as "general assent in modernity: proceduralism, disinterestedness, autonomy, and solidarity."[20] Forman argues that, in the 1960s, closed scientific communities, laboratory teams, and academic organizations bowed out of their influence on the interpretations of scientific problems, and mass media began taking center stage. Building on Forman's observations, Edgar Grande furthermore highlights the processes of "medialization," or the rising influence of media in all social spheres, as well as new forms of communicating knowledge production to the public.[21]

Cold War media representations of scientists and scientific research not only proliferated new forms of knowledge but also were steeped in the tensions, conflicts, and convergencies of different political agendas and policies. Film as a medium, for example, traveled across geopolitical borders and evolved in several different genres to reach various audiences and respond to different ideological and educational purposes. With the idea of genres, we introduce a crucial distinction between three types of nonfictional, specifically scientific films, which are of particular importance to some of the chapters that focus on Germany: science films (*Wissenschaftsfilme*); technical research films (*technischer Forschungsfilme)*, a subgenre of the science film; and educational films (*Lehrfilme* or *Unterrichtsfilme*).[22] Science films were part of the interaction between media and research, and they connected scientific innovations and technological changes with national (and nationalist) agendas. These are films made to support and visualize research on scientific topics and experimental practices. They may or may not be used outside of the lab—e.g., for university or professional education.

Educational or pedagogical films were films made to help teach science subjects in schools, mostly at the primary and secondary level. A subcategory here are the health and hygiene films. Because of their use in the classroom they also had impact as instruments of social policies.[23]

Until recently, scholarly interest has gravitated toward the (geo)political, diplomatic, and military functions of science and scientists more than on media or institutional histories.[24] Science and technology studies of the Cold War have recently turned to "science diplomacy," as a subcategory of "soft power," in order to reveal and situate new high-profile protagonists that operated on both the state-sponsored and supranational levels. Simone Turchetti, on the one hand, has investigated how the promotion of scientific collaboration among national research groups can become a powerful tactic to strengthen relations between nonpolitical actors and to counter political divisions.[25] Alison Kraft and Carola Sachse, on the other hand, have pointed out that scientists are now "emphasizing the 'common language of science' as a means to transcend national and ideological allegiances," which has the potential to constitute "a new kind of transnational, cross-bloc scientific diplomacy involving—and made possible by—elite scientists."[26] Kraft, Sachse, and Turchetti agree that scientists have productively shaped the rising international critique related to environmental issues and climate change, the use of nuclear weapons, and nuclear power, along with other social justice movements.[27]

Building on such key studies, *Science on Screen and Paper* seeks to bolster what has been marginal engagement by scholars with nonfiction, nonmainstream films and the politics and agencies that shaped science and education programs in Cold War Europe. The chapters in the volume interrogate a range of important questions: How do we conceive of the Cold War as shaped by evolving, intertwined cultural and ideological processes instead of as static phenomenon? What can the intersections of scientific research with media representations reveal about international relations and transnational exchanges that took place during the Cold War? What types of organizations, institutions, and protagonists played pivotal roles in connecting the laboratory to the screen or printing press? How did state funding influence science programs and their popularization in the media? How were scientists consulted in view of setting policies? How were scientific breakthroughs and discoveries communicated to the public?

To begin answering such questions, *Science on Screen and Paper* explores historical situations that reveal the process of appropriation, medialization, and popularization of science in various countries via various forms of media. Our objective is twofold: to disclose ways in which scientific knowledge was utilized for knowledge production and ideological purposes during the Cold War, and to explore ways in which media simultaneously shaped scientific discourse and/or experimental practices.[28] Spanning a wide range of media—from PR materials and print journals to educational and science films to children's magazines and television—the contributions in this volume seek not to ignore or marginalize differences between media but rather to focus on intersectional themes and common strategies in representing science. Taking the unique history of each type of media into account, each chapter highlights overlaps and mutual influences among distinct media representations of science with regard to narratives, representations, and aesthetics. The volume thus combines contextualized media archaeology with in-depth media analysis. Anchored in the history of knowledge, film, and media analysis, it aims to begin developing an archive to help understand how moving images represented, produced, impacted, and transformed scientific knowledge and epistemic practices in the Cold War.

The volume is centered on Germany and other European countries. While it thus offers good coverage of Eastern and Western European examples, the important contribution made by chapters on print and audiovisual media in North and South must not be overlooked. In approaching their topics, these fold theories of knowledge production and critical race and postcolonial theories into an interdisciplinary media studies approach. Finally, in our attempt to facilitate comparisons of Cold War approaches to science, media, and knowledge production in different political and temporal contexts, we chose to group the chapters by topic, not by national focus.

Why Do Media Representations of Science Matter?

Media representations have produced knowledge and contested the promise of scientific objectivity, while at the same time foregrounding often contentious relationships between narrative and visual presentations of information. Media representations, as defined in this volume, are images and narratives created in different historical and political contexts of the long twentieth century. Around 1900, methods of visualization, such as filming experiments, increasingly became part of scientific practices in diverse fields, including biological studies of animal and human behavior. In the post-war era, scientists made observations of molecular and biochemical processes, and neurological imaging procedures. In the process, educators, journalists, and filmmakers entered laboratories and scientific institutes and their communities to apply their training and skills to the visualization of scientific work. One aim was to disseminate the knowledge of technology and scientific innovation among wider audiences; another goal was to explore experimental processes and methods for the creation of scientific evidence. As Juliane Scholz's contribution to this volume demonstrates, such collaborations among media, science organizations, and researchers often continued over long periods of time, sometimes surviving war, political crises, and changes of government and ideology before and after 1945. In a similar consideration of longevity, Mario Schulze's chapter tracing the materiality of film demonstrates ways in which the medium was manipulated and recycled to attend to changing political demands during the course of the Cold War.

Science on Screen and Paper foregrounds a variety of agents and audiences involved in science-related media representations, thus building on media studies scholarship by authors such as Stuart Hall, Christian Bonah, Anja Laukötter, Shani Orgad, and Myra Macdonald, to name but a few.[29] Bonah and Laukötter have argued that visual media are not merely "a mirror or expression of the things they depict, (. . .) media and art are endowed with their own distinct, interactive, performative power."[30] Several chapters in this volume, notably those by Darina Volf, Doru Aurel Pop, Sophia Gräfe and Kerrin von Engelhardt, and Ettore Costa, productively engage with underlying power structures, asking how media representations of science and scientists have negotiated concepts of self and other, of authenticity and objectivity, science fiction and propaganda. Power, as Orgad has observed, can emerge through "players (institutions, groups, individuals) [who] increasingly compete for visibility and voice," as well as "through the projection of symbolic representations on to the global media space."[31] Indeed, as Hall has argued, visibility in the media images and narratives becomes "a means of claiming recognition and exercising power."[32] In other words, he continues, representations are not mere images or reflections but also interpretations that frame our understanding of the world and our lived reality. Chapters by Evan Torner, Sarah Stein, and Jan Uelzmann bring Hall's scholarship on uncovering the politics behind the "work" done by representations into dialogue with studies on Cold War media and representation of non-European peoples from the Global South. Like other scholars who have responded to Hall's reframing of media representations and explored them in a global context, these authors study their role in both local and national contexts while accounting for alterity, identity, positionality, and political position.[33]

This volume also advances our understanding of media representations by approaching them in their relationship to science and economic, social, and political power. "Power relations are encoded in media representations," Orgad writes, "and media representations in turn produce and reproduce power relations by constructing knowledge, values, conceptions, and beliefs."[34] This leads us to the idea of representations as social constructs. This fundamental idea, proposed by Myra MacDonald, among others, destabilizes the way in which we see media; from this point of view, media simultaneously cite, depict, reflect, and represent scientific knowledge and discoveries. In other words, the very act of representation—let alone the various genres or types of media representations—often subtly, but nevertheless intrinsically, participates in the reproduction and contestation of existing power relations. While this phenomenon certainly existed well before the Cold War, in this period, as many of the contributions to this volume illustrate, it was amplified and reached a new level of sophistication.

Science on Screen and Paper thus seeks to both open a forum and provide methodologies for developing a critical understanding of the media representations of scientists and scientific knowledge during the Cold War. To this end, the chapters in this volume perceptively foreground the importance of science, nongovernmental institutions, and the changing patterns of production, distribution, and receptions of different media forms, genres, and formats. Doru Aurel Pop's chapter, for example, approaches the significance of scientific advancement and the space race for sci-fi literature through the lens of "anticipation"—a concept completely foreign to Western and Western European science fiction. Sandra Schnädelbach looks at how scientific and pseudoscientific knowledge was disseminated in the 1960s and 1970s, focusing on the example of "suggestology"—a learning technique that originated on the fringe of the European continent and swept up Eastern European audiences via television. Josephine Diecke takes a different approach, positioning her chapter at the intersections of environmental and media studies; she argues that, as scientific concepts became more mediated and popularized, they also entered public ecological and political debates. Sophia Gräfe and Kerrin von Engelhardt trace and describe government institutions that largely focused on the aesthetics and didactics of educational films, seeking the most effective use of cinematic images to optimize learning processes in various social settings.

Research institutions in Cold War Europe developed their own active communications with public stakeholders in order to maintain steady funding for their projects and secure their image as guarantors of scientific progress and modernization. Juliane Scholz's and Mario Schulze's chapters insightfully interrogate these processes. These authors demonstrate that, as research institutions came under public scrutiny, they had to legitimize their need for state funding by explaining and publicizing their scientific inventions and potential applications for new technologies. This trend toward professionalized communication strategies on the part of science institutions engendered internal PR and communications departments, as well as science journalism, which evolved in the 1970s and 1980s—all of which fostered the public reach of their work.[35] To expose the vagaries of making scientific information public knowledge, but also its appropriation for propaganda purposes, Ettore Costa interrogates the rhetoric used in Italian communist media. In addition, Uelzmann's and Stein's chapters in this volume engage with these topics to reveal understudied uses of film, such as West German PR films and French educational films produced for postcolonial Africa. Like Torner, these authors expose and critique the underlying political nature of state-sponsored films in a context of decolonization and persisting Eurocentrism and racism.

Organization and Scope of the Volume

Science on Screen and Paper thus offers a series of in-depth case studies at the intersection of science and media studies as it investigates relationships between scientific achievements, film, television, and institutional and individual actors in science and politics. The outlook is international, covering East and West Germany, Romania, Bulgaria, Italy, and France, but also Guinea, Burkina Faso, the Soviet Union, and the United States. The three sections are organized to group chapters that interact with one another. There are, of course, also important intersections and interconnections between chapters in different sections as well, and we hope that readers will discover even more points of comparison themselves. Part 1, "Institutional and Industrial Contexts," focuses on the appropriation and reuse of scientific films by educational institutions and the dissemination of scientific knowledge from scientists and labs to the wider public. Part 2, "Imaginaries of Self and Other," explores the role that science-fiction media—including feature films, literature, and print media like newspapers and popular science journals—played with respect to science and in Cold War cultural diplomacy. Part 3, "Science in Public and Educational Media," interrogates the educational agendas that informed film, audiovisual, and print media production, focusing in particular on propaganda, Eurocentrism, and disguised racism, and provides a critical engagement with capitalist and socialist states.

Part I: Institutional and Industrial Contexts

The three chapters included in this part focus on specific scientists and film officials who legitimized and popularized film production for research purposes, thus shedding light on institutional and industrial contexts in East and

West Germany and the United States. Two of the chapters focus on the problematic reuse of scientific films in educational and science institutions after World War II and on their transnational movements and distribution. The last chapter in this section focuses on the role of scholarly debates in science publications and print media and recounts how they, in turn, influenced larger political debates on environmental issues; it also analyzes the institutionalization and archival preservation of research films in East Germany.

The volume opens with Juliane Scholz's study of postwar West Germany, focusing on institutional transformations and the appropriation of science films that had been made in the Third Reich. "Whitewashing the Nazi Past: Continuity and Transformation in Scientific Cinematography in Germany, 1934–56" looks at one of the advocates for the recycling of these films, Gotthard Wolf, who—after a career in the Nazi film industry—spearheaded founding the Institute for Scientific Film (IWF) in Göttingen. In 1949, Wolf advocated for distinguishing the political propaganda of Nazi nonfictional *Kulturfilme* (culture films) and commissioned *Unterrichtsfilme* (educational films) from *Wissenschaftsfilme* (science films), which focused mostly on research experiments and therefore, he claimed, had only scientific functions that were free of propaganda. After World War II, Wolf's definition constituted a successful exculpatory strategy in the attempt to absolve himself and his colleagues from political incrimination. By tracing the evolution of science films through their iterations during the Nazi period and in postwar scientific organizations in West Germany, Scholz outlines how science films, specifically the subgroups "technical research films" and "medical films" used in military research and to legitimate "eugenic" policies during World War II, were repurposed during the Cold War for educational purposes.

In "Flow Vis for the Space Race: German Science Films in U.S. Education during the Cold War," Mario Schulze adopts a different vantage point in examining the international reappropriation of research films. He focuses on the US-based National Committee for Fluid Mechanics Film (NCFMF), led by Ascher H. Shapiro, an MIT professor of mechanical engineering. Cosponsored by the Ford Foundation and the National Science Foundation, the NCFMF produced one of the most extensive US educational film programs during the Cold War. Though the films attempted to represent science as a nonideological endeavor, Schulze demonstrates that they were deeply enmeshed in the politics of the Cold War. In creating educational materials for US students and professionals, the NCFMF used film sequences made for military purposes before World War II that were later instrumentalized by the Nazis. Schulze's analysis of the circulation, instrumentalization, and distribution of science films across decades and governments helps us trace the materiality and mediality of scientific knowledge; categories of research and educational film also become more tangible and distinct—both in their ideological implications and for our understanding of how films functioned as political instruments in the Cold War.

Turning our attention to institutional contexts on the other side of the Iron Curtain, Josephine Diecke's "Raw Film Manufacturing: Between Economic Efficiency and Environmental Awareness in East Germany" looks at raw film stock production in the GDR in relation to evolving debates on pollution and environmental issues. The author analyzes professional discussions about manufacturing, environmental pollution, and economic efficiency in the trade journal *Bild und Ton* (Image and Sound), which circulated in the GDR from 1948 to 1992. Despite its finite resources, the East German economy was able to maintain production for the lucrative markets for chemical and film production. Although water pollution and other harmful side effects of chemical processes were public by the 1970s, changes in favor of environmental protection were not enacted until the 1980s. Diecke's important study draws connections between the chemical and film industries while demonstrating that short-term industrial efficiency was ultimately more important to the GDR than the incentive of long-term sustainability.

Part II: Imaginaries of Self and Other

The four chapters in part 2 trace the works of filmmakers and authors who created utopian and dystopian visions of the future, as well as of self and other, within their respective national contexts, in Bulgaria, Romania and the Soviet Union, the United States, and the GDR. These studies demonstrate that science magazines and sci-fi films and novels popularized expert knowledge in the process of legitimizing or critiquing national ideologies, racial biases, and alterity. They alternately endorsed and perpetuated or criticized and scrutinized both Western capitalism and Eastern communisms. Within socialist countries, the lively export of science and sci-fi texts and films and exchange of ideas—on topics ranging from science in the service of peace, to the psychedelic revolution and parascience—did not preclude racial bias and discrimination on screen and paper, however. Critically assessing the transnationalism of such exchanges, the chapters in this section indicate that, however forward-thinking or modern they were, popular media representations of science and sci-fi remained tainted by colonial and racist imaginary.

In "Between Cooperation and Competition: Cold War Imaginaries and Representations of US-Soviet Encounters in Space," Darina Volf argues that cultural and media imaginaries in both East and West oscillated between dreams of productive cooperation and the reality of competitive national displays. Focusing on cinematic representations of cooperation and competition, she demonstrates how fictional films actually influenced space exploration practices and the development of space technologies. Volf analyzes "sociotechnical imaginaries" of space cooperation in three films—one Soviet, one US, and the first East German–Polish coproduction. In envisioning technological innovations, such sci-fi films and television shows presented fantasies of what was technologically attainable and disseminated normative views on what was socially desirable. Over the course of the 1960s, these views helped bolster the idea of space cooperation, which gradually replaced the previously dominant perception of space as an arena for national and systemic competition. Bringing in examples from opposing US and Soviet national media outlets, Volf argues that media coverage contributed to the further spread of imaginaries originating in sci-fi films. As the ideal of space cooperation was articulated as the desirable norm in the mass media, it increasingly formed public expectations of national space programs.

In "Postcolonial Science Heroes in East German Science-Fiction Films," Evan Torner offers a different approach, applying the early sci-fi concept of "science heroes" to four canonical East German science-fiction films. As the socialist bloc embraced antiracist internationalism as an ideology and as an aesthetic strategy, attempts were made in some of these films to represent a post-racist future. Nevertheless, Torner argues, these films reveal the latent whiteness inherent in European socialist culture. The author then traces the ways in which tropes of the sci-fi genre lend themselves to a patriarchal, colonial understanding of science and the cosmos—in keeping with the actual oppressed social circumstances of BIPOC in the GDR. His interrogation of the bias inherent in sci-fi films by Eastern European, socialist partners reveals that these films rely on tokenism and the image of the white savior. In a brief look at the context of German and Eastern bloc sci-fi, as well as the production and reception of the four films, the chapter highlights the understated postcolonial regimes in the films, which confirm a socialist multicultural white mindset.

In "The Culture of Anticipation and Science Fiction in Socialist Romania," Doru Aurel Pop analyzes how Romanian science-fiction cinema was born and evolved under Romania's socialist regime. The popularization of space fiction coincided with the popularization of science and the modernization of Romanian society. Pop adopts an interdisciplinary approach that couples film analysis with an explanation of the political and historical contexts. As a case study, the author explores Ion Popescu-Gopo's science-fiction films made at the state-owned Buftea Film Studios. Moving between text and film, this chapter links the Romanian sci-fi genre to a genre known as *film de anticipație* (anticipation films). "Anticipation" was a notion that Romanian authors borrowed from French literary theory to develop the concept of scientific anticipation literature. Alongside a review of sci-fi literature widely available in socialist Romania, Pop's analysis interprets a larger cultural phenomenon in Romania, which he calls a "culture of anticipation"; cultural expressions of this phenomenon included anticipation almanacs, books popularizing science, and science-fiction novels and comics, all of which influenced social interactions under socialism.

In "(Un)Healthy Tunes: Body, Mind, and Music on Socialist Television," Sandra Schnädelbach focuses on a set of television programs, shown in the GDR in the late 1970s and 1980s, which popularized "suggestology," a pseudoscientific method of knowledge acquisition developed in Bulgaria. Schnädelbach demonstrates that TV broadcasts on "suggestology" had an impact far beyond Bulgaria and became part of an international scientific competition. Building on Soviet psychological research on hypnosis and suggestion, suggestology claimed that it accessed the hidden resources of the brain by means of relaxation techniques and sensory, especially acoustic stimulation. Introduced by a team led by psychotherapist Georgi Lozanov in the mid-1960s, the research field of suggestology and its pedagogical application, suggestopedia, were particularly present in East German science TV shows of the 1970s. In addition to providing a fascinating excursion into the history of Soviet and Eastern European pedagogical research, this chapter argues that suggestopedia touched the core of the Cold War preoccupation with "brainwashing" and found applications in neurology as well as psychology.

Part III: Science in Public and Educational Media

The contributions in part 3 share an interest in how film and media products were employed for educational purposes in different national and international contexts. They point to both the potency and pitfalls of media once harnessed for education beyond national borders. The first chapter explores the ideological underpinnings of representations of the space race in print media put out by the Italian Communist Party (PCI). The second chapter examines the role of science films in GDR education and research, skillfully incorporating discussions taking place in the trade magazine *Bild und Ton* (Image and Sound). Taken together, these chapters demonstrate that, even when scientific achievements were highjacked for ideological purposes, audiovisual and print media invited a differentiated reception in both East and West. The second set of chapters offers a comparative look at PR and educational films about Africa, one made in France and the other in West Germany. Each country engaged in educational and aid projects in Africa and made films about them. The chapters offer very different insights into Western European relations with Africa in the early postcolonial period, however, as France—a former colonizer—and West Germany—a nation with colonial history—had different motives for offering aid. Ultimately, though, neither treated their African counterparts as equals, demonstrating that racism still unified Western European attitudes and media on Africa.

In "Science for Children and Adults: The Space Race in the Publications of the Italian Communist Party, 1957–64," Ettore Costa analyzes communist discourses on the Cold War's space race by reviewing the media coverage in periodicals published by the PCI. His analysis of articles in the children's magazine *Il Pioniere* (The pioneer) and the adult magazine *Rinascita* (Rebirth) reveals how Italian communists understood the space race and the role that science played in their view of themselves and the world. Unsurprisingly, the periodicals celebrated the Soviet Union as the vanguard of scientific progress, which proved the superiority of the socialist system. Interrogating the type of scientific information that was selected and how it was invested with political significance through rhetoric, Costa focuses on the illocutionary and perlocutionary aspects of print articles to uncover the goals of writers; letters from readers reveal the effect of events and articles on the audience. He contrasts a range of ideological tropes and arguments of Soviet origin, which were used by Italian communist authors, against representations of the space race in the rest of Italian and Western media, which relied primarily on ideas and modes of representing scientific knowledge developed in the United States. Costa concludes that the impetus to recenter communist ideology around scientific progress and achievements was linked to the need of post-Stalinist leaders to disentangle communism's associations with repression and exploitation.

In "'Effective Films": Science Film and Education in East Germany," Sophia Gräfe and Kerrin von Engelhardt tackle the crucial position that science and education films occupied within the GDR cinematographic landscape. Gräfe and von Engelhardt trace the evolution of East German pedagogical theory, which drew heavily on Ivan Pavlov and other Soviet theorists. Given the economic challenges facing a nation rebuilding after war, a great deal of attention was spent on using educational films to make the subject matter useful and teaching methods effective. Involved in the idea of effective teaching, of course, was instilling the students with the correct political line and attitudes. Institutions were developed to explore film aesthetics and didactics in order to optimize effective teaching and learning in various settings. This mission, in turn, led to the emergence of a distinct field of research at universities and professional academies, focusing specifically on the educational impact of film. In an effort to encourage more research into this subject, the authors provide a comprehensive account of the East German institutions that were involved, as well as current archives that hold educational and research science films and related matter.

In "German Technology and Education for 'Young Nations': Cold War Politics and Aesthetics of Development Aid in Two West German Governmental PR Films of 1961," Jan Uelzmann analyzes the politics behind the filmed representation of West German (FRG) developmental aid to Guinea. The author points out that prior to the rise in popularity of and access to television, public relations films played a crucial role in the domestic diplomacy of Konrad Adenauer's administration. At the same time, they also played a role in asserting West Germany's place on the world stage, especially in the eyes of the former Western Allies. The chapter focuses on two examples filmed in 1961, one about visiting the FRG's capital city, Bonn, and one about the Foreign Office. Uelzmann's close analysis of these films shows that, in making their case to the predominantly white West Germans, they presented racist and Eurocentric worldviews, advocating for white European supremacy over peoples and places indiscriminately represented onscreen as "Africans" and "Africa." Uelzmann interrogates the progressive rhetoric used in the films, which list the FRG's advanced and successful economics, technology, and educational system to favorably distinguish West Germany from the socialist states (most notably the GDR) that were competing for influence in decolonialized countries. Finally, the author links the Adenauer administration's combination of democratic rhetoric and persistent racist and colonial modes of thinking to continuities in the careers of Nazi-era officials in West German governmental structures.

In the volume's final chapter, "French Educational Films for Africa: The Question of Decolonization in the 1960s and 1970s," Sarah Stein explores the evolution of French film policy for former colonies in the 1960s and early 1970s and the production of educational films to assist in aid projects. The chapter focuses on the case of two films, produced for screenings in Upper Volta (today Burkina Faso) by the International Audiovisual Consortium (CAI), created in 1961 by the French government to provide former African colonies with newsreels and educational films. The films were created within a complex, international production framework. Stein traces the history of this framework, from the early influence of the high-ranking colonial administrator Robert Delavignette, who wrote *Service Africain* in 1946, to France's National Center of Cinematography (CNC) and film diplomacy and decolonization, to the postcolonial moment in the 1960s when CAI made the two films in question with scenes shot in Upper Volta. Leaning on the work of media theorist Eef Masson, Stein's film analysis accounts for the implicit viewer addressed by the film, analyzes the rhetorical means used to attract and motivate viewers, and uncovers the creation of references to the nonfilmic world. In the process, she exposes connections between postcolonial cultural policy initiatives and colonial-institutional ties and presents ways in which these films invited viewers—whether in France or Burkina Faso—to adopt specific Eurocentric behaviors and modes of living.

Conclusion

Science on Screen and Paper extends and complements the existing scholarship on Cold War media representations of science and scientists by looking at parallel period developments in countries in Eastern and Western Europe. To the more familiar fields of Cold War studies and science studies, the chapters in this volume add an overdue comparative dimension by considering, for instance, decades-long trajectories of films produced within scientific contexts, or how media representations in various print journals, PR films, and television were affected by Cold War politics. As editors, we have attempted to give voice to both emerging and well-established scholars from all over Europe in an effort to broaden our shared fund of experience and perspectives. The contributions selected for this volume were accordingly written by scholars based and working in eastern and western, northern and southern Europe, as well as in the United States.

Mariana Ivanova is associate professor for German film and media and the academic director of the DEFA Film Library at the University of Massachusetts Amherst. Previously, she has taught at Rice University and Miami University. Her scholarship focuses on twentieth- and twenty-first-century German and European cinemas and cultures, theories of transnational filmmaking and coproduction, heritage cinema, artistic networks, and cultural mediation. Ivanova's 2019 monograph, *Cinema of Collaboration: DEFA Coproductions and International Exchange in Cold War Europe*, was finalist for the international Willy Haas award. Her research has been published in *The German Studies Review*, *The Historical Journal of Film, Radio and Television*, and in numerous edited volumes on East German and Central European cinema.

Juliane Scholz is a postdoctoral researcher and graduate coordinator at the Leibniz Centre for Contemporary History in Potsdam, Germany. Her research focuses on twentieth-century German, European, and US-American social, cultural, and media history. She studied cultural sciences, media studies and psychology at the University of Leipzig, where she completed her dissertation on the professional histories of screenwriters in the United States and Germany in 2014. She has served as a senior lecturer and postdoctoral researcher at the University of Leipzig and the Max Planck Institute for the History of Science in Berlin. Her recent publications on the social history of the Max Planck Society sheds light on working conditions in the academic field and the professionalization of science communication.

Notes

1. Mainstream cinema and television continue to churn out biopics of scientists who worked during the Cold War. Often considered "box office hits" and nominated for Academy Awards, these films largely employ clichéd Cold War representations of science and scientists and heavily emphasize fictionalized personal stories to enable audiences to identify with the protagonists. Examples in the last ten years include: *Einstein and Eddington* (2008, dir. Peter Moffat, UK; *The Imitation Game* (2014, dir. Morten Tyldum, US); *Hidden Figures* (2016, dir. Theodore Melfi, US); *Charlatan* (2020, dir. Agnieszka Holland, Czech Republic/Ireland/Slovakia/Poland); *Tesla* (2020, dir. Michael Almereyda, US); *Radioactive* (2019, dir. Marjane Satrapi, UK); and, most recently, *Oppenheimer* (2023, dir. Christopher Nolan, US/UK).
2. Some historians argue that the concept of a "Cold War" was linked to the creation of the atomic bomb, from its inception, by scientists who had perceived the possibility of nuclear weapons. For this discussion, see Oreskes and Krige, *Science and Technology in the Global Cold War*; Vowinckel, Payk, and Lindenberger, *Cold War Cultures: Perspectives on Eastern and Western European Societies*.
3. Orwell, "You and the Atomic Bomb." For similar definitions, also see Westad, "The Cold War and the International History of the Twentieth Century."
4. "Dual use" technologies were defined by the US National Research Council as "technologies intended for civilian application that also can be used for military purposes." See Forge, "A Note on the Definition of 'Dual Use.'"
5. Kraus, *Von der Uranspaltung zur Göttinger Erklärung*.
6. See "Pugwash Conferences in Science and World Affairs"; Kraft, *From Dissent to Diplomacy*; Kraft and Sachse, *Science, (Anti-)Communism and Diplomacy*.
7. Nehring, "Remembering War, Forgetting Hiroshima."
8. Van Dongen, *Cold War Science*.
9. On the competing discourses and definitions of both science and scientists as public figures, see Hacking, "What About the Natural Sciences?" On the impact that government-sponsored projects had on scientists' work, see Oreskes, "Science in the Origins of the Cold War."
10. Gordin, *Red Cloud at Dawn*, 29.
11. Westad, *The Global Cold War*, 5.
12. Leffler and Westad, *The Cambridge History of the Cold War*.
13. Oreskes and Krige, *Science and Technology in the Global Cold War*.
14. Oreskes, "Science in the Origins of the Cold War," 17.
15. Jasanoff, *States of Knowledge*.
16. Jasanoff, *States of Knowledge*, 2–3.
17. Jasanoff, "Reconstructing the Past, Constructing the Present"; Daston, "Science Studies and the History of Science."
18. Weingart, "Science and the Media."
19. Weingart, "Science and the Media," 869.
20. Forman, "On the Historical Forms of Knowledge Production," 72.
21. For a detailed discussion of "medialization," see Grande, "Die neue Governance der Wissenschaft."
22. For definitions of "research film" and "science film" as well as the concepts of "reusing" and "recycling" films, see Sattelmacher, Schulze, and Waltenspül, "Introduction: Reusing Research Film."

23. See Schmidt, *Medical Films, Ethics, and Euthanasia in Nazi Germany*.
24. See, e.g., McCauley, *The Origins of the Cold War, 1941–1949*; Wettig, *Stalin and the Cold War in Europe*; Bridger, *Scientists at War*.
25. Turchetti, *Greeting the Alliance*, 5. Also see Roberts, "Science, Peace and Internationalism."
26. Kraft and Sachse, *(Anti-)Communism and Diplomacy*, 5 and 22.
27. Kraft, Nehring, and Sachse, "The Pugwash Conferences and the Global Cold War"; Turchetti and Roberts, *The Surveillance Imperative*.
28. On the twentieth century as the "age of techno-science," see Pestre, "Regimes of Knowledge Production in Society."
29. See, for instance, Orgad, *Media Representations and the Global Imagination*; Gaycken, *Devices of Curiosity*; Mitman and Wilder, *Documenting the World*.
30. Bonah and Laukötter, *Body, Capital, and Screens*, 14.
31. Orgad, *Media Representations and the Global Imagination*, 5.
32. Hall, *Representation*, 1–11.
33. See, for instance, Couldry, Hepp, and Krotz, *Media Events in a Global Age*; Machin and Van Leeuwen, *Global Media Discourse*; Robertson, *Mediated Cosmopolitanism*.
34. Orgad, *Media Representations and the Global Imagination*, 25.
35. See: Höhn, *Wissenschafts-PR*; Kohring, *Die Funktion des Wissenschaftsjournalismus*; Göpfert, "The Strength of PR and the Weakness of Journalism."

Bibliography

Aronova, Elena, and Simone Turchetti, eds. *Science Studies during the Cold War and Beyond: Paradigms Defected*. New York: Palgrave Macmillan, 2016.

Bauer, Martin W., and Massimiano Bucchi, eds. *Journalism, Science and Society: Science Communication between News and Public Relations*. London/New York: Routledge 2007.

Bonah, Christian, David Cantor, and Anja Laukötter, eds. *Health Education Films in the Twentieth Century*. Rochester, NY: University of Rochester Press, 2018.

Bonah, Christian, and Anja Laukötter, eds. *Body, Capital, and Screens: Visual Media and the Healthy Self in the 20th Century*. Amsterdam: Amsterdam University Press, 2020.

Bridger, Sarah. *Scientists at War: The Ethics of Cold War Weapons Research*. Cambridge, MA: Harvard University Press, 2015.

Couldry, Nick, Andreas Hepp, and Friedrich Krotz, eds. *Media Events in a Global Age*. London/New York: Routledge, 2010.

Daston, Lorraine. "Science Studies and the History of Science." *Critical Inquiry* 35(4) (2009): 798–813.

Forge, John. "A Note on the Definition of 'Dual Use.'" *Science and Engineering Ethics* 16(1) (2010): 111–18.

Forman, Paul. "On the Historical Forms of Knowledge Production and Curation: Modernity Entailed Disciplinarity, Postmodernity Entails Antidisciplinarity." *Osiris* 27(1) (2012): 56–97.

Fuchs, Rolf, and Klaus Kroll, eds. *Audiovisuelle Lehrmittel: Methode ihrer Anwendung und ihre Gestaltung*. Leipzig: VEB Fotokinoverlag, 1982.

Gaycken, Oliver. "The Secret Life of Plants: Visualizing Vegetative Movement, 1880–1903." *Early Popular Visual Culture* 10(1) (2012): 51–69.

——. *Devices of Curiosity. Early Cinema and Popular Science*. Oxford/New York: Oxford University Press, 2015.

Gordin, Michael D. *Red Cloud at Dawn: Truman, Stalin and the End of the Atomic Monopoly*. New York: Farrar, Straus, and Giroux, 2009.

Göpfert, Winfried. "The Strength of PR and the Weakness of Journalism." *Journalism, Science and Society: Science Communication between News and Public Relations*, edited by Martin W. Bauer and Massimiamo Bucchi, 215–26. London/New York: Routledge 2007.

Grande, Edgar. "Die neue Governance der Wissenschaft: Zur Einleitung." In *Neue Governance der Wissenschaft. Reorganisation—externe Anforderungen—Medialisierung*, edited by Edgar Grande, Dorothea Jansen, Otfried Jarren, Arie Rip, Uwe Schimank, and Peter Weingart, 15–45. Bielefeld: transcript, 2013.

Hacking, Ian. "What About the Natural Sciences?" In *The Social Construction of What?* 63–99. Cambridge, MA: Harvard University Press, 1999.

Hall, Stuart, ed. *Representation: Cultural Representations and Signifying Practices*. London: Sage, 1997.

Höhn, Tobias D. *Wissenschafts-PR: Eine Studie zur Öffentlichkeitsarbeit von Hochschulen und außeruniversitären Forschungseinrichtungen*. Konstanz: UVK Verlagsgesellschaft, 2011.

Jasanoff, Sheila, ed. *States of Knowledge: The Co-production of Science and the Social Order*. London/New York: Routledge, 2004.

——. "Reconstructing the Past, Constructing the Present: Can Science Studies and the History of Science Live Happily Ever After?" *Social Studies of Science* 30(4) (2000): 621–31.

Kohring, Matthias. *Die Funktion des Wissenschaftsjournalismus: Ein systemtheoretischer Entwurf*. Wiesbaden: VS Verlag für Sozialwissenschaften, 1997.

Kraft, Alison. *From Dissent to Diplomacy: The Pugwash Project during the 1960s Cold War*. Berlin: Springer, 2022.

Kraft, Alison, Holger Nehring, and Carola Sachse. "The Pugwash Conferences and the Global Cold War: Scientists, Transnational Networks, and the Complexity of Nuclear Histories." *Journal of Cold War Studies* 20(1) (2018): 4–30.

Kraft, Alison, and Carola Sachse, eds. *Science, (Anti-)Communism and Diplomacy: The Pugwash Conferences on Science and World Affairs in the Early Cold War*. Leiden/Boston: Brill, 2020.

Kraus, Elisabeth. *Von der Uranspaltung zur Göttinger Erklärung: Otto Hahn, Werner Heisenberg, Carl Friedrich von Weizsäcker und die Verantwortung des Wissenschaftlers*. Würzburg: Königshausen & Neumann, 2001.

Leffler, Melvyn P., and Odd Arne Westad, eds. *The Cambridge History of the Cold War*. 3 vols. Cambridge: Cambridge University Press, 2010.

Leslie, Stuart W. *The Cold War and American Science: The Military-Industrial-Academic Complex at MIT and Stanford*. New York: Columbia University Press, 1993.

Machin, David, and Theo Van Leeuwen. *Global Media Discourse: A Critical Introduction*. London/New York: Routledge, 2007.

McCauley, Martin. *The Origins of the Cold War, 1941–1949*. Harlow: Pearson Education Limited, 2003.

Mitman, Gregg, and Kelley Wilder, eds. *Documenting the World: Film, Photography, and the Scientific Record*. Chicago: University of Chicago Press, 2016.
Nehring, Holger. "Remembering War, Forgetting Hiroshima: 'Euroshima' and the West German Anti-Nuclear Weapons Movements in the Cold War." In *The Age of Hiroshima*, edited by Michael D. Gordin and G. John Ikenberry, 179–200. Princeton, NJ: Princeton University Press, 2020.
Oreskes, Naomi. "Science in the Origins of the Cold War." In *Science and Technology in the Global Cold War*, edited by Naomi Oreskes and John Krige, 11–29. Cambridge, MA: MIT Press, 2014.
Oreskes, Naomi, and John Krige, eds. *Science and Technology in the Global Cold War*. Cambridge, MA: MIT Press, 2014.
Orgad, Shani. *Media Representations and the Global Imagination*. Cambridge/Malden, MA: Polity, 2012.
Orwell, George. "You and Atomic Bomb." *Tribune*, 19 October 1945. Retrieved on 30 August 2023 from https://orwell.ru/library/articles/ABomb/english/e_abomb.
Pestre, Dominique. "Regimes of Knowledge Production in Society: Towards a More Political and Social Reading." *Minerva* 41(3) (2003): 245–61.
"Pugwash Conferences in Science and World Affairs." Retrieved 30 August 2023 from https://pugwash.org/about-pugwash/.
Roberts, Geoffrey. "Science, Peace and Internationalism: Frédérique Joliot-Curie, the World Federation of Scientific Workers and the Origins of the Pugwash Movement." In *Science, (Anti-)Communism and Diplomacy: The Pugwash Conferences on Science and World Affairs in the Early Cold War*, edited by Alison Kraft and Carola Sachse, 43–79. Leiden/Boston: Brill, 2020.
Robertson, Alexa. *Mediated Cosmopolitanism: The World of Television News*. Cambridge/Malden, MA: Polity, 2010.
Rubinson, Paul. *Redefining Science: Scientists, the National Security State, and Nuclear Weapons in Cold War America*. Amherst: University of Massachusetts Press, 2016.
Sattelmacher, Anja, Mario Schulze, and Sarine Waltenspül. "Introduction: Reusing Research Film and the Institute for Scientific Film." *Isis* 112(2) (2021): 291–98.
Schmidt, Ulf. *Medical Films, Ethics, and Euthanasia in Nazi Germany: The History of Medical Research and Teaching Films of the Reich Office for Educational Films / Reich Institute for Films in Science and Education, 1933–1945*. Husum: Matthiesen Verlag, 2002.
Turchetti, Simone. *Greeting the Alliance: The Diplomacy of NATO's Science and Environmental Initiatives*. Chicago: University of Chicago Press, 2019.
Turchetti, Simone, and Peder Roberts, eds. *The Surveillance Imperative: The Rise of the Geosciences during the Cold War*. New York: Palgrave Macmillan, 2014.
Unger, Corinna R. "Cold War Science: Wissenschaft, Politik und Ideologie im Kalten Krieg." *Neue Politische Literatur* 1 (2006): 49–60.
van Dongen, Jeroen, ed. *Cold War Science and the Circulation of Transatlantic Knowledge*. Leiden/Boston: Brill, 2015.
Vowinckel, Annette, Markus Payk, and Thomas Lindenberger, eds. *Cold War Cultures: Perspectives on Eastern and Western European Societies*. New York/Oxford: Berghahn Books, 2012.
Walker, Mark. "Physics, History, and the German Atomic Bomb." *Berichte Zur Wissenschaftsgeschichte* 40(3) (2017): 1–18.
Weingart, Peter. "Science and the Media." *Research Policy* 27 (1998): 869–79.

Weingart, Peter, and Patricia Schulz, eds. *Wissen—Nachricht—Sensation: Zur Kommunikation zwischen Wissenschaft, Öffentlichkeit und Medien*. Weilerswist: Velbrück, 2014.

Westad, Odd Arne. *The Global Cold War: Third World Interventions and the Making of Our Times*. Cambridge: Cambridge University Press, 2007.

———. "The Cold War and the International History of the Twentieth Century." In *The Cambridge History of the Cold War*, edited by Melvyn P. Leffler and Odd Arne Westad, 1:1–19. Cambridge: Cambridge University Press, 2010.

Wettig, Gerhard. *Stalin and the Cold War in Europe: The Emergence and Development of East-West Conflict 1939–1953*. Lanham, MD: Rowman & Littlefield, 2008.

PART I

Institutional and Industrial Contexts

CHAPTER 1

Whitewashing the Nazi Past

Continuity and Transformation in Scientific Cinematography in Germany 1934–56

Juliane Scholz

Introduction

Perhaps the most influential advocate for science films in postwar West Germany (FRG) was the head of the Institute for Scientific Film (IWF), a production company founded in Göttingen in 1956 and dissolved in 2010.[1] Electrical engineer and film official Gotthard Wolf (1910–95) introduced a new definition of scientific films. In a seminal speech in 1949, Wolf advocated that a clear distinction be made between the political propaganda of the National Socialist nonfiction *Kulturfilme* (culture films) and the purpose of establishing *Wissenschaftsfilme* (science films) with purely epistemic functions.[2] In his lecture, Wolf foregrounded the need to advance knowledge and support experimental processes in the laboratory as the main mission of science films.[3] The heterogeneous films under the umbrella of "scientific cinematography" had played an important role in German governmental health and hygiene programs since the 1920s; during and after Nazi rule, from 1933 to 1945, they also served political and ideological purposes. For instance, they justified the scientific foundation of and aided racist Nazi "euthanasia" programs by establishing medical training films to sterilize women; they promoted and supported ballistics and weapons production during World War II; and, after the war, they supported the FRG's redemocratization process and national science programs.[4] In particular, Wolf's personal agency helped to reinstate science-film organizations such as the IWF in postwar West Germany.

In this chapter, I will follow Wolf as one of the key figures involved in these processes. Wolf's career as a film official started in the Higher Education Department of the Reich Office for Educational Films (RfdU), established in Nazi Germany in 1934. In 1940, the RfdU was expanded and transformed into the Reich Institute for Film and Image in Science and Education (RWU). The RWU played a pivotal role under the Nazis, both centralizing the production of a range of instructional and educational films for classroom use and starting to produce and collect highly specialized science films. Wolf held leading positions in these organizations as of 1939. Among other things, he produced top-secret and technical research films for the Wehrmacht and other contractors, including research organizations such as the Kaiser Wilhelm Society (KWS).

After World War II, his recurring statements on science films continued to shape the definition and institutionalization of the genre in the FRG. Most importantly, Wolf's characterization of scientific cinematography as a purely objective research tool implied that the films—showing experimental practices and intended to support the advancement of scientific knowledge—had essentially been an apolitical medium all along.[5] What was left out of this definition was the past political function of science films, especially their ideological application and utilization in racist medical programs as well as war-related industries supporting military purposes. This decontextualization of science films was part of the FRG's continuing effort to circulate these older films, a distinct practice that has been described as reusing or recycling the material.[6]

Historians like Sarine Waltenspül, Anja Sattelmacher, Mario Schulze, and Ulf Schmidt have shed light on the reuse of science films, Wolf's problematic engagement in war-related and ballistic research as of the 1940s, and the whitewashing of science film organizations.[7] Building on their research, I explore how, as of the early postwar years, such science films inevitably became part of academic curricula in the FRG and were used for technical applications in non-war-related industries and as scientific instruments in academic education, benefiting from the promotion of leading researchers at the Kaiser Wilhelm Society and its successor organization, the Max Planck Society (MPS).

To illustrate the RWU's role and its participation in the Nazi effort to "Aryanize" Jewish film property, I scrutinize the case of the confiscation of the medical film collection of the German Society for Scientific Film (DEGEWI) in 1935 and 1936, showing how medical training films were unlawfully incorporated into the RWU. With the start of World War II and under Wolf's leadership, science films and their personnel and institutions became involved in military research by expanding the Technical Research Film Unit (TF Unit). After analyzing the alliances forged between researchers, the film producers and the state under the Nazi dictatorship, I explore how science films' functions were adapted, repurposed, and whitewashed at the start of the Cold War.

In this chapter, I thus explore continuities in institutional history as well as personnel and structural consistencies between 1934 and 1956. As Gotthard Wolf's professional career highlights the continuities between the Nazi RWU and the FRG's IWF, I scrutinize his vital professional networks, which proved immensely useful during the denazification period and were also supported by the MPS.

Science Film Organizations in Germany

In *The Shape of Spectatorship*, Scott Curtis suggests that scientific cinematography played a pivotal role during the experimental stage of cinematography.[8] Neither the impact of scientific cinematography on epistemic practices nor the histories of the German institutions involved have been studied in depth after 1920, a period when the production of science films thrived in Germany and the government saw educational health films as key in its effort to communicate hygiene and health programs to the wider population.[9]

After World War I, several small private companies, such as the Medical Film Archive of the Universum Film AG (UFA), dealt with physical injuries, disabilities, and neurotrauma inflicted by the war. Another small private company was the Medical Cinematographic University Institute,[10] which was established in 1922 and, as we shall see, later forced to merge with the RfdU under the Nazis, became one of the first examples of centralizing production and gathering governmental control over valuable film stocks.[11] Subsequently, the government centralized parts of the educational and science film production, which resulted in a specialized film division for the coordination of medical film production.[12]

As Ulf Schmidt has shown, the strategy of distancing the RWU from any political crime and characterizing educational and science film as apolitical under the Nazi dictatorship established a powerful narrative after 1945.[13] Science films were not only repurposed as scientific tools for visualizing epistemic practices, as Wolf stated in 1949, however; they were also used in service of various political, educational, and military functions over time.[14]

The RfdU/RWU and Its Higher Education Department, 1934–45

The launch of the Reich Office for Educational Films in 1934 clearly marked the centralization of educational and science film production and distribution by Nazi Germany.[15] As Malte Ewert explains, the RfdU was a semiprivate organization under the auspices of the Reich Ministry for Science, Education, and National Culture (REM).[16] At first, it had relative political autonomy because it did not belong to the Reich Film Chamber (Reichsfilmkammer) and therefore was not subject to the direct control of propaganda minister Joseph Goebbels.[17] The RfdU maintained an independent, internal board of reviewers within the REM for some years, which mostly consisted of teachers, some of whom even rejected Nazi propaganda films as inappropriate for school curricula.[18]

The RfdU also built a vast collection of educational and, later, science films, and its film categories were published annually in a growing film catalog,[19] with films categorized by discipline in "series" designated by a specific letter and number.[20] Educational and instructional films were organized in the F-series, science films and academic training were in the C-series, while the B-series remained an eclectic collection of "technical science films." In 1941, the new AR-series was introduced, which mainly included propagandistic instructional films on 16 mm reels for use in schools.

The Higher Education Department of the RfdU was established in 1935 to produce science film stock, foremost in medicine and later in disciplines such as physics and biology. These films were distributed to the twenty leading research universities, nine technical colleges, and sixteen other higher-education organizations in Germany.[21] The head of the Higher Education Department was physician and psychologist Kurt Gauger (1899–1959), a Nazi Party member from 1933 on, who managed day-to-day operations under founding RfdU director Egon von Werner. Gauger expanded the educational film program to include ideological content and fostered cooperation with leading German scientists.[22]

In 1940, the RfdU was expanded and reorganized as the Reich Institute for Film and Image in Science and Education, or RWU. The formation of the RWU marked the beginning of centralized state control of the production, distribution, and collection of scientific films. In the same year, the Wehrmacht sponsored its Higher Education Department with 26,800 Reichsmarks (RM). With the outbreak of World War II, the Wehrmacht declared the department to be essential for the war effort and entertaining the troops.[23] As of September 1939, the RfdU began organizing the screening of over four hundred productions for Wehrmacht troops all over Europe.[24] The newly founded Special Technical Research Film Unit, or TF Unit, fulfilled contracts to film top-secret weapon tests and other military research.[25]

The RWU also produced and distributed racist and propagandistic medical films, with the support of leading research organizations and scholars. Surgeons, neurologists, and psychiatrists actively participated in National Socialist euthanasia programs.[26] These films also legitimized the forced sterilization of approximately four hundred thousand German citizens, including people with disabilities, people of color, LGBTIQ persons, sexual abusers, and the so-called "Rhineland Bastards" and other "racially inferior" persons.[27] The forced sterilization of these groups was carried out with medical and legal justification, including the sterilization law.[28] The RWU catalog lists, for instance, a three-part film on sterilization titled *Unfruchtbarmachung der Frau* (C-15, C-16, C-17, *Sterilization of Women*), directed by Prof. Georg August Wagner. As the director of the Women's Clinic at Berlin's Charité hospital from 1928 to 1945,[29] Wagner conducted X-ray and radium castration on men and women, which caused severe physical and psychological pain and damage to his patients. By the end of World War II, five thousand people—mainly women—had died as a result of such drastically and recurrently forced surgical interventions.[30]

The RWU films not only displayed, taught, and normalized similar invasive procedures, which were part of the National Socialist heredity and "eugenic" programs, but also actively supported Nazi racist "racial hygiene" policies by providing the necessary medical training for physicians.[31] It is important to note that, even after the war, RWU sterilization films (distributed by the IWF) continued to circulate in German universities throughout the 1950s and 1960s mainly because forced sterilization was not legally perceived as a crime until the Rome Statute of the International Criminal Court in 1998 marked it as a crime against humanity.[32] Until then, victims of National Socialist eugenic programs were not officially recognized as victims of the Nazi dictatorship in Germany.[33]

Confiscation of DEGEWI Medical Film Collection, 1935–37

Kurt Gauger was appointed as the head of the Higher Education Department in 1935 and, the following year, became the RfdU's chief operating officer. Under his leadership, science films were considered to be of ideological as well as scientific significance. An extension of this was the "Aryanization" (i.e., expropriation) of Jewish film companies, such as the forced takeover of the most valuable medical film collection in Germany, that of the DEGEWI.[34]

As "one of the loudest proponents" of "a new German psychotherapy," Gauger was inspired by Carl Gustav Jung's idea of "political medicine" and was eager to build a medical and science film collection.[35] Gauger had studied medicine and psychology with the financial backing of Kurt Ludwig Richard Zierold, a conservative nationalist. In 1934, Zierold was appointed RfdU supervisor by the Department of Public Education.[36] In this role, he undermined the authority of RfdU founding director August von Werner and lobbied for Gauger to take over. Gauger was indeed appointed as president of the reorganized RWU in 1941, although, thanks to Zierold, he had been informally in charge of the RfdU long before his official promotion.[37]

The DEGEWI case provides a key insight into the RfdU's unscrupulous tactics in developing the portfolio of medical science films that later became the core collection of the Higher Education Film Department. Gauger orchestrated the DEGEWI takeover along with influential political officials—such as Kurt Zierold, Alfred Pudelko, and officials at the Gestapo—and, most notably, with the help of leading eugenics expert Hans Nachtsheim. In 1941, Nachtsheim was then appointed director of the Kaiser Wilhelm Institute for Anthropology, Human Heredity, and Eugenics, which provided the expert medical knowledge for ethnic cleansing and racist population planning.[38]

Shortly after his initial appointment in 1935, Gauger tried to instigate the confiscation of the DEGEWI, basing the move on racist laws for excluding Jewish businesses, such as the 1933 Law on the Seizure of Assets of Enemies of the People and the State.[39] Amid the ongoing "Aryanization" of film companies, the confiscation of the DEGEWI would be yet another example of the nationalization of the film industry by the Nazis.

The DEGEWI was founded as a private enterprise by Alexander von Rothe (1880–1932) in 1922.[40] Von Rothe had earlier experimented with filming surgical methods and invented so-called X-ray cinematography,[41] using something called the Rothe Device (*Rothscher Apparat*).[42] The films were immensely successful because they enabled medical instructors to present surgical techniques to their students multiple times.[43] Von Rothe initially financed his company with support from his patients and the Ministry of Education. His surgical films and recording techniques drew the attention of neurologists, psychiatrists, and gynecologists from Italy, Austria, Germany, and Czechoslovakia. Soon, von Rothe's surgical films and the application of the apparatus became an integral part of the academic medical curriculum,[44] and his X-ray camera became a preferred device for demonstrating complex surgery procedures to medical students in training.[45]

In the mid-1920s, von Rothe's brother Thorben founded a company for the distribution of science films under the auspices of a publishing house. He published a bimonthly journal and the accompanying newsreel, *Medical Film Week* (*Medizinische Filmwoche*), which was widely distributed at German universities.[46] Part of their success came with the move of Rothe's institute to Berlin's famous Charité hospital in 1927. Despite partial government funding, the institute maintained the status of a private business.[47] Von Rothe's distributor went bankrupt in 1930,[48] and two years later, the entity known as the Charité-based Medical Cinematographic Institute was officially dissolved by the Prussian government and renamed DEGEWI.[49]

The DEGEWI medical film collection was attractive to Gauger because of its scope, international recognition, and academic value. The precious film collection would also bring prestige to the RfdU's newly established Higher Education Department he was heading. By 1935, the DEGEWI medical film archive already contained 36,738 meters of 16 mm and 35 mm film, and the reels had been exported worldwide, as indicated by shipping lists.[50] In contrast, the RfdU was known as a film contractor rather than a producer of original science films and owned only very heterogeneous film stock.[51] Ulf Schmidt maintains that this wide distribution network and the accompanying patents constituted the main incentives for the takeover.[52]

The legal confiscation of the DEGEWI in 1936 began when Max Liebermann, who had been responsible for its international distribution rights since 1933, was forced to resign. The film stock and subsequent copyrights were transferred to Otto von Lans, a former admiral,[53] who insisted on the low price of 85,000 RM for the film stock and 20,000 RM for the international distribution rights.[54] During the negotiations, von Rothe and Liebermann secretly tried to secure the film negatives, which were still stored in Berlin, by transporting them to Vienna and then taking them abroad.[55] Gauger had already gotten wind of this plan from cameraman Viktor Trinkler, who had worked for von Rothe at the Charité. It was Trinkler who had earlier denounced Liebermann, who was a Jew and who, according to the Nazi Nuremberg Laws, could no longer conduct any business; Gauger later employed Trinkler as a cinematographer at the RfdU, perhaps as a reward for this denunciation.[56]

Gauger warned the Reich Film Chamber and alarmed officials in the REM about von Rothe and Liebermann's plan to transport the DEGEWI film stock to Austria. In the fall of 1935, Gauger also employed what he called "elastic methods" to convince new owner von Lans to agree to the merger of the DEGEWI with the RfdU, exercising immense political pressure on him with the help of high-ranking SA and SS members. In a letter to Alfred Pudelko (1899–1981) dated September 1935,[57] he introduced himself as a fellow Nazi Party member and described the DEGEWI case in detail, which he had now "personally monitored for months." Gauger claimed that "the Jew" Liebermann had bought the bankrupt DEGEWI in 1931 for a pittance and had thus "tricked the former Prussian education ministry." Gauger claimed that the DEGEWI was illegally withholding the film stock from the government.[58] He also slandered new DEGEWI holder von Lans, calling him a stand-in ("dummy for a Jew") and then reporting him to the Gestapo in a November 1935 meeting.[59]

Pudelko was head of the Nazis' Central Institute for Education and Teaching, a leading pedagogue, and the editor of the journal *Race Education as a Teaching Foundation* (*Rassische Erziehung als Unterrichtsgrundsatz*).[60] He was a high-ranking official at the REM, in the Nazi Party since 1932, and became an SS member in 1933.[61] Gauger used Pudelko's important position and his strong connections to the SS and Gestapo to force the confiscation of the DEGEWI stock and its transfer to the RfdU.[62]

Gauger also engaged the leading Nazi eugenics specialist Hans Nachtsheim, who had collaborated with the DEGEWI on the eighty-eight-minute film *Erbkranke Kaninchen* (Rabbits with hereditary defects, 1934). The film's main message was that animals who appeared physically healthy could secretly carry genetic defects and, therefore, had to be prevented from breeding. Of course, Nachtsheim linked his findings to humans and the 1933 Law for the Prevention of Offspring with Hereditary Diseases (Gesetz zur Verhütung erbkranken Nachwuchses). The film saw several re-releases and was an effective multiplier, as Alexander von Schwerin points out.[63] Nachtsheim served as a key eyewitness in support of Gauger and Trinkler's allegations and accused von Rothe of misleading him into believing that the DEGEWI was a state-owned institution.[64] He also reported that von Rothe had fraudulently used the letterhead of the Medical Cinematographic University Institute at the Charité.[65] Nachtsheim and his assistant then testified that they had seen Liebermann at a film screening at the Charité and that the company was still in "Jewish hands."[66]

On 16 October 1935, Gauger, Nachtsheim, Friedrich Curtius (head of the hereditary pathology department at the Charité), and Eugen Schwarz (an employee at the RfdU's Higher Education Department) concluded that, as it was by now state property, the DEGEWI film stock legally belonged to the RfdU.[67] Subsequently, the film collection and its distribution rights were classified as being of high public and political interest, and the film reels were considered at risk of being "abducted" abroad.[68]

Unable to withstand the intensified political pressure, von Lans and von Rothe eventually agreed to the merger. The DEGEWI collection was bought for the unreasonably low price of 10,000 RM and incorporated into RfdU holdings on 1 December 1935.[69] Approximately two hundred DEGEWI films were thus usurped by the Higher Education Department. To ensure that film stock was appropriated immediately, Gauger also took over DEGEWI offices at the Charité on behalf of the RfdU in November 1935.[70] The official "merger agreement" was signed on 3 January 1936.[71] In 1937, the international film distribution rights were bought for 20,000 RM.[72]

Most DEGEWI medical films were incorporated into the B- and C-series of the RfdU collection.[73] Only 13 films of the B-series and 122 of the C-series were original RfdU productions; the remaining two-thirds were original DEGEWI productions.[74] As Wolfgang Tolle has pointed out, the confiscation of DEGEWI films (including 332 negatives and 330 positives), of numerous domestic and international rights for copying and distribution, and of valuable equipment were all essential for the establishment of the Higher Education Department.[75] While most films by Jewish directors remained in the expanded RfdU collection, these films were not distributed, and the names of the directors were erased from corresponding index cards and title frames.[76] The RfdU profited immensely from the confiscation of Jewish film property, later re-editing many DEGEWI films to serve its own propaganda purposes and using the technical equipment for secret assignments.[77]

Secret Military and Industrial Applications: The TF Unit, 1939–45

In 1939, immediately after the outbreak of World War II, the Technical Research Film Unit was established by recruiting staff from the Higher Education and Technical Departments of the RfdU.[78] New collaborations with the Wehrmacht's Development and Testing Department became part of the RfdU's restructuring into the RWU, and additional funding from medical education was crucial amid war shortages.[79] Led by the engineer Gotthard Wolf, who probably worked full time for the new unit from 1939 onward, the TF Unit was classified as important to the war effort (*kriegswichtig*).[80] Wolf and his team of cinematographers, engineers, cutters, scientists, and technicians formed a mobile film crew that traveled throughout Germany; they worked on civil and industry-related projects but were also contracted for top-secret assignments by the Wehrmacht. The TF Unit specialized in what were called technical research films, in sectors such as weaponry, ballistics and their use, and the diagnosis of wartime bone fractures using X-ray film techniques.[81]

The birth of the TF Unit was a troubled one because Wolf and Gauger were initially unsuccessful in finding state sponsorship. In November 1939, Gauger inquired if the KWS would finance a new Technical Research Film Unit at the RfdU with 200,000 RM.[82] In a letter to the KWS, Gauger and Wolf stressed the importance of the TF Unit for the war effort and described proposed film projects as important to the Wehrmacht and war-related industry.[83]

Since Wolf had already collaborated with several KWS institutes on science film projects in the early 1930s, he also contacted KWS general secretary Ernst Telschow with the proposal.[84] In his request to Telschow, Wolf clearly defined his idea of science films and distinguished them from educational films used for high school instruction. The latter, he argued, blatantly displayed tropes of National Socialist propaganda, as movies on German geography and folklore started to show swastikas and other well-known Nazi symbols and underlying messages on the "Volkskörper" and "Volksgesundheit" (the nation's body and health). In contrast, Wolf clarified, science films would ultimately build evidence based on research, provide exact scientific methods, and be useful in industrial applications.[85]

Technical research films in scientific contexts did, in fact, cleave quite closely to the experimental method. The new slow-motion and time-lapse cameras were used frequently, and motion and time became increasingly important for both technical and stylistic design. Slow-motion, close-up, and time-lapse cameras could, for example, provide insights on chemical, biological, and physical flows, movements, and reactions.[86] For example, the science film *Elektrenkephalographie* (C-552, Electroencephalography, 1942), which the RWU produced with Alois E. Kornmüller (1905–68), director of the Kaiser Wilhelm Institute for Brain Research, was used to illustrate brain waves and explain cramps and seizures.[87] Kornmüller's institute conducted research on the brains of killed German air force men and contributed extensively to the military's wartime research on central nervous system damage. The institute's own Research Center for Brain, Spinal Cord, and Nerve Injuries (Forschungsstelle für Hirn-, Rückenmark- und Nervenverletzte) also produced over four hundred science films on war-induced brain injuries and possible surgical methods. After 1945, this film collection was handed over to the Institute for Film and Image in Science and Education (FWU), which was declared the RWU's successor in 1950.[88]

In addition to medical films, RWU collaborations with research institutes, including technical research films on weapons engineering, became more integrated into the military-industrial complex. These demonstrated the direct application of war-related research using the support of visual media. In their letter seeking support for the TF Unit, Gauger and Wolf itemized planned film projects that were essential for the German war effort, including, for instance, films about the rolling trajectory of automobile tires, the combustion process in engines, and heat flows in small rooms.[89] Although Wolf and Gauger's request to the KWS did not secure funding right away, several coproductions with KWS research institutes were realized in the following years. In 1943, for example, the Kaiser Wilhelm Institute for Coal Research hired the TF Unit to produce two science films to address problems with chemical reactions in coal-coking processes.[90]

With the expansion of Nazi Germany's war effort and an ongoing personnel shortage, the TF Unit was declared high priority with the support of the Wehrmacht and was commissioned to work on projects, including biological warfare projects, for the Reich Aviation Ministry.[91] With long-term sponsorship from both the REM and the Wehrmacht, Wolf was at last able to stabilize RWU revenues during World War II.[92] Later, the Reich Office for Economic Expansion (Reichsstelle für Wirtschaftsaufbau) also cosponsored the unit.[93]

Most of the TF Unit's top-secret films were produced at the Wehrmacht High Command and its military research facilities located in Hillersleben, Kummersdorf, Rügenwalde, and Peenemünde on the Baltic coast, among other places.[94] One such top-secret project employed time-lapse and microcamera recordings to study persistent problems with the MG 42 machine gun.[95] RWU official reports only included civilian or merely scientific projects, but internal plans indicated the TF Unit's filming locations, which could easily be linked to military bases and research facilities.[96] However, after World

War II, only two clearly military films made by the TF Unit appeared in the public RWU catalog: *Ballistische Untersuchungen* (C-96, Ballistic studies, 1937, dir. Joachim Kömmnick), with twenty-three copies sold by 1940 and Hubert Schardin's *Dum-Dum Wirkung* (C-229, Dumdum effect, 1938).[97] The RWU work report of 1942 coined the corporation and commissioned work for the military vaguely as war-related special tasks ("Kriegsonderaufgaben").[98]

Over the course of World War II, the TF Unit's personnel expanded; in early 1940, Wolf was the only listed employee, but in 1943 there were thirty-five employees, including five cameramen, one technician, and six staff researchers. In 1944, the unit also took on seventeen Italian forced laborers.[99] Reporting to RWU president Gauger, the TF Unit assignments remained top secret.[100] In 1944, Erich Schneider, head of the Army Ordinance Office for the Technical Development of Arms and Weapons, praised the TF Unit's work.[101] Schneider's Wehrmacht office became the main sponsor for the TF Unit, contributing to salaries and housing costs, and assisting the unit's employees to move out of Berlin and relocate during the war.[102]

Whitewashing Science Film Organizations and Personnel, 1945–56

At the time of Germany's capitulation in May 1945, the RWU had gross assets of nearly 22 million RM, and though there were orders for its dissolution, the organization was instead transformed. Most assets listed in the official reports were generated by revenue from technical and copying services and from aerial photos taken for the Wehrmacht. No evidence remained of the World War II military projects.[103]

Wolf claimed that the TF Unit had provided for the advancement of technical science films for industrial sectors and had been forced into cooperation with the Wehrmacht.[104] As a result, RWU departments and personnel were able to set up new organizations in both East and West Germany in the initial postwar years. These organizations were characterized by strong structural and personnel continuities.[105] The former Educational School Film Department relocated to Hamburg and was reestablished in 1946. In June 1946, the Soviets established their own Central Office for Educational Screenings (Zentralschulbildstelle), which was also declared to be the successor of the RWU's Educational Film Department.[106] In 1950, the RWU itself was moved to Munich, renamed the Institute for Film and Image in Science and Education (FWU), and declared the RWU's legal successor.[107]

Kurt Gauger was initially interned, then released in mid-1946 due to poor health. His 1948 denazification report indicated that he had not engaged in the production of propaganda films.[108] He argued that the RWU could in fact be considered a "political oasis" where educational and teaching films were supposedly "safeguarded" from political control—a myth that Schmidt later dismantled.[109] In a personal statement, he downplayed his political role and even labeled himself an "opponent of the Nazi regime" because he had once hired a communist. He received his denazification certificate. He even requested financial compensation for losing his position at the RWU and demanded a lifelong pension payment of 72,000 Deutsche Mark (DM); this was not granted, but he later sued the FWU and was awarded compensation of 32,500 DM instead.[110] Gauger opened a clinic for psychosomatic therapy near Uelzen and specialized in treating war veterans with suggestion therapy until 1952. He continued to screen RWU medical films in his facility.[111]

Several evaluations of the RWU film collection in fact nurtured the myth of a safeguarded political oasis. When the British and American officials assessed the RWU film collection in 1945 and again in 1946, however, they did not have access to most films in the Higher Education Department's collection, nor did they review the propaganda films or racist medical films. Most of the RWU film reels had been destroyed during the bombing or later thrown out of bunkers and probably dumped in a lake near Berlin-Tegel.[112] The Allies eventually whitelisted 744 of 847 RWU educational films. Starting in summer of 1945, the Audio-Visual Aids Commission of UNESCO's Conference of Allied Ministers of Education reviewed the RWU educational and teaching films; the purpose of this initiative, coming from Czech and Slovakian members, was to reuse German films for teaching in the liberated countries. Ultimately, their *Report on the German Educational Film* (1946), which was conducted by the British Film Institute with viewing panels of teachers and both technical and academic experts, reviewed 175 films mainly from the BF and some of the F-series (instructive films for high schools and science films). This was only a fraction of the approximately nine hundred films produced by the RWU. Only a small number of explicit propaganda films (mainly the history section) were blacklisted; however, 13 percent were regarded as tendentious and not propagandist in approach. Another 10 percent showed Nazi flags and needed editing before reuse.[113] Most of the medical and surgical films, including the sterilization and brain surgery films discussed earlier, were not reviewed by the commission and were later transferred into the science film catalog for medical training distributed by the FWU and IWF, which occurred into the late 1950s or possibly longer.[114]

This successful whitewashing of RWU personnel and its film collection and the evolving Cold War confrontation between East and West made it possible for Gotthard Wolf to embark on a second career.[115] After receiving a work permit after the war, Wolf prevented a full incorporation of the Higher Education Department into the FWU, which was controlled by US officials,[116] and pulled strings to secure several RWU higher education films to establish his own science film organization.[117] In 1945, he tentatively named it the Institute for Film and Images (Institut für Film und Bild) and later established it in Göttingen. The institute finally achieved independent status on 1 April 1956, when the name was changed to the Institute for Scientific Film (IWF).[118]

Former friends and influential science managers, such as Ernst Telschow and Kurt Zierold, aided him in setting up an independent institute in Göttingen, in the state of Lower Saxony, in the British Occupation Zone. The aim was to establish it as an influential science hub for postwar West Germany, mainly producing science films. In respect to this, Wolf contacted Zierold—who had become the head of the Higher Education Department of the government of Lower Saxony—about the possible reorganization of a science film institute. The two met several times in Göttingen, where Zierold had been appointed as executive vice president of the Emergency Association of German Science (Notgemeinschaft der deutschen Wissenschaft) in 1949. Wolf also reached out to Ernst Telschow and Nobel Laureate Otto Hahn, who was the first president of the Max Planck Society (MPS) in Göttingen, the successor of the Kaiser Wilhelm Society.

Wolf planned to cooperate with the MPS and was eager to form a scientific film institute within it. He therefore corresponded with its general secretary Ernst Telschow several times between October 1945 and 1948.[119] Telschow was willing to continue his working relationship with Wolf, but the idea of founding a science film institute was rejected by the MPS senate.[120] After Wolf secured political and moral support from Adolf Grimme, Lower Saxony's cultural minister, the MPS granted Wolf's team housing and office space as well as film studio spaces in the former wind tunnels in the MPS research institutes in Göttingen.[121] The IWF, one of the RWU successor organizations, emerged in Göttingen through the advocacy of old friends like Telschow and Zierold and the help of the British regional government in Hanover.[122] At this point nobody knew or talked about Wolf's engagement in military research, even though we can assume it was an open secret and part of the reason the MPS did not incorporate the IWF in its midst. Wolf's extensive network provided the necessary technical equipment.[123] And, in 1950, Wolf's former colleague and TF Unit member Willi Mohaupt supplied him with more film reels from the RWU collection, mainly from the BF-, LF- and F-series.[124]

Although Wolf's attempt to establish a science film institute within the MPS had failed,[125] the collaboration between the MPS and the IWF expanded into a productive and, as MPS president Otto Hahn remembered, "symbiotic work relationship."[126] Telschow, Zierold, and other MPS directors, such as behavioral scientist Konrad Lorenz, remained active members of the IWF advisory board.[127] The cooperation between the IWF and the MPS, by now the leading organization for basic nonapplied research, resulted in approximately ninety coproductions with several Max Planck Institutes by 1960 and numerous short films for the mammoth IWF project of an archive of filmed motion: the *Encyclopaedia Cinematographica*.[128] Other commissioned works for the MPG involved long-take interviews with leading scientists or microcinematography of biological and chemical experiments. In 1956, the first issue of the *Bulletin of the Institute for Scientific Film* (*IWF Bulletin: Mitteilungen des Instituts für den Wissenschaftlichen Film*) was published. In the cover article, Wolf recapped the IWF's history reaching back twenty years to RWU times and his early vision of supporting science through science films.[129]

Interestingly, the IWF was financially and structurally founded and reinforced at the same time as competing institutions in East Berlin, including the Central Institute for Film and Image in Teaching, Science, and Education (ZFB), founded in 1950, and the DEFA Studio for Popular Science Films (DEFA Studio für Populärwissenschaftliche Filme).[130] Both the FWU in Munich and the East German ZFB took over RWU *Bildstellen* (image centers in provinces for educational film).[131] In 1954, the government of the German Democratic Republic (GDR) incorporated the ZFB under an umbrella organization called the Central Institute for Teaching Materials (Zentralinstitut für Lehrmittel).[132]

At the outset of the Cold War, competing science policies on either side of the Iron Curtain impacted the reorganization of scientific cinematography.[133] Wolf was determined to build a strong, internationally recognized West German organization and oppose GDR efforts to partake in the growing educational film sector. Ultimately, the increased ideological competition during this period and the founding of the GDR's competing ZFB brought Wolf new funding opportunities. In 1955, Wolf wrote to Telschow stressing the discrepancy between science film funding in East and West Germany. He pointed out that East Germany had dedicated 900,000 D-Mark (West German currency) in annual funding from the state and had already produced ninety-six science films in two years, whereas the IWF had released seven films with funding of only 35,000 D-Mark.[134] Wolf warned Telschow about the IWF's fading international reputation and the possible Soviet domination in the field, as the USSR was spending twenty times more than West Germany on educational and science films.[135]

Wolf also strategically organized a meeting of the International Scientific Film Association (ISFA) in Göttingen in early 1954.[136] Later that year, East German counterparts demonstrated the high quality of their science films at the ISFA conference in Rome.[137] The ISFA was Wolf's ticket to international recognition. In 1952, he and French astro-photographer Jean Dragesco created the first film section dedicated to science films in the ISFA.[138] At the same time, Wolf used the restored international reputation of the IWF to secure further funding from the government for the organization.[139]

In 1955, the IWF finally received generous continuous state funding and became part of the Königsstein Agreement. The agreement among the West German federal states addressing multiple issues on subsidy research was first introduced in 1949.[140] This success was followed by a green light for a new IWF building in Göttingen in 1956, which opened a year later and can be seen as the high point of reinstitutionalizing science film production in the FRG. Consequently, the IWF became one of the most important science film producers in the world, with a targeted portfolio of technical research films and scientific films, while the FWU in Munich concentrated on educational and instructive films for high schools.

On 10 October 1961, the RWU was officially dissolved after a long legal liquidation process, and its assets—including 76,678 D-Mark, educational and science films, and their distribution rights—were officially transferred to the FWU and IWF even though they had already been working with the film collection for years.[141] Amid the raging Cold War competition between the East and West, scientific cinematography was redefined, with an emphasis on science films being most useful as research tools and part of epistemic practices to support researchers. The IWF continued to cooperate with universities and research organizations such as the MPS until its official dissolution in 2010. The FWU concentrated on instructive films for high schools and educational programs with the reorganization of hundreds of *Bildstellen* after its establishment in Munich in 1950. It's first film was the thirteen-minute animal movie *Quick, das Eichhörnchen* (Quick, the squirrel, 1952, dir. Heinz Sielmann), which was awarded the Bundesfilmpreis—the federal film award.

Conclusion

This chapter explores the emergence of German science films, along with related organizations and personnel, from the 1930s through the 1950s. The story necessarily involves confronting the ideological uses to which such films and science film production were put and continuities before and after 1945. To illustrate the entanglement of German science programs and politics, the chapter focuses on the forced merger of the DEGEWI medical film collection into the Reich Office for Education, Science, and National Culture (RfdU) under the Nazis.

The chapter reveals that the DEGEWI film stock was forcefully merged with that of the RfdU. This was orchestrated by Kurt Gauger—later president of the RWU—and was only made possible thanks to crucial support from Germany's leading geneticist, Hans Nachtsheim, later director of the notorious Kaiser Wilhelm Institute of Anthropology, Human Heredity, and Eugenics. After confiscating the medical film stock from its Jewish owner, Gauger built a dedicated Higher Education Department within the RfdU and concurrently strengthened the idea of a scientific film collection. It shows that the medical film collection, which became known around the world, involved one of the many forced takeovers of Jewish-owned businesses during the National Socialist dictatorship. The confiscation of film stock and rights and the seizure of production companies were covered up as "mergers" and orchestrated as part of the ongoing "Aryanization" of "Jewish film property" even before the Nuremberg racial laws became effective.

This chapter also traces the institutional history and the multiple reuses of science films, which exemplify the multiple levels of cooperation in science, military, industry, and politics that existed as of 1933 and especially during World War II. Ideological and political agendas were not always acknowledged but were rather incorporated within assignments that supported war-relevant industries, such as coal mining, agriculture, clinical surgery, and ballistics. During World War II, the Wehrmacht's need for research further engendered the secret TF Unit, led by Gotthard Wolf, Gauger's protégé. The archival findings presented in this chapter contradict Wolf's postwar claims—for instance, that most projects conducted by the TF Unit were civilian in nature. On the contrary, TF Unit projects substantially supported warfare and military research and were primarily sponsored by the Wehrmacht and other government institutions. The process of whitewashing these science films was epitomized by the new definitions and meanings that Wolf offered in 1946—namely, that the films were apolitical, epistemologically objective, and regulated by scientific research methods. They were redefined as timeless vehicles. These definitions underpinned Wolf's strategy to reorganize the IWF, with the help of the RWU film collection, and concurrently give this highly problematic film collection a fresh start with a blank slate.

One of the chapter's major insights occurs in the discussion of the important role played by influential scientists and scientific institutions, such as the MPS, before and after 1945. The eleven years from 1945 to 1956 are not only striking for structural and personnel continuities in science film production but were also crucial for the reshaping and remodeling of German science film organizations, largely through the advocacy of Gotthard Wolf. Wolf's long postwar career as the head of the IWF resulted from his strategy of recontextualizing and redefining science films and their roles while withholding facts and misrepresenting the truth in order to fit the IWF story into the founding narrative of West Germany's reconstruction and redemocratization. In championing the idea of objective scientific practices, Wolf aligned his efforts with the whitewashing strategies of other science organizations that also chose to remain silent about their military and industrial research under the Nazis.

The rebuilding of West and East German science film organizations after 1945 was indeed a complex reconciliation process. It ultimately reinstated the authority of leading scientific film experts, researchers, and cultural officials who had gained professional standing and power under the Nazi dictatorship. A similar, albeit differently framed pattern of whitewashing science film personnel and organizations became evident in both East and West Germany as part of their growing Cold War competition.[142] After the immediate postwar period (1945–49), former second-tier bureaucrats and technocrats, in particular, as well as science and educational film specialists were reinstated in both German states.

Juliane Scholz is a postdoctoral researcher and graduate coordinator at the Leibniz Centre for Contemporary History in Potsdam, Germany. Her research focuses on twentieth-century German, European, and US-American social, cultural, and media history. She studied cultural sciences, media studies and psychology at the University of Leipzig, where she completed her dissertation on the professional histories of screenwriters in the United States and Germany in 2014. She has served as a senior lecturer and postdoctoral researcher at the University of Leipzig and the Max Planck Institute for the History of Science in Berlin. Her recent publications on the social history of the Max Planck Society shed light on working conditions in the academic field and the professionalization of science communication.

Notes

1. "Science films" is an umbrella term for a rather heterogeneous group of films. Typically low-budget productions made in smaller quantities, they are not intended for commercial distribution and predominantly target a narrow audience of scholars, educators, and experts. "Science film" is a direct translation of the German *Wissenschaftsfilm* and "research film" of the more technical and specialized term *Forschungsfilm*. On the problems of translation and meaning of science and research film, see: Sattelmacher et al., "Introduction"; Waltenspül und Schulze, "Follow the Films"; Gaycken, *Devices of Curiosity*.
2. *Kulturfilme*—culture, or cultural advertising films—mostly accompanied feature films in cinemas and were heavily used for Nazi propaganda after 1933. See Zimmermann, "'Auferstanden aus Ruinen.'"
3. Wolf, "Der wissenschaftliche Film," 3.
4. Schmidt, *Medical Films*, 135–42.
5. Wolf, *Der wissenschaftliche Film*, 7.
6. Waltenspül und Schulze, "Follow the Films"; Schmidt, *Medical Films*, 130–32.
7. Waltenspül, "Reusable and Nonreusable Films," 342–51; Sattelmacher et al., "Introduction," 291–98.

8. Curtis, *The Shape of Spectatorship*, 1–17.
9. Schmidt, *Medical Films*, 42.
10. Medizinisch-kinematographisches Universitäts-Institut.
11. Laukötter, *Sex—richtig!* 14–20; Schmidt, *Medical Films*, 46–47.
12. Schmidt, *Medical Films*, 46.
13. Schmidt, *Medical Films*, 129–31.
14. Wolfgang Tolle, *RWU*; Gotthard Wolf, "Die Aufgaben"; Große, *Filme für die Wissenschaft*.
15. von Keitz, "Die Kinematographie," 463–88.
16. The Reichsministerium für Wissenschaft, Erziehung und Volksbildung, unofficially referred to as the Reichserziehungsministerium, or REM. For a comprehensive study on the REM, see Nagel, *Hitlers Bildungsreformer*.
17. See Mohaupt's printed letters, in Ewert, *Die Reichsanstalt*, 473–75.
18. See the review and rejection of *Der Aufmarsch der Millionen* by the internal board, in *Mitteilungsblatt der Reichsstelle für den Unterrichtsfilm*, 2 August 1934, Bundesarchiv (Barch), R/169/5.
19. Barch-Farch, "Filme."
20. RWU, Übersicht, 1942.
21. For the history of the RfdU/RWU, see Tolle, *RWU*, 25–41.
22. Born in Stettin, Gauger first joined the anti-bolshevist *Freikorps* paramilitary group and then the SA and NSDAP in 1933. He served as a lieutenant colonel (*Obersturmbannführer*) in the medical corps of the SA marines. Schmidt, *Medical Films*, 82–88.
23. Revenues of RWU 1940, Barch, R/169/11; Budget RWU 1943/44, Barch, R/169/13.
24. See *Mitteilungsblatt der Reichsstelle für den Unterrichtsfilm 1934: Gutachten der Reichsstelle*, 2.8.1934, Barch, R/169/5.
25. Große, *Filme*, 7.
26. Weindling, "Genetik und Menschenversuche"; Offermann, "Die biopolitische Produktion."
27. Nazis and other conservatives used the derogatory term "Rhineland Bastards" to describe BIPOC children who were believed to have been fathered by French Army personnel of North African descent stationed in the Rhineland area during the French occupation of the Rhineland after World War I. See Roos, "The Race to Forget."
28. Eugenic specialists like Herrmann Muckermann and Ernst Rüdin were commentators who helped draft the law's first version. From 1935 on, the law strictly criminalized abortion. See "Gesetz"; Bock, *Zwangsterilisation*.
29. Schmidt, *Medical Films*, 142.
30. Schmidt, *Medical Films*, 141–43.
31. Wagner, "Die Technik."
32. Zeidman, *Brain Science*, 692.
33. See the Deutscher Bundestag, *Opfer von NS-"Euthanasie" und Zwangssterilisation als Verfolgte des Nationalsozialismus anerkennen – Aufarbeitung vorantreiben*.
34. Schmidt, *Medical Films*, 88–101.
35. See Budde, *Kurt Gauger*; Kühn, *Unterrichtsfilm*, 261–62.
36. Schmidt, *Medical Films*, 20.
37. Schmidt, *Medical Films*, 69, 130.
38. von Schwerin, *Experimentalisierung*, 21–26.
39. *Gesetz über die Einziehung volks- und staatsfeindlichen Vermögens vom 14. Juli 1933*, in *RGBl*, 479.

40. Wolf, "Das Institut," 5.
41. Laukötter, *Sex—richtig!* 144.
42. The camera position, above and facing the patient on the operating table, could be controlled by the surgeon with a foot pedal.
43. For an evaluation of Rothe's surgical recording process, see Prof. A. Miete, TU Berlin: 2.5.1922, Barch, R/109/I 5759.
44. Hay, *Photographisches Praktikum*.
45. von Rothe, list of medical experts and professors using method, 1927: Barch, R/109/I 5759.
46. von Rothe, report on the Medical Cinematographic Institute, 1927, Barch, R/109/I 5759.
47. von Rothe, 1927, Barch, R/109/I 5759.
48. von Rothe, 1927, Barch, R/109/I 5759.
49. Friedland, "'. . . doch erscheint in seiner Denkschrift,'" 33.
50. List of DEGEWI films, July 1935, Barch R/169/15, fol. 96–98.
51. Note by Alexander Herzfeld on the planned merger of DEGEWI to Gauger and von Werner, 21 October 1935, Barch, R/169/15, fol. 37–38.
52. Schmidt, *Medical Films*, 93–94.
53. Schmidt, *Medical Films*, 88.
54. Schmidt, *Medical Films*, 92.
55. Schwarz to Reichsfilmkammer, 16 October 1935, Barch, R/169/15, fol. 41–43; Schwarz to Zierold, report on the ownership of the Medical Cinematographic Institute, 27 July 1935, Barch, R/169/15, fol. 92-94; Schwarz on DEGEWI at Reich Film Chamber, 15 October 1935, Barch, R/169/15, fol. 44–46.
56. Trinkler earned 500 RM including rent for a flat. After 1945, his widow gathered RWU support in the form of a small pension of 60 D-Mark, which was paid until 1961. Appendix RfdU budget list 1939/40, Barch, R/169/11; presentation on the liquidation status of the RWU, 3 July 1958, Barch, R/169/20.
57. Gauger to Pudelko 16 September 1935, Barch, R/169/15, fol. 71.
58. Schwarz, presentation on DEGEWI at Reich Film Chamber, 15 October 1935, fol. 44–46; Schwarz to Reich Film Chamber, 16 October 1935, Barch, R/169/15, fol. 41–43.
59. Gauger and von Werner to Fehlis, 23 November 1935, Barch, R/169/15, fol. 32.
60. Harten et al., *Rassenhygiene als Erziehungsideologie des Dritten Reichs*, 219–20.
61. Pudelko was a trained pedagogue volunteered for the SS in 1933 and made a rapid political career advancement at SS Sicherheitsdienst (Security Service of the Reichsführer-SS), Rassenpolitisches Amt der NSDAP (Nazi Party Office of Racial Policy) and the Rasse- und Siedlungshauptamt (Race and Settlement Headquarters). During the war he was member of Waffen-SS and specialist for education in occupied Norway. Harten et al., *Rassenhygiene als Erziehungsideologie des Dritten Reichs*, 219–20.
62. Schmidt, *Medical Films, Ethics and Euthanasia*, 73–74.
63. von Schwerin, *Experimentalisierung*, 194–197; Film: *Erbkranke Kaninchen (Eine erbliche Nervenkrankheit: Schüttellähmung). Ein Lehrfilm von Hans Nachtsheim*, re-release 1938, Archiv der Max-Planck-Gesellschaft (AMPG), 6. Abt., Rep. 1, F 116.
64. Schwarz to Reich Film Chamber, 16 October 1935, Barch, R/169/15, fol. 41–43.
65. Note by Rottenburg/Schwarz on a phone call with Nachtsheim, 18 September 1935, Barch, R/169/15, fol. 89.

66. Cited from a note by Schwarz on DEGEWI, 16 September 1935, Barch, R/169/15, fol. 90–91; note, 18 September 1935, Barch, R/169/15, fol. 89.
67. Schwarz to Reich Film Chamber, 16 October 1935, Barch, R/169/15, fol. 41–43.
68. Schwarz to Reich Film Chamber, 16 October 1935, Barch, R/169/15, fol. 41–43; report on the Medical Cinematographic Institute by von Rothe, May 1927, Barch, R/109/I 5759.
69. Schmidt, *Medical Films*, 93.
70. Note by Gauger, 21 November 1935, Barch, R/169/15, fol. 34.
71. Agreement RWU and DEGEWI, 3 January 1936, Barch, R/169/15, fol. 16–17.
72. Note by Schwarz, 24 September 1937, Barch R/169/15, fol. 10.
73. Annotated DEGEWI Film List, Barch, R/169/15, fol. 105–6, 241–47, 354–60.
74. For a comprehensive list of the RWU B- and C-series films, including number of sold copies, see Tolle, *RWU*, 124–36.
75. Tolle, *RWU*, 119.
76. Mommsen-Spiess to Schwarz, 22 July 1934, Barch, R/169/15, fol. 3–4.
77. List of new productions using acquired material, Barch, R/169/15, fol. 354.
78. RWU budget reports 1936–1941, Barch, R/169/11.
79. Wolf, Gotthard, NSDAP parteiliche Erhebung 1939, Barch, R/9361/I 3991.
80. Schmidt, *Medical Films*, 280.
81. Schmidt, *Medical Films*, 127.
82. Note by Ernst Telschow, 1 February 1939, AMPG, 2. Abt., Rep. 1A,1043.
83. Gauger to Reich Office for Economic Expansion, 29 November 1939, AMPG, 2. Abt., Rep. 1A, 1043.
84. Telschow (KWS) to Wolf (RfdU), 23 February 1940, AMPG, 2. Abt., Rep. 1A, 1043.
85. Sattelmacher, "Shuffled Zeppelin Clips."
86. Also see the chapter by Mario Schulze in this volume, analyzing the visualization of flow research films.
87. List of IWF and MPG coproductions 1932/33–1959, 118–23, AMPG, II. Abt., Rep. 70 Nr. 1016.
88. Schmuhl, "Hirnforschung und Krankenmord," 591.
89. Report, "Der Wehrwichtige Forschungsfilm," 1939, AMPG, 2. Abt., Rep. 1A, 1043.
90. Minutes of advisory board meeting, 25 February 1943 in Berlin, AMPG, 1. Abt., Rep. 1A, 1044.
91. Work report, Technical Research Film Department, 1942, Barch, R/169/9; Reichsanstalt, Übersicht, 1942, 6–11.
92. RWU budgets 1940, Barch, R/169/11.
93. Work report, Technical Research Film Department, 1942, Barch, R/169/9.
94. List of business trips TF Unit, R/169/9.
95. Work report, Technical Research Film Department, from 1 April to 30 September 1942, Barch, R/169/9.
96. Work reports, TF Unit, 1942, Barch, R/169/9.
97. Tolle, *RWU*, 126 and 129.
98. Reichsanstalt, *Übersicht*, 1942, 11.
99. Employees included: cinematographers Karl Koch and Schmid; assistant cinematographers Phillip and Grimm; former head of department "Schriftum" Willi Mohaupt; scientists Schicha, Mende, Mense, Joachim Rieck, and Schweitzer; and several technicians. Work Report TF Unit, Barch, R/169/9; Personnel list, TF Unit, Barch, R/169/6; List of business trips TF Unit, R/169/9.

100. Interim report department I, 1944, Barch, R/169/1.
101. Schneider Erich, über die Bedeutung der Forschungsfilmarbeit der Reichsanstalt für die wehrtechnische Entwicklung, 26 June 1944, printed in Barch, R/169/1.
102. Interim report department I, 1944, Barch, R/169/1.
103. These assets were worth 2.1 million DM after the 1948 currency reform. Tolle, RWU liquidation, 10 October 1961, Barch, R/169/18.
104. In reference to his interviews with Wolf: Ewert, *Die Reichsanstalt*, 109.
105. Waltenspül and Schulze, *Follow the Films*, 240.
106. Zabel, "Die Lehrmaschine."
107. Große, *Filme*, 6–7.
108. Entnazifizierungs-Hauptausschuss Hannover, 11.10.1948, Barch, R/169/16.
109. Schmidt, *Medical Films*, 129–34.
110. Letter from Kreuz to Bundesministerium für Wirtschaft 22.7.1953, Barch, R/169/16; Tolle, presentation on the liquidation status of RWU, 3 July 1958, Barch, R/169/20.
111. Schmidt, *Medical Films*, 283.
112. Schmidt, *Medical Films*, 276.
113. For the full report, see appendices in Ewert, *Geschichte*, 315–79.
114. List of *RWU/FWU* educational films in Kühn, *Unterrichtsfilm*, 267–308.
115. Schmidt, *Medical Films*, 282.
116. Große, *Filme*, 7–9.
117. Sachse, "'Whitewash Culture.'"
118. Wolf, "Die Aufgaben."
119. Note by Wolf 1948, AMPG, 2. Abt., Rep. 102, Nr. 209, fol. 95; note from Telschow, Unterbringungsmöglichkeiten für das Institut für Film und Bild, AMPG, 2. Abt., Rep. 102, Nr. 207, fol. 452.
120. List of IWF and MPG coproductions 1932/33–1959, 118–23, AMPG, II. Abt., Rep. 70 Nr. 1016.
121. 3. Sitzung des Wissenschaftlichen Rates der Max-Planck-Gesellschaft, 4 October 1950, 10, AMPG, 2. Abt. Rep. 62, 3.WR.
122. Telschow to Research Control Branch, Economics Division, Carl H. Nordstrom, 23 April 1948, AMPG, 2. Abt., Rep. 102, Nr. 207, fol. 472.
123. Rieck to Purchase 24 September 1947, AMPG, 2. Abt., Rep. 102, Nr. 207, fol. 481.
124. Große, *Filme*, 11. Born in 1909, Mohaupt studied chemistry and geology and was head of the Schrifttum Department as of 1937. That year, he also joined the NSDAP. He started working with the TF Unit in 1943. Kühn, *Unterrichtsfilm*, 262.
125. Letter from Wolf to Telschow 2 October 1945 and 19 January 1946, AMPG, 2. Abt., Rep. 102, Nr. 207, fol. 500–502.
126. See Otto Hahn's speech at the inauguration of the IWF building in 1961, AMPG, Abt. III, Rep. 14 A, 6242 and Film list KWS/MPS and RWU/IWF 1932–1960, AMPG, 2. Abt., Rep. 70 Nr. 1016, fol. 118–22.
127. Protocol IWF extended advisory board meeting, 16–17 May 1955 in Bochum, AMPG, 2. Abt., Rep. 102, Nr. 207, fol. 43–53.
128. Wolf, *Der wissenschaftliche Film*, 6–13.
129. Wolf, "Die Aufgaben," 1–2.
130. Hackbarth and Günter, "Foreword."
131. For the institutional history of East Germany's science and educational program, see Hortzschansky, *Das deutsche Zentralinstitut*.

132. Also see the chapters, in this volume, by Josephine Diecke and by Sophia Gräfe and Kerrin von Engelhardt for more about the role of science films in the GDR.
133. Also see Mario Schulze's chapter in this volume, about Cold War recycling of research film footage.
134. Wolf to Telschow, 22 September 1955, AMPG, 2. Abt., Rep. 102, Nr. 207, fol. 4.
135. Wolf to Telschow, 22 September 1955, AMPG, 2. Abt., Rep. 102, Nr. 207, fol. 4–10. (Also see: list of soviet science films, GDR medical film index, *IWF* advisory board memorandum)
136. Report on *ISFA* congress in Göttingen, 8–9 April 1954, AMPG, 2. Abt., Rep. 1a, 187–96; Internationale Tagung des Films, Göttinger Presse, 7 April 1954, AMPG, 2. Abt., Rep. 102, Nr. 207, fol. 201.
137. Report on Wolf's business trip to Rome, 4 November–1 December 1954, AMPG, 2. Abt., Rep. 102, Nr. 207, fol. 117–21.
138. Tosi, *Cinematography*, 28.
139. Walther Gerlach to Otto Hahn, 5.7.1954, AMPG, 2. Abt., Rep. 102 Nr. 207, fol. 133.
140. Note by MPS general administration (Pfuhl), 24.11.1955, AMPG, 2. Abt., Rep. 102, Nr. 209, fol. 62.
141. Tolle, RWU liquidation, 10 October 1961, Barch, R/169/18.
142. Also see the chapter by Sophia Gräfe and Kerrin von Engelhardt in this volume.

Bibliography

Bock, Gisela. *Zwangssterilisation im Nationalsozialismus: Studien zur Rassenpolitik und Frauenpolitik*. Opladen: Westdeutscher Verlag, 1986.

Budde, Christian. *Kurt Gauger (1899–1957): Psychotherapeut, Filmschaffender, Literat*. Vol. 5. Schriften zur Medizingeschichte. Hamburg: Verlag Dr. Kovač, 2014.

Bundesarchiv-Filmarchiv (BArch-FArch). "Filme der Reichsanstalt für Film und Bild in Wissenschaft und Unterricht (RfdU/RWU) Bestandsübersicht Filme." [Overview of the holdings of the RdfU/RWU.] Berlin, 2005. Retrieved 10 August 2023 from https://www.bundesarchiv.de/DE/Content/Downloads/Finden/Film/rwu-filme.pdf?__blob=publicationFile.

Curtis, Scott. *The Shape of Spectatorship: Art, Science, and Early Cinema in Germany*. New York: Columbia University Press, 2015.

Deutscher Bundestag. *Opfer von NS-"Euthanasie" und Zwangssterilisation als Verfolgte des Nationalsozialismus anerkennen—Aufarbeitung vorantreiben*. Berlin: 2022. Retrieved 10 August 2023 from https://dserver.bundestag.de/btd/20/024/2002429.pdf.

Ewert, Malte. *Die Reichsanstalt für Film und Bild in Wissenschaft und Unterricht (1934–1945)*. Vol. 15. Schriften zur Kulturwissenschaft. Hamburg: Kovač, 1998.

———. *Neue Dokumente zur Geschichte der Schulfilmbewegung in Deutschland: Berlin und Britische Besatzungszone 1945/46*. Schriftenreihe Schriften zur Kulturwissenschaft 19. Hamburg: Kovač, 1998.

Friedland, Alexander. "'. . . doch erscheint in seiner Denkschrift die Bedeutung des klinischen Films für den Unterricht allzustark betont': Zur Geschichte des Medizinisch-kinematographischen Instituts der Charité 1923–1931." *Medizinhistorisches Journal* 52(2/3) (2017): 33.

Gaycken, Oliver. *Devices of Curiosity: Early Cinema and Popular Science*. Oxford/New York: Oxford University Press, 2015.

"Gesetz zur Verhütung erbkranken Nachwuchses: Vom 14. Juli 1933." *Reichsgesetzblatt* 1(88) (1933): 529–31.

Große, Werner. *Filme für die Wissenschaft: Die Epoche des wissenschaftlichen Films in Göttingen*. Göttingen: Buchverlag Göttinger Tageblatt, 2012.

Hackbarth, Doris, and Günter Schulz. "Foreword." In *DEFA 1946–1964: Studio für populärwissenschaftliche Filme (und Vorläufer)*. Berlin: Henschel, 1997.

Harten, Hans-Christian, Uwe Neirich, and Matthias Schwerendt. *Rassenhygiene als Erziehungsideologie des Dritten Reichs: Bio-bibliographisches Handbuch*. Berlin: Akademie Verlag, 2006.

Hay, Alfred. *Photographisches Praktikum für Mediziner und Naturwissenschaftler*. Berlin: Springer, 1930.

Hoffmann, Kay, and Peter Zimmermann. "Von der Gleichschaltungs- und Proaganda-These zur differenzierten Erforschung dokumentarischer Genres." In *Geschichte des dokumentarischen Films in Deutschland. Band 3 "Drittes Reich" 1933–1945*, edited by Kay Hoffmann and Peter Zimmerman, 45–56. Stuttgart: Reclam, 2005.

Hortzschansky, Werner. *Das deutsche Zentralinstitut für Lehrmittel, seine Aufgaben, Organisation und Arbeitsweise*. Schriftenreihe des Deutschen Zentralinstituts für Lehrmittel 11. Berlin: Volk und Wissen, 1957.

Kühn, Michael. *Unterrichtsfilm im Nationalsozialismus: Die Arbeit der Reichstelle für den Unterrichtsfilm/Reichsanstalt für Film und Bild in Wissenschaft und Unterricht*. Memmendorf: septem artes, 1998.

Laukötter, Anja. *Sex—richtig! Körperpolitik und Gefühlserziehung im Kino des 20. Jahrhunderts*. Göttingen: Wallstein Verlag, 2021.

Nagel, Anne C. *Hitlers Bildungsreformer: Das Reichsministerium für Wissenschaft, Erziehung und Volksbildung 1934–1945*. Frankfurt am Main: S. Fischer, 2012.

Offermann, Stefan. "Die biopolitische Produktion 'lebensunwerter' Subjekte im Rahmen der 'Aktion T4'—Eine Re-Lektüre von 'Ich klage an.'" *Sudhoffs Archiv* 97(1) (2013): 25.

Reichsanstalt für Film und Bild in Wissenschaft und Unterricht. Übersicht Über Die Arbeit Der Reichsanstalt Für Film Und Bild in Wissenschaft Und Unterricht. Stand Vom 1. April 1942. Berlin, 1942.

———. Übersicht Über Die Arbeit Der Reichsanstalt Für Film Und Bild in Wissenschaft Und Unterricht. *Stand Vom 1. Januar 1944*. Berlin, 1944.

Roos, Julia. "The Race to Forget? Bi-racial Descendants of the First Rhineland Occupation in 1950s West German Debates about the Children of African American GIs." *German History* 37(4) (2019): 517–39. https://doi.org/10.1093/gerhis/ghz081.

Sachse, Carola. "'Whitewash Culture': How the Kaiser Wilhelm/Max Planck Society Dealt with the Nazi Past." In *The Kaiser Wilhelm Society under National Socialism*, edited by Susanne Heim, Carola Sachse, and Mark Walker, 373–99. Cambridge: Cambridge University Press, 2009.

———. "Grundlagenforschung: Zur Historisierung eines wissenschaftspolitischen Ordnungsprinzips am Beispiel der Max-Planck-Gesellschaft (1945–1970)." In *"Dem Anwenden muss das Erkennen vorausgehen": Auf dem Weg zu einer Geschichte der Kaiser-Wilhelm-/Max-Planck-Gesellschaft*, edited by Dieter Hoffmann, Birgit Kolboske, and Jürgen Renn, 215–39. Max Planck Research Library für the History and Development of Knowledge Proceedings 6. Berlin: epubli, 2014.

Sattelmacher, Anja. “Shuffled Zeppelin Clips: The Flight and Crash of LZ 129 Hindenburg in the Archives.” *Isis* 112(2) (2021): 352–60.
Sattelmacher, Anja, Mario Schulze, und Sarine Waltenspül. “Introduction: Reusing Research Film and the Institute for Scientific Film.” *Isis* 112(2) (2021): 291–98. https://doi.org/10.1086/714823.
Schmidt, Ulf. “Der medizinische Forschungsfilm im ‘Dritten Reich’: Institutionalisierung, politische Funktion und ethische Dimension.” *Zeitgeschichte* 28(4) (2001): 200–213.
———. *Medical Films, Ethics and Euthanasia: The History of the Reich Office for Educational Films / Reich Institute for Films in Science and Education*. Husum: Matthiesen Verlag, 2002.
Schmuhl, Hans-Walter. “Hirnforschung und Krankenmord. Das Kaiser-Wilhelm-Institut für Hirnforschung 1937–1945.” *Vierteljahrshefte für Zeitgeschichte* 50(4) (2002): 559–609.
Schulze, Mario, and Sarine Waltenspül. “From Images of Lines to Images of Particles: The Role of the Film Camera in Flow Visualization.” In *Image Evolution: Technological Transformations of Visual Media Culture*, edited by Lars C. Grabbe, Patrick Rupert-Kruse, and Norbert M. Schmitz, 161–86. Marburg: Büchner, 2019.
Tolle, Wolfgang. *RWU: Reichsanstalt Für Film Und Bild in Wissenschaft Und Unterricht*. Berlin, 1961.
Tosi, Virgilio. *Cinematography and Scientific Research*. Paris: UNESCO, 1977.
von Keitz, Ursula. “Die Kinematographie in der Schule: Zur politischen Pädagogik des Unterrichtsfilms von RfdU und RWU.” In *Geschichte des dokumentarischen Films in Deutschland. 3 “Drittes Reich” 1933–1945*, edited by Kay Hoffmann and Peter Zimmermann, 463–88. Stuttgart: Reclam, 2005.
von Schwerin, Alexander. *Experimentalisierung des Menschen: Der Genetiker Hans Nachtsheim und die vergleichende Erbpathologie, 1920–1945*. Geschichte der Kaiser-Wilhelm-Gesellschaft im Nationalsozialismus, Bd. 10. Göttingen: Wallstein, 2004.
Waltenspül, Sarine. “Reusable and Nonreusable Films: From Ballistic Films to the Encyclopaedia Cinematographica.” In *Isis* 112(2) (2021): 342–51.
Waltenspül, Sarine, und Mario Schulze. “Follow the Films: Reuses of a Research Film; Biography, Recycling, Whitewashing, Appropriation and Palimpsesting.” In *Images on the Move: Materiality—Networks—Formats*, edited by Olga Moskatova, 231–58. Bielefeld: Transcript Verlag, 2021.
Wagner, Georg August. “Die Technik der Unfruchtbarmachung.” In *Die psychiatrischen Aufgaben bei der Ausführung des Gesetzes zur Verhütung erbkranken Nachwuchses: Mit einem Anhang Die Technik der Unfruchtbarmachung; Klinische Vorträge im erbbiologischen Kurs*, edited by Karl Bonhoeffer. Berlin: Karger, 1934.
Weindling, Paul. “Genetik und Menschenversuche in Deutschland 1940–1950: Hans Nachtsheim, die Kaninchen von Dahlem und die Kinder vom Bullenhuser Damm.” In *Rassenforschung an Kaiser-Wilhelm-Instituten vor und nach 1933*, edited by Hans-Walter Schmuhl, 245–74. Geschichte der Kaiser-Wilhelm-Gesellschaft im Nationalsozialismus 4. Göttingen: Wallstein, 2003.
Wolf, Gotthard. “Der wissenschaftliche Film.” Presented at the Jahrestagung der Hochschulfilmreferenten, Göttingen, 27 October 1949.
———. “Das Institut für den Wissenschaftlichen Film: 25 Jahre wissenschaftliche Filmarbeit.” In *Zwei Jahre Jugendvertretung: Festschrift zur Einweihung des Neubaues* für das Institut für den wissenschaftlichen Film, edited by Institut für den wissenschaftlichen Film, 5–40. Göttingen: Institut für den wissenschaftlichen Film, 1961.

——. *Der Wissenschaftliche Dokumentationsfilm und die Encyclopaedia Cinematographica*. Heidelberg: Springer-Verlag, 1967.
——. *Der wissenschaftliche Film in der Bundesrepublik Deutschland*. Born-Bad Godesberg: Inter Nationes, 1975.
——. "Die Aufgaben Des Instituts Für Wissenschaftlichen Film." *Mitteilungen Des Instituts Für Den Wissenschaftlichem Film* (1) (1956): 1–3.
Zabel, Nicole. "Zur Geschichte des Deutschen Pädagogischen Zentralinstituts der DDR: Eine institutionsgeschichtliche Studie." Chemnitz, 2009.
——. "Die Lehrmaschine und der Programmierte Unterricht—Chancen und Grenzen im Bildungswesen der DDR in den 1960er und 1970er Jahren." *Jahrbuch für Historische Bildungsforschung*, no. Schwerpunkt Maschinen (2014): 123–52.
Zeidman, Lawrence A. *Brain Science under the Swastika: Ethical Violations, Resistance, and Victimization of Neuroscientists in Nazi Europe*. Oxford: Oxford University Press, 2020.
Zimmermann, Peter. "'Auferstanden aus Ruinen': Zur Wirkungsgeschichte von Kulturfilm und Wochenschau nach 1945." ["Revived from ruins": Toward a history of culture films and weekly newsreels after 1945.] In *Geschichte des dokumentarischen Films in Deutschland. 3 "Drittes Reich" 1933–1945*, edited by Kay Hoffmann and Peter Zimmermann, 691–711. Stuttgart: Reclam, 2005.

CHAPTER 2

Flow Vis for the Space Race

German Science Films in US Education during the Cold War

Mario Schulze

Introduction

The National Committee for Fluid Dynamics Film (NCFMF) was an association of ten—all white male—professors from various engineering departments in the United States, under the leadership of Ascher H. Shapiro, a professor of mechanical engineering at MIT.[1] Supported by the Ford Foundation and the National Science Foundation (NSF), it produced one of the most extensive educational film programs in US history.[2] By 1969, the NCFMF had released twenty-two 16 mm sound films, as well as one hundred thirty-three Super 8 cassettes of silent films ranging from two to four minutes in length.

The NCFMF films are prime examples of the countless science education films that reached US classrooms and lecture halls in the late 1950s and 1960s. Many of these films were deeply enmeshed in the political and scientific fault lines of the Cold War. These films promulgated the idea of science as an integral and nonideological endeavor independent of political influence. At the same time, they were distributed globally and instrumentalized politically in a cultural and psychological war between the West and the Eastern Bloc. Such entanglements, however, are not easy to see in the films. Traditional film analyses such as those developed for narrative films, which often view the film as a text, are of little help here. Instead, methods are needed that look at films in their institutional contexts, in their alliances with other media, and in the ways in which they are used and handled. Scholars within film history occupied with early film and nonfiction films—including educational films, science films, or industry films—have developed such methods and put distribution, circulation, and the concrete practices of screening and reception at the center of their analysis.[3] In the first half of this chapter, I apply these methods to a corpus of films so far unexplored in scholarly literature; the literature has barely touched upon educational films about physics and fluid dynamics, with no mention of the flow visualization films or, in short, flow vis films produced by the NCFMF at all.[4]

In the second part of the chapter, I attempt to extend these methodologies toward a history of science by proposing an approach to the NCFMF films that centers on footage that originated in laboratory contexts and served research purposes. For example, the NCFMF film *Vorticity* (1961, dir. Jack Churchill, USA) twice reuses footage that was produced in neither the Watertown film studio where the film was made nor even in the United States. These clips instead originate from a pre–Cold War 1920s film by German physicist and engineer Ludwig Prandtl, who is often credited as one of the founders of modern aerodynamics, and by his assistant Oskar Tietjens. Just under three minutes of the forty-four-minute film in fact consist of footage from Prandtl's research film, *Production of Vortices by Bodies Travelling in Water* (1927, dirs. Oskar Tietjens, Ludwig Prandtl, Germany).

Why do these filmic reuses merit attention when they, at first sight, seem to be so negligible? A visual citation of film in film is nothing special. Rather, as Claudia Blümlinger writes, the "use of found footage is in a sense inscribed in the medium through montage and reproducibility."[5] But while a wealth of examples of the reuse of historical film material in new works can be found in film and art history,[6] the historical literature on educational films includes almost no specific example of such found footage appropriation, let alone reuses of recordings that originally served as laboratory material for research.[7] In the history of art films, appropriations are often analyzed as an expression of visual memory that creates a relationship between remembering and forgetting, integrating and excluding, reactivating and discarding. Applying this focus on appropriation to educational films in the sciences promises insights into the modalities of memory in science, the history of functionalizing laboratory material in society and education, and the technical, political, and social circumstances of image circulation in general.

While the analysis of circulation, instrumentalization, and distribution of films presented in the first half of this chapter is somewhat detached from the actual content and narration of the films, the focus on found footage in Cold War educational films establishes a connection between production, narration, and use of the films. These reuses of film footage can therefore be informative sources, both for a historical epistemology that sets out to trace the historicity, materiality, and mediality of our knowledge, and for a film and media history interested in the technical, political, and epistemic uses of particular technologies.[8] In the process, the categories of research film and educational film become more tangible and distinct from each other when the focus is placed on the reuse of footage in educational films that was originally created in a laboratory for research purposes.

US Science Education during the Cold War

The NCFMF films were made possible by a plethora of new educational programs and funding in the United States.[9] Especially after the successful Soviet launch of *Sputnik* in 1957, the US government pumped hundreds of millions of dollars into revising school and college curricula, mainly through the National Defense Education Act. In particular, the relatively young NSF launched a series of curriculum programs in physics, biology, chemistry, mathematics, and many other subjects. MIT's Physical Science Studies Committee (PSSC), which produced new educational material in physics for high school students, served as a pioneer and model in this endeavor.[10] The PSSC not only produced Ascher Shapiro's first film but also served as a key model for the educational film program that Shapiro later established, the NCFMF. In 1960, Shapiro started to work on an NCFMF course of study, which followed the PSSC's organizational structure, financing, and orientation, but was mainly geared toward college and university students. In the process, Educational Services Incorporated (ESI) was installed as the administrative umbrella of the PSSC and NCFMF and also became responsible for a variety of other curricula initiatives.[11] The NCFMF relied on ESI's human and physical resources, most notably the film directors and film studio in Watertown, Massachusetts, near the MIT campus.

Within the framework of these programs, scientists from elite North American universities, rather than educators, developed new textbooks, sourcebooks, classroom equipment, and films. In cooperation with a wide range of government agencies, including the intelligence services, these new teaching materials were distributed, first nationally, and then internationally. The new science education programs, promoted by the US government in the face of the *Sputnik* crisis, were primarily designed for recruiting scientists that could work in military research programs. The military and US industry had been complaining about an increasing shortage of scientifically trained personnel for some time. Furthermore, the military success of the almost mythically charged, large-scale US research initiatives of World War II—like the Manhattan Project's development of nuclear weapons—had shown how closely scientific research was linked to possible military superiority.[12] It thus seemed that the domestic recruitment of new scientists would be strategically decisive in the Cold War, and educational programs were one of the many means to ensure this recruitment. The curricula were therefore intended to motivate more students to study science and deepen their knowledge of topics that were crucial to contemporary US military research.

On another level, the new US curricula were also part of a cultural and psychological battle between East and West over the definition of the role of research and science in society. The Communist Party in the Soviet Union was pivoting away from "bourgeois science" in favor of a science that cleaved to the philosophy of dialectical materialism and would lead to immediately useful results; the US government, in contrast, promulgated the idea of science as an allegedly apolitical, nonideological enterprise. Framing science in general, and physics in particular, as apolitical was a highly paradoxical strategy, however, which the contested distinction between applied research and basic research is crucial to understanding. As the world wars had shown, applied research was at the core of the technical developments that helped to colonize the air and waters and could be decisive in any upcoming war. However, when understood as a tool to tackle our fundamental understanding of natural and other phenomena, basic research, in general, and physics, in particular, also represented rationality, freedom from ideology, and the freedom of science. This seemingly apolitical science thus simultaneously served as a symbolic weapon in countering totalitarian regimes and leading the world to democracy.[13] In turn, the new science education programs were welcome to paint a specifically Western picture of a rational and nonideological science and, at the same time, to de-emphasize its relevance for modern warfare. Therefore, the task for US science education in the Cold War was multidimensional—or, in other words, ambivalent: recruiting scientists for war-related research and representing physics as an apolitical discipline were supposed to fit together as one, noncontradictory whole.

The productions of the PSSC exemplify how recruitment and the political goal of presenting science as apolitical could come together in the films produced for US science education. The PSSC physics course was launched in 1960 and quickly became the most widely used program for teaching physics in the world.[14] In addition to textbooks, a manual for experiments, specially designed laboratory apparatus for schools, a sourcebook for teachers, and a series of over fifty small, in-depth paperbacks, the PSSC also produced over fifty films.

As a local initiative of white, male physicists who aspired to reform physics education, the personal goals of the PSSC scientists were not necessarily congruent with the comprehensive plans of the US government, however.[15] Founded in 1956 by a group of MIT professors, led by Jerrold Zacharias, the PSSC physicists wanted to present physics as a "method of thinking."[16] The new course was intended to convey an image of scientific work that corresponded to what Zacharias and his colleagues saw as the essence of their discipline: empirical-experimental observation, methodologically guided conclusions, and a thought process based on ingenuity and initiative.[17] This approach stood in sharp contrast to the "life adjustment" curricula, which had dominated school pedagogy in the United States from 1945 to the mid-1950s and which were designed to introduce useful facts that students could relate to their everyday lives. By emphasizing the methods and ideas of basic research, the PSSC physicists also hoped to counteract anti-intellectual and racist resentment in the United States, as well as McCarthyist anti-communism.[18] In the climate of irrational attacks against science in the 1950s, a new science pedagogy seemed to present a suitable means to regain trust in science and scientific research. Finally, the strategy to focus on basic research must also be seen as a response to the role played by the sciences in World War II, for example the Manhattan Project and the Nazi rocket program. In contrast, fundamental research and basic didactics were to stand for a science that cannot be corrupted and cannot be roped into politics.

Educational films were regarded as an ideal means of demonstrating to high school and college audiences how basic science supports and justifies its scientific claims in practice. The first annual report of the PSSC emphasized their central importance to the overall project, stating: "The films are expected to set the tone and level for the entire course."[19] In their films, Zacharias and his colleagues often staged experiments and subsequently drew clear and sometimes surprising logical lines between the empirical data obtained from the experiments and physical theories. For example, Shapiro begins his PSSC movie on the *Fluid Dynamics of Drag* (1960, dir. Quentin Brown, USA) with Galileo's famous experiment testing whether a ball or a feather falls faster. Repeating the same experiment in a vacuum in which they both indeed fall with the same velocity, Shapiro concludes that the invisible and impalpable air exerts the resistance, or a drag, that is the central subject of the film. Moreover, all PSSC films were intended to show "real" scientists tackling equally "real" problems. The aim was to counteract popular stereotypes of scientists—be it the mad scientist that dominated public perception after World War II, as epitomized by Peter Sellers in *Dr. Strangelove* (1964, dir. Stanley Kubrick, USA), or the socially inept scientific immigrant or ivory tower dweller with a disembodied mind and possibly communist ideas. As one of the directors recalls, the rule was "no beards, no accents."[20] Instead, the scientist was to be depicted in the films as a normal, active, and occasionally fallible American poster boy, who liked to watch a science-fiction movie or, even better, a western in his spare time.[21]

The filmic means employed by the PSSC and NCFMF responded to media developments of the 1950s, as television rose to become the number one mass medium in the United States. In 1965, 93 percent of American households owned a television set that played an average of 5.5 hours per day.[22]

The new medium was aesthetically and technically inscribed into the films. The introductory words of the anchorman, the deliberately chosen one-studio atmosphere, in which most experiments were conducted directly in front of the camera(s), and the small number of interludes, anchored to the studio by voice-over, are all evocative of a television setting. The PSSC and NCFMF films seem now like a mixture of news program, morning show, and television documentary. Furthermore, the new format of educational television was apparently an inspiration for these educational movies: in the United States, educational television programs were broadcasted from 1948 on.[23]

Born in 1916, Ascher Shapiro fit the image of the "American poster boy" perfectly. He grew up in humble circumstances in Brooklyn and devoted his free time to sports and playing the flute. Before attending MIT, he was on the boxing team at the City College of New York. In his obituary, written by his colleague Ronald Probstein, this sport is credited for the course of his stellar university career: "Ascher's training as a fighter was ideal for someone who chose to spend his life at M.I.T."[24] From 1935 until his retirement, Shapiro studied and conducted research at MIT with few interruptions. Like many others in the PSSC, Shapiro was part of a generation of engineers for whom military research was the most obvious field of activity. During World War II, he headed a navy laboratory that developed torpedo engines. After the war, he worked for NACA, the predecessor of NASA, and the Atomic Energy Commission. In 1948, he was part of a study group at MIT that became known as the Lexington Project, which advanced the development of atomic-powered aircraft.[25]

In spite of their wartime careers in military and industrial research, the PSSC physicists saw themselves as independent scientists, and their personal agendas did not necessarily coincide with that of the US government. Nonetheless, the courses they developed in the 1950s and 1960s were still compatible with governmental goals on two levels. First, the PSSC lecture course program radically deviated from traditional US physics curricula. The physics of refrigerators and internal combustion engines was replaced by the fundamental physics of waves, atomic energy levels, and particle movements. This focus on more sophisticated and theoretical research brought into focus the knowledge that formed the basis for contemporary large-scale military-industrial research in the United States. Second, the course insisted on a US conception of science, based on investigation, experimentation, and consistency, that could be contrasted with the science of the Soviet Union, cast as state-controlled puppetry. Intended to foreground the independence and integrity of US science, the PSSC lecture course implicitly distinguished itself from the conception of science and its teaching east of the Iron Curtain, which was considered firmly embedded in Soviet propaganda. There is a certain contradiction in pretending to be apolitical while simultaneously espousing freedom of research as a political goal, however. While the course developed by the PSSC physicists may have been an act of resistance to the anti-science sentiments of the McCarthy era, it was also consistent with the government's agenda to use scientific freedom as a cultural trope in the Cold War.

The Politics of Education and Film

Functionalizing film as a political instrument has a long tradition in the United States and elsewhere. Since the 1930s and by the time the 16 mm format was fully established, some countries already had robust networks for distributing instructional films that could serve governmental goals.[26] Eventually, World War II made mass media instruction a shared social experience, wherein films served to educate, train, persuade, sell, inspire, or indoctrinate.

In the immediate postwar period, numerous organizations across various sectors of society—from UNESCO to the American College of Surgeons—entered the film business, leading to an even more rapid growth in audiovisual educational offerings.[27] In the United States, the 1950s through the 1970s marked the heyday of educational film production, with more than ten thousand new films produced each year.[28] The spectrum of productions ranged in content from live recordings of medical operations to instructions on how to behave during atomic bomb attacks to pedagogical films about educational film production. Targeted audiences ranged from preschoolers to PhD students around the world. These postwar films were intended to serve the benevolent purposes of education; at the same time, many of them used and modified propaganda techniques that had been employed in the films produced during the war. The implicit strategy of the US government in the 1950s and 1960s was to support educational initiatives that directly or indirectly propagated a vision of society and science that could be functionalized for governmental purposes. That is to say, the politico-military goals of using film as propaganda in wartime did not simply disappear after the war; instead, education policy became a means to increase governmental influence.

Following in the tradition, dating back to the 1920s, of using educational film for political purposes, the PSSC and NCFMF can thus be seen as organizations that functionalized film to form recruitable subjects domestically and, on an international scale, used film to promote a certain form of nonideological Western science. In 1956, the PSSC received its first generous grant from the NSF in the amount of $300,000. ESI, the administrative umbrella of the PSSC and the NCFMF, subsequently pursued a policy of national and international expansion.[29] The priority was national curriculum development, but ESI also worked with numerous foreign institutions that expressed interest in the lecture courses and wanted to adapt them for their countries with the help of private or semiprivate US international relations agencies, such as the

Ford Foundation or the Asia Foundation. Both these foundations were openly subject to US policy and were not only funded in part by CIA money, but they were also headed by directors closely associated with the US intelligence agencies.[30] PSSC and NCFMF projects illustrate the intimate relationship between national agenda, state power, and globalization during the Cold War era. Not only was scientific and technological knowledge a key national asset in the competition between the two blocs, but providing this knowledge in teaching programs, in turn, also legitimized the global influence claimed by the United States.

Estimates from the late 1960s claimed that at least half a million people had watched the NCFMF movies in the United States alone. All major US universities and colleges—as well as industrial companies from General Motors to Whirlpool, military research facilities, and civil institutions—had bought or rented the films.[31] In other words, the films became the blockbusters of US engineering studies. The Super 8 film cassettes that the NCFMF distributed from the mid-1960s onward further increased the dissemination of the visual materials and expanded the number of recipients. Mounted on cassettes and shown on special projectors, they eliminated the time-consuming and material-invasive fumbling with film reels on hard-to-reach gears; these so-called "Magi cartridges," each containing fifty feet of Kodachrome Super 8 film, could simply be inserted and started with a single push of a button. These cassettes were especially useful for small groups and individual instruction. They could be checked out like books from numerous libraries, along with a projector, or could be viewed right at the library.[32] The NCFMF distributed the cassettes in close cooperation with the Technicolor film production company, which in turn was aiming to develop new markets, since educational films were a way of advertising the private use of the new 8 mm technology.

Along with other newly developed curricula during the 1960s, Shapiro's films were distributed worldwide. Sales are documented in at least twenty-two countries.[33] In other words, these educational films appeared in numerous classrooms, lecture halls, and boardrooms, shaping the patterns of thought and vision of an upcoming generation of scientists and engineers while implicitly promoting a politically charged vision of science in the Cold War conflict.

Iconic Excerpts: NCFMF Use of the Göttingen Recordings

From 1925 until around 1932, Ludwig Prandtl and his assistant Oskar Tietjens had filmed images of alluminium particles moving in water channels around objects (like cylinders, airfoils, or so-called bluff bodies) in their hydrodynamic laboratory at the Kaiser Wilhelm Institute for Fluid Mechanics Research (KWI) in Göttingen. In 1927, Prandtl compiled a collection of these lab recordings for the prestigious Wilbur Wright Memorial Lecture at the Royal Aeronautical Society in London. This compilation film, *Production of Vortices of Bodies Travelling in Water*, features mesmerizing images of white dots and stripes moving around simple geometric figures, like circles and ellipses. The flow patterns seen in the movement of the white particles visually convey the central ideas of Prandtl's highly influential boundary layer theory.

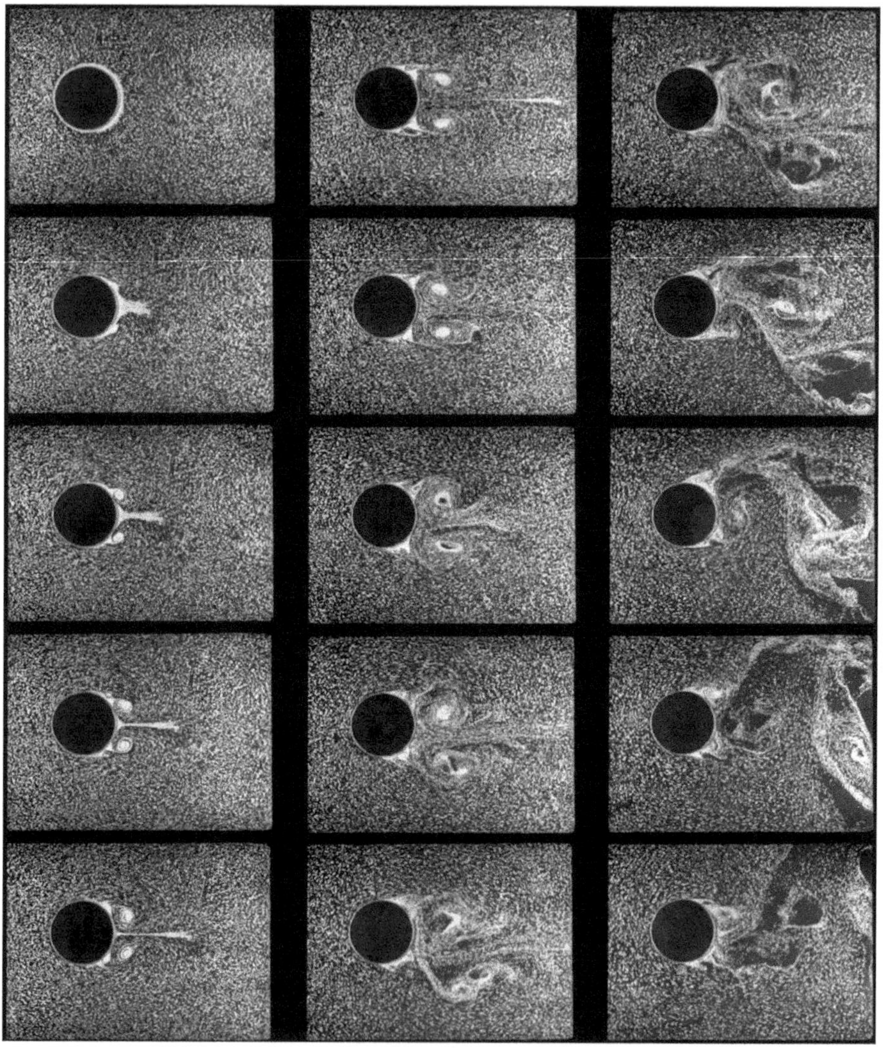

Figure 2.1. Vertically arranged film frames from *Production of Vortices by Bodies Travelling in Water* (1927, dirs. Oskar Tietjens, Ludwig Prandtl, Germany). Screen captures and montage by Mario Schulze. © Archive of the Max Planck Society

Prandtl's film had an eventful history even before excerpts were reused in the NCFMF film *Vorticity*.[34] He showed it at many conferences and took it on a world tour, presenting it at many universities along the way, until it was stolen on the platform of the Detroit train station in 1929. In 1936, the film was published by the Reich Office for Educational Films (RfdU), along with some of Prandtl's other lab recordings, and became one of the most widely used university teaching films in Nazi Germany. After the war, it was republished by the Institute for Scientific Film (IWF) in Göttingen.[35] But it was only after Prandtl's death in 1953 that his footage found its new application in the educational and scientific landscapes of the United States during the Cold War. The first three NCFMF films—*Vorticity*, *Fluid Dynamics of Drag*, and *Flow Visualization* (1963, dir. Jack Churchill, USA)—all include footage from Prandtl and Tietjens's film. Additionally, seven of the 8 mm cassettes issued by the NCFMF consist exclusively of film recordings from Prandtl's lab.[36]

It remains unknown how Shapiro got the film material from Prandtl's lab. Shapiro and Prandtl had not been in contact, or, at least, no correspondence has been preserved. Nevertheless, it can be deduced from the reused scenes and their frames that Shapiro had 35 mm film material from Prandtl's lab at his disposal, not just a 16 mm rental copy of *Production of Vortices* from the postwar IWF version, with cropped edges. Shapiro may have had material on hand that Prandtl left behind during his 1929 lecture tour through the United States. However, it is more likely that the footage from Göttingen came to MIT via Shapiro's colleague in the Department of Mechanical Engineering, Jacob P. Den Hartog. After emigrating to the United States from Holland in the 1920s, Den Hartog had worked in the Mechanics Department at Westinghouse Electric and Mechanical Company, where he met Oskar Tietjens, who had been jointly responsible for filming *Production of Vortices*. Subsequently, Den Hartog took a sabbatical and went to Göttingen to spend a research year at Prandtl's institute. He also translated the textbook *Applied Hydro- and Aeromechanics*, which Tietjens had published based on Prandtl's lectures, into English.[37] In Göttingen, he made his own films with spinning cylinders in water channels.

Inquiries by MIT students at the Göttingen Institute in the 1930s also show that the film *Production of Vortices* must have been part of the curriculum as early as that time.[38] The film was the exception rather than the rule for US engineering curricula of the 1930s and 1940s, however. Even though educational film production in the United States had been expanding more and more since the 1920s, no major educational film initiative existed in the field of fluid mechanics. This marks the niche that Shapiro and his educational film program intended to fill.[39]

It was highly unusual that Shapiro reused recordings made at another lab. NCFMF (and PSSC) films very rarely used footage shot outside of the Watertown studio. For Shapiro, the clips from Prandtl's film were not just fillers but instead an important source and inspirational model, and he created his own film program in the tradition of Prandtl. In the preface to his book *Shape and Flow*, which he calls "the film in print," Shapiro explicitly writes: "I am indebted . . . to the late Ludwig Prandtl for classic film sequences now thirty years old as well as the inspiration of his style."[40] Shapiro declares a kinship between Prandtl's and his own films that goes beyond the mere use of some sequences. He does not specify which "style" he is referring to—perhaps the immediate clarity of the images, or their sober transtemporal aesthetics. But although he does not explicitly relate to a particular concept of "style" here, he does grant Prandtl's recordings the status of images that introduced a specific way of looking at fluid flow—a way of looking that he tried to copy and promote.[41] He thus constructs a visual genealogy along filmic lines and elevates Prandtl's footage to the status of a "classic" as if it were an artist's work rather than lab footage whose epistemic value was initially rather precarious.

Prandtl was a central reference in the fluid mechanics of the 1950s and 1960s. Publications in respected journals of the time, such as the *Journal of Fluid Mechanics* and *Physics in Fluids*, show that Prandtl's concepts and scientific findings continued to shape and guide discussions in engineering during the 1960s, and not only because of the numerous figures and theorems that bear his name. Many discussions in fluid dynamics research dealt in a narrower sense with the set of questions that Prandtl had raised, such as the onset of a flow instability or fully developed turbulence. Furthermore, specific phenomena that his film visually highlighted are addressed in engineering debates, such as the special case of boundary layer development at rotating cylinders.[42] Prandtl was also omnipresent in the standard textbooks on fluid mechanics of the time, not least due to his own textbook, which was reissued in English by the American Association of Engineering Societies in 1957.[43]

The integration of Prandtl's recordings into Shapiro's films thus represents the consolidation and canonization of certain scientific images and approaches. By the time Shapiro reused the clips in his educational films, the movement of particles in a water channel in a German laboratory of the 1920s, which had once been so open to interpretation by Prandtl and the other scientists in his lab, had become the clearly defined and thoroughly researched facts of fluid movement. The precarity of the production of these fluid flow processes, the uncertainty of their evaluation, and the randomness of their transmission—in short, the historicity of the facts—is lost in the reappropriation. In other words, Shapiro finalized the transformation of Prandtl's film from an interpretable research film for experts to an educational film conveying established information with instructive purpose.

The citation itself brought the film images an iconic status, comparable to other repeatedly reprinted scientific photographs, such as Ernst Mach's projectile photographs or the photographs with which Prandtl first substantiated his boundary layer theory.[44] With their reuse in the NCFMF films and 8 mm cassettes, the moving image from Göttingen became a tradition of depicting fluid processes and now assured the visual conditioning of the flow vis community. Seen in this light, a history of seeing and showing in fluid dynamics shines through in this example of reuse.

Although Shapiro seems to have quoted the flow visualization images from Prandtl's lab as a means to construct a patrilineal line connecting him to Prandtl, he in fact integrates the reused material into his films in a way that obscures the context of their production. Prandtl is never explicitly mentioned in Shapiro's film. Acknowledgment only appears in the credits, listed as "Scenes from 'Production of Vortices' by L. Prandtl," along with visual material from NACA, or its successor NASA, and a poem by meteorologist L. F. Richardson. The appropriation of the foreign footage in *Vorticity* works as follows: first, we see Shapiro using pencil drawings of an airfoil and formulas of the Kelvin theorem to introduce the phenomenon of rotation-free flow. Shortly thereafter, Prandtl's shots of an airfoil in the water channel are shown, accompanied by Shapiro explaining the scientific experiment that we are witnessing via voice-over that refers to the previously introduced Kelvin theorem. Finally, the film pauses, and the frame from Prandtl's film is overlaid with graphics based on the pencil drawings. In this way, the visual integration of

Figure 2.2. Vertically arranged film frames from *Vorticity* (1961, dir. Jack Churchill, USA). © Education Development Center

Prandtl's film recordings succeeds seamlessly, enabling Shapiro to explain the theorem without mentioning Prandtl at all. The next theoretical concept to be introduced, Helmholtz's vortex theorems, is again illustrated with the footage from Prandtl's films, and again narrative and visual integration renders the footage an imperceptible part of Shapiro's educational film.

It seems as though any reference to the production location or context of the images produced by Prandtl at the Kaiser Wilhelm Institute would have introduced an intolerable rupture in Shapiro's educational films. The gap between the hydrodynamic lab named after the last German emperor and the post-McCarthy-era film studio on the US East Coast would have been irreconcilable. Had the lab and its technicians not been offscreen in Prandtl's film, Shapiro could not have reused it. The new curricula of the 1950s and 1960s were supposed to depict science as a nonideological, nonpolitical endeavor reaching for the basic laws of the universe. A forthright homage to a German physicist and his lab, which later conducted pivotal research for the Nazi war of extermination, would have ruined this illusion.[45]

Teaching and Selling Flow: The Pervasive Power of the Flow Film

The history of this reappropriation suggests that Shapiro was guided by a logic of iconicity, canon formation, and patrilineal descent. Prandtl's film was iconic in that it served as a model for the fluid motion image and was therefore essential to any educational program on fluid dynamics. But how can the iconic status of the Göttingen lab shots be reconciled with the complicated politics of recruitment and proclaimed freedom of science in the postwar United States? Is there a connection between the fact that the NCFMF films constituted a comprehensive educational program instrumentalized by the US government and that recordings from a 1920s German lab played a programmatic role in those same films? Can the two central findings of this chapter, thus far, be linked to a common denominator?

Shapiro's own reflections on the task and function of educational films help explain how he thought lab recordings would increase the persuasive and mobilizing power of his finished educational film. Shapiro repeatedly described his films as "demonstration experiment films."[46] The medium of film, he argued, guarantees a special participation in the shown experiment that other media cannot provide. This use of visuals of the experiment would ultimately bridge the gap between abstract theoretical models and concrete phenomena. Such a mediation between abstraction and tangibility was, according to Shapiro, the central challenge of educational reform; after all, Cold War curricular reforms specifically called for teaching more fundamentals, even as these became increasingly abstract.

The clips from Prandtl's film helped Shapiro interweave theoretical modeling and experiments in his films. In *Fluid Dynamics of Drag*, for example, Shapiro introduces students to the abstract boundary layer theory, which is fundamental to fluid dynamics, using footage from *Production of Vortices*. In doing so, he also implicitly imparts some fundamentals of contemporary military research, insofar as basic knowledge about stall on the airfoil was essential for the construction of airplanes and missiles. While the films were generally highly didactic and accurate, Shapiro intended the presentation of experiments to be gripping. Therefore, it was the captivating qualities of the clips from Prandtl's lab that provided a link between science education, armament research, and fluid dynamics. The moving images of flow experiments figuratively enveloped the complex and dangerous realities of national armament research and recruitment in a seductive cloak.

It is not definitively known whether or not Shapiro himself was aware that the films were supposed to mobilize more students for military research and were disseminated as a kind of propaganda for a superior Western science. However, fulfilling such US government goals depended on the success of selling educational films and thus on the appeal of the educational product. Though there is no archival evidence that Shapiro reflected on the political applicability of his films, he worked hard to transform the promise of the scientific process into a consumer product. Political applicability and Shapiro's concern for the commercial exploitation of his products therefore overlapped. Shapiro, always concerned about current sales figures, worried about keeping diverse promotional materials for the NCFMF films up to date. In a letter to the Encyclopedia Britannica Educational Corporation, which took over distribution and administration from the ESI in 1966, Shapiro baldly says, "I would like to see both Encyclopedia Britannica and the MIT Press make some money."[47]

Several aspects of the NCFMF films can be interpreted as skillful science marketing. First, the aestheticized images of fluid dynamics experiments were particularly cleverly used, as described exemplarily on the basis of the clips from Prandtl and their visual integration into the film *Vorticity*. Second, the PSSC film *Fluid Dynamics of Drag* appeared in tandem with a book, the snappily titled *Shape and Flow*. In this aforementioned "film in print," Shapiro explains the theoretical concepts presented in the film in more detail and invokes, as already quoted, Prandtl's "style." Third, the book's cover quotes *Production of Vortices*, with the scene of boundary layer separation reduced to a pictogram of flow lines. This cover was designed by Robert Flynn, widely known for his collaboration with the popular jazz label Impulse! Records. Finally, the NCFMF (like all other ESI film programs) worked with talented film directors who knew how to appeal to mass audiences. For example, Jack Churchill, who showcased his skills directing *Vorticity* and *Flow Visualization*, was later hired as the co-cinematographer of the 1977 Hollywood film *ABBA: The Movie* (dir. Lasse Hallström, USA).

In collaboration with his production company, Shapiro thus developed a carefully designed and marketable product that appealed to the desires of modern consumers in the 1960s. The aforementioned collaboration with Technicolor on 8 mm cassettes also speaks to commodification. Shapiro recognized the footage from Prandtl's laboratory as a potential attraction that could satisfy a desire for the visual tangibility of mathematical complexities. He then helped to edit the footage into a product that multiplied the impact of the original lab material.

Later in his career, Shapiro again reused the footage from Prandtl's lab when he produced a video course in fluid dynamics at MIT in the 1980s, jumping on the bandwagon of the latest technical innovation in moving images. This video course too incorporates nearly all of the scenes from the Göttingen footage, which are scattered throughout the thirty-nine videotapes that were available in all half-inch or three-quarter-inch video formats common at the time of their release. The course was not intended for lecture halls and classrooms but for private study and the boardrooms of private and public companies. The format corresponds to a filmed lecture in which Shapiro explains the basics of fluid mechanics with a blackboard and overhead projector, with footage of experiments interspersed sporadically.

The commodification and canonization of the flow films is even more clearly apparent for this video course. The promises made in the promotional material are grandiose: "A new video course . . . teaches the concepts, principles and practical applications that apply to all technologies . . . old and new."[48] In addition, discounts provide incentives: purchasing the entire collection of thirty-nine tapes for $17,355 potentially saves the buyer $2,145. The stirring visual quality of the recorded experiments remains a central selling point: "dramatic film footage," as the brochure puts it.

The video course also makes the process of canonizing science images evident. In it, Shapiro uses numerous images from the *Album of Fluid Motion* by Milton van Dyke, the most prominent scientific atlas of fluid dynamics, first published in 1982. Like other scientific atlases, this one provides a canon of images that "define the most significant research subjects of a discipline."[49] Shapiro made a comparable selection for his video course, but with moving images. In fact, in the introduction to his atlas, Van Dyke explicitly apologizes for not being able to locate the "beautiful photographs that Prandtl used to illustrate his papers and books."[50] Shapiro had Prandtl's film still at hand, however. While his film programs turned the footage into products with himself as the anchorman, Prandtl's film material retained a central, canonical position.

Figure 2.3. Vertically arranged film frames from *MIT Video Course Fluid Dynamics* (1984, Ascher Shapiro, Cambridge MA: MIT/CAES). Screen captures and montage by Mario Schulze. © Video Course: MIT Libraries

Conclusion

Today, the NCFMF films are freely available on YouTube. They rank high when searching for a number of key fluid mechanics terms. There is no critical engagement with the films on YouTube, not to mention any assessment of their origin. In outlining the historical background of their production and distribution, I have argued that these films were meant to serve as a "cold weapon" in the not-really-cold Cold War. These educational films were linked to the goals of recruiting more scientists and propagating a specifically Western way of doing science.

By focusing on the appropriation of particular footage by Shapiro in his NCFMF films, this chapter explores and complicates the story behind a specific set of educational films. Shapiro's films inscribed historical film sequences from a pre–World War II German hydrodynamics laboratory into the visual memory of many postwar students and professionals in the United States and other countries. A close examination of this reuse of Prandtl's film *Production of Vortices*, in particular, opens a new, critical perspective on the aesthetic, ideological, and material networks shaping educational films in fluid dynamics during the Cold War. Because the prewar lab recordings depict the flow experiment in an aesthetically gripping way—while concealing any reference to where and by whom they were made and what they were used for—they presented an ideal material and stylistic model for postwar educational films that aimed to depict science as a nonideological, time-transcending, rational endeavor. By interweaving the aesthetics of flow with the basic knowledge of military research, Shapiro's films were able to grasp their viewers and interest them in a field of knowledge that promised economic, military, and scientific-ideological superiority during the Cold War.

Shapiro's found footage appropriation exemplifies how fluid mechanics, the medium of film, and Cold War knowledge regimes came together in these educational films. Shapiro's films thus constitute a case study of revolutionary changes in media culture that took place during the Cold War. The global expansion of communications media and their use as powerful tools in education made it possible for the visual to become an everyday form of circulating knowledge. Albeit, the film recordings from Prandtl's laboratory contributed only a small, though important, element to educational films on fluid dynamics. But in its very marginality, this reappropriation reveals something about both the Cold War and knowledge circulation, if only that laboratory experiments and the phenomena they uncover can, depending on their medialized form, persist in quite unexpected ways in the sciences and beyond.

Mario Schulze is a researcher and curator based in Basel. He completed a PhD in cultural analysis at the University of Zurich and has held postdoctoral positions at the Collegium Helveticum (Zurich), the Humboldt University Berlin, the Zurich University of the Arts and the Universities of Lüneburg, Basel, and Lucerne. His transdisciplinary research focuses on the visuality, mediality, and materiality of knowledge. His 2017 book, *Wie die Dinge sprechen lernten: Eine Geschichte des Museumsobjektes 1968–2000* [*How Things Learned to Speak: A History of the Museum Object, 1968–2000*] discusses how the conceptualization of objects in cultural history exhibitions has changed. His work has appeared in journals such as *Isis*, *Representations*, and *Berichte zur Wissenschaftsgeschichte*, as well as in various edited volumes. His film *unlearning flow* (codirected with Christoph Oeschger and Sarine Waltenspül) was selected for several international festivals.

Notes

1. This chapter expands upon my article on reusing film in the 2021 *Isis* Focus Section on "Research Film." See Schulze, "Mobilizing Moving Images."
2. For a list of NCFMF and other educational film programs, see National Science Foundation, *Released Textbooks*, 54–58.
3. For an overview, see Zimmermann, "Analyse nicht-fiktionaler Filmformen." Also see Juliane Scholz's chapter in this volume.
4. Rudolph, *Scientists in the Classroom*, 96–97.
5. Blümlinger, *Kino aus zweiter Hand*, 9. Translation mine.
6. Wees, *Recycled Images*.
7. There is only scant literature on the appropriation of scientific films in art films. See: Gaycken, "'Beauty of Chance'"; Blümlinger, *Kino aus zweiter Hand*, 228–37. Hanna Rose Shell reflected on Marey's reuse of film material in an article and a corresponding film: Shell, "Cinehistory."
8. On the concept of reuse in science films, see Sattelmacher, Schulze, und Waltenspül, "Introduction: Reusing Research Film." For another way to trace the political usages of media technology, see Josephine Diecke's chapter in this volume.
9. Some excellent studies are devoted to different aspects of this educational reform. The argumentation in this chapter relies mainly on Rudolph, *Scientists in the Classroom*, and Simon, "The Transnational."
10. Rudolph, *Scientists in the Classroom*, 84.
11. "A PROGRESS REPORT 1956–1962: The Physical Science Study Committee and Educational Services Incorporated." In PSSC Records, MC 626, Box 1, MIT Department of Distinctive Collections, Cambridge MA. (Hereafter, MIT Archives.)
12. Moore, *Disrupting Science*.
13. See Wolfe, *Freedom's Laboratory*.
14. For some numbers on the PSSC, see Simon "The Transnational," 314. Also, "A Partial List of Teachers Using the PSSC Course." In PSSC Records, MC 626, Box 1, MIT Archives.
15. See Rudolph, *Scientists in the Classroom*, 113–36.
16. Walter Michels, "Committee on High School Physics Teaching: High School Physics Texts." In MIT Oral History Program, MC 602, Box 1, MIT Archives.
17. See Rudolph, *Scientists in the Classroom*, 135.
18. See Rudolph, *Scientists in the Classroom*, 43–47.
19. Jerrold Zacharias, "Movie Aids for Teaching Physics in High Schools" (15 March 1956). In MIT Oral History Program, MC 602, Box 1, MIT Archives. Because of the perceived special suitability of films for the message of physicists, films even played a leading role in the first PSSC funding application: "First Annual Report of the Physical Science Study Committee." In PSSC Records, MC 626, Box 1, MIT Archives.
20. "Interview with Jack Churchill." In MIT Oral History Program, MC 602, Box 1, MIT Archives, 47.
21. I did not find a single nonwhite person and only one female physicist, Dorothy Montgomery, in the fifty-nine films of the PSSC course. For the mad scientist as a visual topos, see Frayling, *Mad, Bad and Dangerous*. For the new type of the creative, and passionate scientist that emerged in the 1960s, see te Heesen, "'The Way of the Scientist.'" For biographies of the PSSC film presenters, see "Biographies of PSSC Film Personalities." In Shapiro Papers, MC 387, MIT Archives.

22. Hilmes, *Only Connect*.
23. Graf and Elwanger, "Entwicklung, Funktion, Präsentationsformen und Texttypen."
24. Probstein, "Ascher Shapiro."
25. J. L. Livesey, "Degree Congregation Professor Asher Shapiro," 13 July 1973. In Shapiro Papers, MC 387, MIT Archives.
26. Waller, "Projecting the Promise of 16mm, 1935–45." For the German case, also see Juliane Scholz's chapter in this volume.
27. Ostherr, "Health Films, Cold War," 103. For another example of soft diplomacy via films, see Jan Uelzmann's chapter in this volume.
28. Orgeron, Orgeron, and Streible, *Learning with the Lights Off*, 52.
29. See Simon, "The Transnational," 317–26.
30. The literature on the cultural Cold War and the CIA is wide-ranging; see, for example, the work of Gilbert Scott Smith, Richard J. Aldrich, and Peter Coleman. For information on the Ford and Asia Foundations, see Saunders, *The Cultural Cold War*, 138–39.
31. Letter from Charles Conn (NCFMF) to Alfred Borg (EDC), 24 January 1968. And Newsletter No. 4 of the NCFMF. Both in Shapiro Papers, MC 387, Box 12, MIT Archives.
32. Correspondence from 1968. In Shapiro Papers, MC 387, Box 12, MIT Archives.
33. Letter from Charles Conn to Alfred Borg (NSF), 24 January 1968. In Shapiro Papers, MC 387, Box 12, MIT Archives. Unfortunately, the twenty-two countries are not specified, and it is therefore not possible to say whether they were also distributed to Eastern Bloc countries.
34. My colleague Sarine Waltenspül and I are in the process of compiling the history of these lab recordings. Our publications on the history of this particular film to date include: Schulze and Waltenspül, "Follow the Films!" and "From Images of Lines." A book is currently in progress.
35. For the history of the IWF, see Juliane Scholz's chapter in this volume and Waltenspül, "Reusable and Nonreusable Films."
36. Shapiro, *Illustrated Experiments*, 172, 182, and 207–9.
37. Prandtl and Tietjens, *Applied Hydro- and Aeromechanics*.
38. Letter from Edgardo N. Accinelli to Prandtl, 20 March 1939. In Prandtl Nachlass, I/44/209, Archive of the Max Planck Society, Berlin.
39. There was one other major NSF-funded film program working with fluid dynamics; it was out of the University of Iowa and led by Hunter Rouse. See Letter from Hunter Rouse to the Course Content Improvement Section of the NSF, 22 November 1965. Provided by the IIHR-Hydroscience & Engineering, University of Iowa.
40. Shapiro, *Shape and Flow*, xv.
41. Style has been an often-theorized topos in the history of science, mostly in reference to Ludwig Fleck. For an overview, see Heumann, *Gegenstücke*, 119.
42. For Prandtl's scientific legacy, see Eckert, *Ludwig Prandtl*, 343–50.
43. Prandtl and Tietjens, *Applied Hydro- and Aeromechanics*.
44. Hoffmann, "Bildmaterial," 156–60. Ernst Mach was an Austrian physicist and philosopher famous for, among other things, his contributions to the study of shock waves.
45. See the role of KWI in fluid mechanics during World War II in: Epple and Schmaltz, "Göttingen: Das Max-Planck-Institut."
46. Shapiro, "Motion Pictures for the Engineering Sciences," 475; and Shapiro, "Motion Pictures of Fluid Mechanics Phenomena."
47. Letter from Shapiro to John Field (EBEC) and Rebecca McGovern (MIT Press), 18 July 1972, in Shapiro Papers, MC 387, Box 12, MIT Archives.

48. "MIT Video Course Fluid Dynamics." Undated anonymous leaflet. In Shapiro Papers, MC 387, Box 12, MIT Archives.
49. Daston and Galison, *Objectivity*, 17 and 402–3.
50. Van Dyke, *An Album of Fluid Motion*, 6. Nevertheless, this statement is somewhat surprising. After all, new editions of Prandtl's textbooks were also published in the 1980s.

Bibliography

Blümlinger, Christa. *Kino aus zweiter Hand: Zur Ästhetik materieller Aneignung im Film und in der Medienkunst*. Berlin: Vorwerk 8, 2009.

Curtis, Scott. *The Shape of Spectatorship: Art, Science, and Early Cinema in Germany*. New York: Columbia University Press, 2015.

Daston, Lorraine, and Peter Galison. *Objectivity*. New York: Zone Books, 2007.

Eckert, Michael. *Ludwig Prandtl—Strömungsforscher und Wissenschaftsmanager: Ein unverstellter Blick auf sein Leben*. Berlin/Heidelberg: Springer, 2017.

Epple, Moritz, and Florian Schmaltz. "Göttingen: Das Max-Planck-Institut für Dynamik und Selbstorganisation." In *Denkorte: Max-Planck-Gesellschaft und Kaiser-Wilhelm-Gesellschaft; Brüche und Kontinuitäten 1911–2011*, edited by Peter Gruss and Reinhard Rürup, 137–45. Dresden: Sandstein, 2010.

Frayling, Christopher. *Mad, Bad and Dangerous: The Scientist and the Cinema*. London: Reaktion Books, 2002.

Gaycken, Oliver. "'Beauty of Chance': Film ist." *Journal of Visual Culture* 11 (2012): 307–27.

Graf, Ludwig. "Entwicklung, Funktion, Präsentationsformen und Texttypen der Telekolleg- oder Akademie-Sendungen." In *Medienwissenschaft*, edited by Joachim-Felix Leonhard et al., 3:2369–83. Berlin/Boston: De Gruyter, 2002.

Heumann, Ina. *Gegenstücke: Populäres Wissen im transatlantischen Vergleich (1948–1984)*. Vienna: Böhlau, 2014.

Hilmes, Michele. *Only Connect: A Cultural History of Broadcasting in the United States*. 4th ed. Belmont: Wadsworth, 2014.

Hoffmann, Christoph. "Bildmaterial: Bearbeiten, Auswerten, Verbrauchen." In *Einwegbilder*, edited by Inge Hinterwaldner, 155–74. Leiden: Wilhelm Fink, 2016.

Latour, Bruno. "Visualisation and Cognition: Drawing Things Together." *Knowledge and Society: Studies in the Sociology of Culture and Present* 6 (1986): 1–40.

Moore, Kelly. *Disrupting Science: Social Movements, American Scientists, and the Politics of the Military*. Princeton, NJ: Princeton University Press, 2008.

National Science Foundation. *Released Textbooks, Films and Other Teaching Materials*. Washington, DC: US Department of Health, Education & Welfare, 1968.

Orgeron, Devin, Marsha Orgeron, and Dan Streible. *Learning with the Lights Off*. New York: Oxford University Press, 2012.

Ostherr, Kirsten. "Health Films, Cold War, and the Production of Patriotic Audiences: The Body Fights Bacteria." In *Useful Cinema*, edited by Charles Acland and Haidee Wasson, 103–22. Durham, NC/London: Duke University Press, 2011.

Prandtl, Ludwig, and Oscar Tietjens. *Applied Hydro- and Aeromechanics: Based on Lectures of L. Prandtl*. New York: Dover, 1957.

Probstein, Ronald. "Ascher Shapiro, 1916–2004." In *Biographical Memoirs*. National Academy of Sciences, 2014. Retrieved 15 February 2021 from www.nasonline.org/memoirs.
Rudolph, John L. *Scientists in the Classroom*. New York: Palgrave, 2002.
Sattelmacher, Anja, Mario Schulze, and Sarine Waltenspül. "Introduction: Reusing Research Film and the Institute for Scientific Film." *Isis* 112(2) (2021): 291–98.
Saunders, Frances Stonor. *The Cultural Cold War: The CIA and the World of Arts and Letters*. New York: New Press, 2001.
Schulze, Mario, and Sarine Waltenspül. "From Images of Lines to Images of Particles: The Role of the Film Camera in Flow Visualization." *Yearbook of Moving Image Studies* 4 (2019): 166–91.
——. "Follow the Films! Reuses of a Research Film." In *Trajectories of Images*, edited by Olga Moskatova. Bielefeld: transcript, 2021.
Schulze, Mario. "Mobilizing Moving Images: Reusing a German Flow Film of the 1920s for US Science Education in the Cold War." *Isis* 112 (2021): 361–69.
Shapiro, Ascher H. *Shape and Flow*. New York: Anchor Books, 1961.
——. "Motion Pictures for the Engineering Sciences." *Journal for Engineering Education* 52 (1962): 474–77.
——. "Motion Pictures of Fluid Mechanics Phenomena." *Bulletin for Mechanical Engineering Education* 2 (1963): 61–65.
——. *Illustrated Experiments in Fluid Mechanics*. Cambridge, MA: MIT Press, 1972.
Shell, Hanna Rose. "Cinehistory and Experiments on Film." *Journal of Visual Culture* 11 (2012): 288–306.
Simon, Josep. "The Transnational Physical Science Studies Committee: The Evolving Nation in the World of Science and Education (1945–1975)." In *How Knowledge Moves: Writing the Transnational History of Science and Knowledge*, edited by John Krige, 308–42. Chicago: University of Chicago Press, 2019.
Star, Susan, and James Griesemer. "Institutional Ecology, 'Translations' and Boundary Objects: Amateurs and Professionals in Berkeley's Museum of Vertebrate Zoology, 1907–39." *Social Studies of Science* 19 (1989): 387–420.
te Heesen, Anke. "The Way of the Scientist, Ikonografie der Unmittelbarkeit." In *Gesichter der Wissenschaft*, edited by Christian Vogel and Sonja Nökel, 161–77. Göttingen: Wallstein, 2019.
Van Dyke, Milton. *An Album of Fluid Motion*. Stanford, CA: Parabolic Press, 1982.
Waller, Gregory A. "Projecting the Promise of 16mm, 1935–45." In *Useful Cinema*, edited by Charles Acland and Haidee Wasson, 125–48. Durham, NC/London: Duke University Press, 2011.
Waltenspül, Sarine. "Reusable and Nonreusable Films: From Ballistic Films to the Encyclopaedia Cinematographica." *Isis* 112 (2021): 342–51.
Wees, William Charles. *Recycled Images: The Art and Politics of Found Footage Films*. New York: Anthology Film Archives, 1993.
Wolfe, Audra J. *Freedom's Laboratory: The Cold War Struggle for the Soul of Science*. Baltimore: Johns Hopkins University Press, 2018.
Zimmermann, Yvonne. "Analyse nicht-fiktionaler Filmformen." In *Handbuch Filmanalyse*, edited by Malte Hagener and Volker Pantenburg, 1–14. Wiesbaden: Springer VS, 2017.

CHAPTER 3

Raw Film Manufacturing in East Germany

Between Economic Efficiency and Environmental Awareness

Josephine Diecke

Introduction

In 1981, the German Democratic Republic (GDR) prohibited the publication of Monika Maron's novel *Flugasche* (*Flight of Ashes*) because it tackled the environmental crisis, a matter the ruling Socialist Unity Party of Germany (SED) insisted did not exist.[1] The plot centers around the visit of the protagonist, a young and ambitious journalist named Josefa Nadler, to a town referred to as "B." Her mission is to cover the planned construction of a new power plant and to celebrate the remarkable achievements of the socialist working class. However, the living conditions in "B" turn out to be unexpectedly exceptional, but not in the way she had anticipated:

> These smokestacks like cannon barrels aimed at the sky shooting their charges of filth at the town day in day out and night after night: not with a roar, no, but quietly, like snow that falls slowly and gently, that stops up drainpipes, covers roofs where the wind blows without waves. In summer it swirls through the air, dry black dust that flies into your eyes. . . . Only strangers stand still and rub the soot from their eyes. The population of B. walks through the town with squinting eyes; you'd think they were smiling.
>
> And these fumes could serve as road signs. Please go straight ahead to the ammonia, then turn left at the nitric acid. When you feel a stabbing pain in your throat and bronchial tubes, turn around and call the doctor, that was sulphur dioxide.[2]

Everyone who grew up in the GDR, from the 1950s onward, would have immediately identified that the town referred to as "B." was none other than Bitterfeld. Bitterfeld served as the epicenter of East Germany's chemical industry and its interconnected industrial hubs. This industrial complex was responsible for the production of various chemical products, including film stock. The production of raw film materials in Germany predates the establishment of the GDR in 1949. The facilities in Berlin had already become insufficient to meet the growing demand for raw film stock as early as 1909. It was during this time that Agfa (Aktien-Gesellschaft für Anilin-Fabrikation) inaugurated a new factory in Wolfen, which subsequently became known as the Wolfen Filmfabrik. Between 1925 and 1945, Agfa significantly expanded its influence across the nation as a member of IG Farben (I. G. Farbenindustrie AG), a chemical conglomerate with an extensive supply network comprising multiple production sites throughout Germany.

The Bitterfeld-Wolfen region was thus known for the production of highly sensitive black-and-white film, as well as for color film processes, such as the Agfacolor lenticular process, Agfa's Pantachrome film, and, in particular, the first chromogenic negative-positive Agfacolor process.[3] The Filmfabrik's leading edge in color film shrank after World War II when, on 20 April 1945, Western Allied forces took control of the Wolfen plant, confiscating all relevant photochemical production equipment. In their reports, the American Field Information Agency, Technical, the British Intelligence Objectives Subcommittee, and the Combined Intelligence Objectives Subcommittee disclosed Agfacolor's formula, which paved the way for the subsequent development of color film processes worldwide.[4] So-called modern color film technologies—such as Fujicolor, Gevacolor, and Orwocolor—thus emerged and spread in an age of rapidly growing chemical industries and then persisted under different political systems in East and West for most of the second half of the twentieth century.[5]

Although an unknown amount of technical and personnel infrastructure was temporarily dismantled or permanently destroyed in 1945, production at the Filmfabrik in Wolfen started up again almost immediately after the war and continued to draw on the knowledge and technology accumulated during the Weimar Republic and Nazi periods. Because the plant was located within the Soviet occupation zone, which became the GDR in 1949, East Germany was always able to rely on domestically manufactured raw film stock for photographic and moving images. In addition to supplying East Germany's state-owned DEFA film production company (founded in 1946), however, it also supplied every other socialist Eastern Bloc country, first and foremost the Soviet Union. While the common *Klassenfeind* (enemy of the working class) lurked in the West, the increasing scarcity within the economy exacerbated tensions in the East. The decision for the Wolfen Filmfabrik to cease production of Orwocolor color negative paper in 1967, while Czechoslovakia (CSSR) abandoned color negative film production, serves as an illustrative example of how a burgeoning economy of scarcity, compounded by the GDR's obligation to supply other Eastern Bloc nations, fostered tensions among socialist states.[6] The GDR sought to offload a technological challenge through the German-Czech division of labor within the Council for Mutual Economic Assistance (RGW), even though this conflicted with the global adoption of the negative-positive process and the increasing demand for colored photographic paper in nonsocialist countries. The ensuing struggle to coordinate production between the Wolfen film factory and Fotochema, the Czech photographic paper factory, exacerbated the problem, ultimately impacting the sale of Wolfen's color negative films.

Around Wolfen and Bitterfeld, the impact of such Cold War competition was borne by the people living and working in the self-proclaimed *Arbeiter-und-Bauern-Staat* (the workers' and peasants' state) of the GDR. A document issued by the Abteilung Grundstoffindustrie des ZK der SED on 19 January 1977 to inform the secretariat of the Central Committee of the SED highlights a crucial aspect of the complex dynamics within the Eastern Bloc during that era.[7] It underscores the challenges stemming from economic constraints, international commitments, and disparities in access to resources. In reference to the consultation conducted on the further development of scientific and technical collaboration between the photochemistry sectors of the USSR and the GDR within the economic framework of Assofoto, it is noted that the implementation of this fundamental direction in the development of Soviet photochemistry has encountered a range of challenges.[8] For instance, due to the reluctance of leading capitalist firms to share the latest knowledge and their insistence on exorbitant pricing, the complete importation of all necessary equipment for establishing closed production lines has not been successful.[9] Consequently, a substantial effort was required in terms of in-house contributions to ensure product quality, given that some of the raw materials and auxiliary substances included in the formulations are either unavailable in the USSR or cannot be used due to labor and environmental protection considerations.[10]

The GDR's struggle to maximize production while cutting corners on investments in modern and eco-friendly technologies had the most palpable impact on the general population. This impact was evident in the escalating pollution levels, the release of toxic wastewater, and the widespread presence of exhaust emissions that polluted the environment. These issues began to receive more focused attention toward the end of the 1970s with the establishment of the first grassroots environmental groups.[11] As we shall see, the production of film materials, such as negatives and positives, was closely tied to environmental considerations in East Germany. In this chapter, I explore how and to what extent the entanglement of production and environmental issues was addressed in contemporary scientific and technological debates about film production. Articles in the East German film tech trade journal *Bild und Ton* (Image and Sound) provide the material needed to explore this question.

Bild und Ton *under the Magnifying Glass*

The monthly film and photographic journal *Bild und Ton* circulated in the GDR and across its national borders from 1948 to 1992. As Sophia Gräfe and Kerrin von Engelhardt point out in their chapter in this book, the journal's scope—"[serving] as a resource for both amateurs and professionals"—indicated both its importance and its reach to those interested in the latest photographic and cinematographic developments. The journal was distributed by a series of East German publishing houses over time: Deutscher Filmverlag (1948–52), Henschelverlag (1952–68), VEB Fotokinoverlag Leipzig (1969–89),[12] and, briefly after German reunification, by Fotokinoverlag (1990–91) and Medienverlag Wilhelmshorst (1992). During the Cold War, *Bild und Ton* acted as a flagship for technological progress and the growth of the GDR's socialist society.

The journal was established in 1948 by founding editor Albert Wilkening who, as a leading figure at the DEFA Film Studios, was familiar with the practices and peculiarities of the state-owned production model as well as official political guidelines that called for moderation or secrecy when it came to criticizing the government. Some other authors who published in *Bild und Ton* also worked for state-owned companies—either under the DEFA umbrella, such as the film labs (VEB DEFA Kopierwerke) and the Central Office for Film Technology (DEFA-Zentralstelle für Filmtechnik), or for the raw film production plant in Wolfen (VEB Filmfabrik Wolfen). Even if the work ethic and personal points of view of these authors were in conflict with political mandates—for instance, in the case of environmental pollution—their professional opinions must thus be seen in relation to their public function as representatives of the GDR.

The aim of this chapter is to explore the extent to which technical discourses referred to or incorporated increasingly pressing environmental issues. To do this, I investigate how scientific discourses in *Bild und Ton* functionalized debates around environmental terminology as a means to discuss and promote technological developments in the film, photographic, and television industries. My approach is informed by a media historical perspective and interdisciplinary research techniques from media archaeology and from science and technology studies.[13] Using discourse analysis, paired with a social-constructivist approach, I augment the argument that photochemical films have been refined within a complex network of social groups, artifacts, problems, and solutions. This social construction of technology (SCOT) perspective was laid out in detail in 1987 by Trevor Pinch and Wiebe Bijker in "The Social Construction of Facts and Artifacts." According to SCOT, technological innovations are negotiated in relation to the problems or needs of "relevant" actors and can only be successful if these problems are solved, or if it is at least suggested that they are.[14] It is therefore important to stress the role of social groups and interests that public discourse framed as relevant—such as scientists, technicians, and political figures—or irrelevant—such as environmentalists or skeptical users.

In this chapter I will analyze the problems—hereafter also referred to as needs—that were included in and excluded from East German scientific debates on environmental concerns in the photographic and film industry.[15] I will focus on technical discussions in *Bild und Ton* to assess whether environmental problems—such as the clearly perceivable toxic fumes of Maron's B.—were mentioned at all, and what issues the journal's authors actually confront instead of avoiding.

As I will show, the first epistemological heralds of environmental awareness can be discerned in professional debates in *Bild und Ton* as of its inception in the late 1940s—even though environmental movements only came into the public eye in the 1970s. The relevant debates revolved around problems facing the film stock industry, such as material scarcity and environmental pollution, and possible solutions, including preservation and recycling.

The Janus-Faced Environmental Awareness of the GDR

As recent historiographical (re)evaluations have found, environmental protection was highly valued in official GDR discourse; but there was a disconnect between this and actual practices.[16] Contrary to today's persistent narrative of the Eastern Bloc as a coalition of environmental sinner states, historians such as Julia Obertreis and Tobias Huff have discovered that quite comprehensive environmental legislation—including for the protection of animals, plants, and nature reserves—existed under Soviet occupation as of 1945. The regulations went back to the Nazi-era *Reichsnaturschutzgesetz* (Reich Conservation Act, 1935) and the Ordinance on the Protection of Wild Plants and Nonhunting Wild Animals (1936).[17] Then, on 4 August 1954, the GDR's legislature passed a Law for the Conservation and Preservation of Domestic Nature (*Gesetz zur Erhaltung und Pflege der heimatlichen Natur*) and, on 14 May 1970, a Law on the Planned Shaping of Socialist Land Culture in the German Democratic Republic (*Gesetz über die planmäßige Gestaltung der sozialistischen Landeskultur in der Deutschen Demokratischen Republik*); East Germany thus became the second European country, after Sweden, to establish an environmental protection law. Importantly, as Obertreis notes, this must be placed in the context of Cold War East-West rivalry, as the socialist regimes apparently hoped to score points in this area against Western countries.[18] In sharp contrast to these positive political achievements, however, what was actually taking place around East German industrial zones and cities, such as Bitterfeld-Wolfen, Leuna, Cottbus, and Berlin, was dramatic environmental damage, as captured by Maron in her novel. Although the Ministry of Public Health introduced emission values for forty-eight substances—including sulfur dioxide—as early as 1968, East German companies were not obliged to comply with them because there were no fixed emission limits.[19] Instead, in 1973, individual thresholds were issued according to a plant's location. But even if these limits were not met, companies could still pay a sort of ecology tax; this, however, was politically predetermined. Pricing of environmental penalties were thus mainly set in the interest of economic incentives for the planned economy, as the bankruptcy of domestic companies had to be avoided.[20] Obertreis and Huff see both economic competition with Western countries and the priority given to economic growth plans as the root of disregard for existing environmental laws. Because short-term plan fulfillment took precedence, plant officials considered long-term investments in new filtration and treatment technologies unimportant.[21] This attitude, in turn, did not correspond to the self-imposed communist claim to use science for the benefit of immediately useful results, as Mario Schulze describes in his chapter, but delayed innovations that would have depended, in case of doubt, on scientific knowledge and resources from the West.[22]

As of its first issue in 1948, *Bild und Ton* regularly reported on topics related to the DEFA film laboratories (*Kopierwerke*) since its first issue in 1948. For instance, in "Das Filmkopierwerk" in 1957, chemist Wolfgang Brune reported on the societal, technical, and economic significance of film labs in general; in 1966 and 1969, the technical director of the DEFA Kopierwerke focused on developments at DEFA's film laboratories.[23] Other early articles described the chemical components involved in film production and processing.[24] For instance, in 1957 Sigurd Mansfeld published "Sulfate. Sulfite. Sulfide," an extensive paper on the various occurrences of sulfur in photographic chemistry.[25] In this article, he described operational processes and areas of application as well as the reactions and availability of sulfur, emphasizing its low occurrence in the Earth's crust and the resulting shortage; however, he made no connection between the use of resources, environmental effects, and possible chemical pollution.

In the following years, articles about color film development concentrated mainly on the need for compatibility in relation to financial and technical limitations. "Better, faster and cheaper!"—as Albert Wilkening summarized it in

1961—became the predominant theme in articles about different film processing techniques, as evident in articles by Robert Hardonyi (1963) and Keiler et al. (1972) about developing film using liquids, sprays, and pastes.[26] Expense and rationalization shaped every discourse on improved methods, focusing in particular upon minimizing developing time by increasing temperature to achieve both a "higher economy" with respect to resources and a guaranteed stable, safe, and high-quality technical procedure and product.[27] This prioritization of productive "effectiveness" was reflected in other contemporary articles and areas of research as well, as Gräfe and van Engelhardt point out with regard to audiovisual educational resources that "were designed not only to be effective but to purposefully direct the learning process."[28]

Through the 1960s, therefore, there was no public debate in *Bild und Ton* about investing in renewable energy or recycling materials, as it did not meet financial needs or create conflicts with Cold War enemies. Only in the early 1970s—after the adaptation of the GDR's Law on the Conservation and Preservation of the Environment—did *Bild und Ton* also start considering topics related to environmental issues. In 1974, the translation of an article by W. Sujew, a member of the Soviet Union's Academy of Sciences, which had been originally published in the Soviet journal *Sotsialisticheskaya industria* (Socialist Industry), was made available to East German readers.[29] Although he did not reflect on the film industry's own environmental impacts, in this paper Sujew stressed the role of audiovisual media as tools for researching pollution. Although pollution was addressed, the connection to East German film processing was fully neglected in *Bild und Ton*, or only mentioned if in regard to making financial improvements. The restrained environmental commitment of the East German government and *Bild und Ton* thus continued under the new leadership of Erich Honecker, who took over as the leader of the SED and East Germany in 1971. According to Huff, although measures were pursued early on to identify the environmental impact and establish legal resolutions during the era of Walter Ulbricht, the situation increasingly deteriorated when Honecker took office as head of the GDR state and SED general secretary in 1971, because his *Einheit von Wirtschafts- und Sozialpolitik* (Economic System of Socialism) favored consumer good supply over environmental protection.[30]

Excruciatingly honest assessments of living conditions nevertheless did emerge in the GDR, although they were kept secret, as documents of the Ministry for State Security (Stasi) show. On 26 September 1988, a year before the fall of the Wall, the district administration of the city of Halle forwarded an information paper to the Stasi, which reveals a different perspective on the relationship between the film industry and environmental protection. The document addressed the situation with regard to pollution, plant safety, and production reliability at the synthetic fiber division of the Wolfen Filmfabrik (FCK), thereby confirming what Monika Maron described in *Flugasche* and the SED was trying to conceal from people:[31]

> The environmental impact of the production of synthetic fibers at FCK can be broken down into: air pollution, with the main pollutants being sulfur dioxide, hydrogen sulfide and carbon disulfide; and wastewater and waste products. In addition, workers at FCK production facilities and the population of nearby residential areas . . . are subject to a strong odor nuisance, caused by organic substances such as amines, mercaptans and nitrate compounds.[32]

The document further stated that large amounts of sulfur dioxide, hydrogen sulfide, and carbon disulfide had been emitted into the air unfiltered, heavily contaminating local rivers and water bodies. A particularly well-known example of such contamination was Grube Johannes, a lake located in the south of Wolfen that was given the nickname Silver Lake (*Silbersee*).[33] Such pollution created a series of consequential problems, such as: impairment of production during film manufacturing; equipment damage due to corrosion; migration of workers to other parts of the GDR; an increase in the number of people with long-term illnesses and recognized occupational diseases; a two-to-three-times-higher incidence of bronchitis in children and disturbances in their physical development; a reduction of overall life expectancy by three to five years; and the loss of the recreational value of the region's natural features.[34]

Documents like this prove that, by the end of the 1980s, even the Ministry for State Security was aware of how bad living conditions were. Yet, despite the evident interest in internally informing officials of environmental problems, these findings were kept secret under classified status until after German unification and the end of the GDR. In the meantime, however, ecological protections had become an important issue for environmental movements, whose supporters protested against the state's overexploitation of nature and resources. Huff and Obertreis point out that there were East Germans concerned about the state of the nature that surrounded them from the start. Although environmental groups were eventually formed within the SED and other parties, many more people were involved in social mass organizations, churches, and universities; even as environmental issues started becoming more public and pressing in the 1970s, however, environmental activists always remained a minority. Furthermore, like other nongovernmental movements in the GDR, they did not agitate for a change of government or economic system; in fact, they were probably less political than, for example, East German human rights or peace groups. Because environmental groups functioned as a sort of melting pot of people with different interests (including the preservation of monuments and cultural assets) as well as of different political stripes (including those with nationalist tendencies), the environmental groups served as a catalyst for the emergence of a non-state-structured political landscape in the GDR.[35] The growth of public awareness and concern for environmental issues over the course of the 1970s and 1980s is also reflected in issues of *Bild und Ton* during this period, in which environmental concerns mutated into a high-profile community issue.

Collaborative Efforts and Resignations in Film Laboratories

It was when *Bild und Ton* started to report on international conferences and councils of socialist states, as of the early 1970s, that concern about environmental damage through film processing were addressed in the journal. For instance, in their conference report on the Second Scientific and Technical All-Union Conference in 1975, Götz Pollakowski—at DEFA's Main Office for Film Technology—and Doris Boetzel—from the DEFA film labs—wrote: "An important problem is the prevention of pollution. We must work harder to find less harmful substances (than color developers and bleach baths, especially), to regenerate and reuse processing solutions, and to purify wastewater."[36] In their statement, Pollakowski and Boetzel created a clear connection between the significance of processing chemicals and the motivations behind recycling them. They did so by highlighting the adverse effects of air pollution and untreated wastewater on human health. One year later, in 1976, the GDR's Ministry of Culture and the GDR Film Industry's Scientific-Technical Advisory Board organized the first Council of Film Laboratory Technicians in Socialist States,[37] which gathered film laboratory directors and technical managers from socialist countries, such as Bulgaria, Mongolia, Poland, and Romania. In this forum, a regular exchange was established, for instance on problems concerning process stabilization, process control, and the reuse of photochemical baths including developer, bleach, fixer, and stabilizer.[38] Every year, Johann-Albrecht Keiler wrote a report on the discussions of the Council of Film Laboratory Technicians in Socialist States for *Bild und Ton*. In his second report, in 1978, he noted that the participants addressed issues of water pollution caused by waste products from film laboratories.[39] Similar considerations took more and more time in the following councils, as well.[40]

Generally speaking, environmental discussions appeared at a higher rate after 1975. Keiler and Götz Pollakowski, in particular, systematically analyzed phases of film processing—in their case, the bleaching process—in a series of articles starting in 1980.[41] Their goal was to either present alternatives to existing chemical solutions or provide instructions for reducing water consumption in the intermediate washing process. In the final article in this series, health protection and work safety issues were also put on the agenda.[42]

Physicist Klaus Dittrich, also from DEFA's Main Office for Film Technology, presented new ways to measure chemicals and advised limiting exposure to toxic dust, liquids, and gases.[43] *Bild und Ton* articles also revealed a gradual shift away from scientific debates primarily focused on economic affordability, and toward an increasing overlap of financial and environmental concerns. Instead of only looking at the bottom line, a new perspective was emerging as pollution came to be seen as a threat to individual and collective safety, a risk with potentially serious environmental, medical, and financial consequences.

While environmental concerns were increasingly addressed by authors like Pollakowski, Keiler, and Dittrich in the 1980s, the analysis and handling of wastewater was an issue of particular importance. In a three-part series published in 1983, Pollakowski turned his attention to the risk factors involved in the discharging of film lab wastewater.[44] He began his article by emphasizing the importance of ecological considerations, before continuing with the positive economic side effects of mitigation: "Not only environmental protection increasingly draws the attention of film processors to chemicals entering the wastewater with processor overflows and the washing water, however; it is also questions of economic efficiency, as the reuse or multiple use of baths is of great interest when the cost of chemicals rises."[45]

Aside from Pollakowski's problematizing of artifacts and components involved in film processing, what is refreshing about this excerpt is the order of importance in which the motives for environmental protection are presented. Since the loss of financial and material resources would threaten economic as well as environmental stability, whenever both problems shared the same solution, Pollakowski stressed both the ecological and the economic advantages. He thus explained that the efficient use of chemicals for economic reasons would also be highly conducive to efforts to reduce environmental hazards.[46] He refrained from prioritizing one or the other and moreover emphasized that economic and environmental protection did not need to be mutually exclusive.

Pollakowski's references to previous studies done by competitors such as Kodak and Agfa-Gevaert made it obvious that the East German discussion on the environment was quite delayed.[47] His articles resemble observation from a distance, as his summaries of the status quo could not provide new results by themselves. This ambivalent attitude can be understood as a result of East Germany's years-long postponement of conflict management between scientific evidence, on the one hand, and a political wait-and-see attitude, on the other, that, in many other written sources, was obscured by "a prevailing rhetoric of success, motivated by the fulfillment of the grand plans and loyalty to the party line."[48] In the 1980s, this situation was reinforced by the introduction of secrecy levels so high that even scientists could not submit, let alone verify, actual emission data.[49]

Silver Recovery as All-Purpose Weapon

Parallel to these rather cautious comments on the environmental impacts of chemical processing in the film industry, *Bild und Ton* published concrete proposals to solve the problem of material loss during the processing of photochemical film, including the recovery of silver. In particular, scientific discussions among film lab employees about silver consumption in the film industry broke through into headline questions such as Pietsch and Epperlein's "Does Silver Halide Photography Have a Future?" and the anonymously published "How Much Silver is Left?"[50] These titles illustrate the skepticism that has been associated with the use of silver in the history of the GDR's raw film industry. Silver was highly important to various departments and products in the photochemical industry, as silver ions are the basis for both black-and-white and chromogenic photographic and moving image processing. While the role of silver in film stock production was part of discussions about film labs, *Bild und Ton* also published numerous articles devoted exclusively to the topic from an economic point of view. In order to save water, chemicals, and time, researchers and engineers sought to optimize and rationalize the entire film processing sequence. Kurt Meyer presented the recycling of baths via closed-loop circulation as one approach for maintaining control over and minimizing the use of materials.[51] According to Joachim Epperlein and Susanne Spigiel, this not only saved money but also reduced some of the environmental impact.[52]

Attempts were made to market the process of retrieving silver from fixer baths. As early as 1949, *Bild und Ton* reported on the cost of losing the undeveloped silver in fixer baths and, in the same article, explained various methods of recovering unused silver to amateur filmmakers.[53] While financial savings were presented as the main motivation, the shortage of raw materials was also considered. Subsequent pieces on silver use in photography included a 1951 article by Grupinski and Schreiber and a three-part series by Sigurd Mansfeld in 1957.[54] Grupinski and Schreiber emphasized that the process of silver recycling had long been the norm in large companies and argued that it would be helpful for amateur photographers as well. In his series, Mansfeld then addressed amateur filmmakers directly, describing how they could process their films without losing precious silver.

The method of silver recovery helped the photochemical industry close the gap in their finite resources.[55] As political scientist Volker von Prittwitz has pointed out, recycling systems—including in chemical companies—generally represent ecological as well as economic optimization,[56] and closed-loop resource systems have a long history, particularly in shortage economies.[57]

By the end of the 1960s, the East German film and photo industry was faced with a dilemma. As the purchasing power of GDR citizens had increased, so had the number of amateur filmmakers; as a result, the demand for silver had increased, which in turn caused the cost of the valuable metal to rise.[58] The solution for film producers was once more to recover as much silver as possible. In 1969, an anonymous author published another possible solution to the use of silver in the film industry, which was to increasingly transition from black-and-white to color:

> . . . Only with black-and-white film material is part of the silver used for the layers lost. With color film material, on the other hand, the silver halide serves only as a medium for the formation of the dye image. [As a result,] with color emulsions not only unexposed, but also exposed silver halide components are dissolved out of the layers during processing and thus recovered. With color reversal, negative and positive images no longer contain silver. So, if the processing of color images continues to be restricted to commercial laboratories . . . the photochemical industry will cease to be a bottomless pit of silver consumption, to the extent that color dominates black-and-white photography and cinematography.[59]

Finally, in addition to the serendipitous reduction in silver loss that accompanied the increasing use of color film, the industry eventually discovered a further reason to reduce silver consumption: environmental protection. As we have seen, from the mid-1970s onward, environmental concerns increasingly entered discussions about film production. In 1978, for instance, Epperlein and Spigiel concluded that the processing of classic silver-based photographic materials consumed a great deal of water and would lead to environmental wastewater pollution.[60] In 1980, Dittrich and Grüßner reported that the free silver ions in wastewater had a toxic effect on microorganisms, inhibiting the biological degradation of waste products.[61] Simultaneously, however, two 1981 articles on silver by Dittrich and Grüßner open by addressing economic advantages and the growing need of consumers for information technologies in particular while leaving out uncomfortable disadvantages for the environment.[62]

Four years after arguing that economic and ecological measures to deal with wastewater were compatible, Goetz Pollakowski also backpedaled, shifting the focus from ecological back to economic concerns and almost pitting the two against each other when it came to silver recovery:

> In addition to the ecological problems faced by companies, the economy is also playing an increasingly important role. A wide range of measures are intended not only to protect the environment, but also to ensure that the chemicals used are handled more economically wherever possible. This is achieved by recycling processing baths, using chemicals multiple times, recovering expensive components from processing solutions, and efforts to use water—which has not been regarded as a source chemical until now—as a raw material that is becoming scarce.[63]

Pollakowski's emphasis on water scarcity was further reflected in a 1988 article on photography and silver by Epperlein and Liebscher.[64] Both articles repeated the same message about sensible management of material goods and waste through thriftiness and material economy.

> Moreover, starting in the 1970s, a notable emphasis on maximizing silver recovery from previously developed film stocks emerged, extending to film preservation facilities. Notably, the significance of silver had a unique impact on the National Film Archive of the GDR (SFA). One of the most radical and financially motivated endeavors by the government to turn the archive into a profitable entity was a study conducted in the early 1980s. This study focused on the quantity of silver stored on the shelves in the form of archived film productions. Former SFA Director Wolfgang Klaue described this extraordinary case as follows: When the price of silver on the world market had risen once more and a serious material bottleneck arose in the GDR, a study was made of how many tons of silver might be extracted from the film materials stored in the archive. Before washing them off, [the idea was that] all films would be transferred to video. This would have been economically very lucrative. It took some effort on the part of the archive [SFA], HV Film [the GDR's Central Film Administration], and the Ministry of Culture to ward off the request.[65]

Ironically enough, the environmentalists' goal to improve the ecological situation coincided with film preservationists' mission to safeguard the GDR's cultural heritage. Air, water, sulfur, and silver—the same components that brought photochemical film to life—also worked against its durability and had to be controlled by archival institutions, and mastering material components and storage conditions was thus paramount.

Despite the abovementioned concerns and events, and although *Bild und Ton* had asked "How Much Silver Is Still Left?" as early as 1976, the photo industry continued to rely on silver.[66] Epperlein and Spigiel summarized three essential reasons for the continuing use of silver in photochemical industries: first, the advanced level of scientific research on silver and the assured high level of production; second, the high photosensitivity of silver halides and other favorable photographic properties; and third, the system-stabilizing power of established equipment and technology for silver halide photograph and film processing.[67] In other words, silver-based photography had proven to be reliable for more than 150 years and satisfied manufacturers' need for stability as well as compatibility with existing procedures and equipment; continuing with silver was thus still less expensive than abandoning the technology.

It is presumably for this reason that Pietsch and Epperlein—employees at the Filmfabrik Wolfen—stated:

> Does silver halide photography have a future? It has a future, scientifically, technically, and economically. The forecasts for the scientific and technical development of silver halide photography are mostly optimistic.[68]

However, as a sort of simultaneously exerted countermeasure, manufacturers had also been delving into imaging processes that used less or no silver—mostly for their application in X-ray photography, which uses the most silver of all photographic processes—as well as diazo films (direct positive contact films) for the graphics industry and technical reproductions.[69] Interestingly, *Bild und Ton* also reported on silver-free and low-silver technologies from capitalist companies such as Kodak and DuPont.[70] As Tobias Huff notes, the contest between socialism and capitalism was not decided by the quality of water, air, or even life but by the assortment of goods on department store shelves.[71] The SED knew this and, after 1971 at the latest, consistently geared its policies toward this knowledge: all other demands had to stand back, including environmental policy. That economic issues ultimately trumped environmental concerns reveals one of the most complex aspects of environmental action: being aware of ecological damage does not necessarily result in consistent and adequate action.

Conclusion

The review of environmentally related discussions in *Bild und Ton* throughout this chapter underscores the pervasive influence of chemical usage and finite resource constraints on every aspect of the GDR's scarcity-driven economy, including cinema and photography. These industries represented lucrative markets at the intersection of the chemical and entertainment sectors, and under the guidance of political incentives, the film-stock industry prioritized ensuring a consistent supply of film and photographic materials throughout the GDR's existence. This commitment is reflected in contemporary articles discussing film lab practices and the role of material components, such as water and silver in film development.

The case of silver recovery serves as a testament to how scarcity and a do-it-yourself culture profoundly shaped both private and professional life within the planned socialist economy, ultimately influencing the GDR's landscape. While recycling processing baths and individual components, like silver, became standardized practices that persisted for decades, the detrimental side effects of these industrial processes were recognized early on. However, serious environmental protection discussions did not gain momentum until the 1980s, and even then, they were primarily framed in economic terms. During the Honecker era, "saving the environment" often served as a marketing strategy to achieve greater economic gains by minimizing resource usage. Short-term industrial production efficiency typically took precedence over long-term sustainability considerations. For environmental activists, as well as those involved in film production and archiving, safeguarding and preserving natural and national resources constituted a daily struggle against political regulations and rationalization efforts. Individual initiatives, such as Pollakowski's investigation into the risks of film lab wastewater discharge and the SFA's commitment to preserving East German film heritage, played indispensable roles in safeguarding the quality of photochemical film production, preservation efforts, and the overall environmental well-being of nearby communities.

By the late 1980s, environmental concerns began to intersect with other issues and contributed as a multiplier effect to the 1989 uprising against the SED state. While it may not have been the primary trigger for the GDR's downfall, environmental awareness became an integral part of East German civil society's efforts to reshape the country's political landscape.[72] However, as Huff points out, the post-reunification discourses in the West after 1990 often presented the GDR as "having failed ecologically," further emphasizing the discrepancy between political actions and societal efforts.[73]

Questions pertaining to resource availability and environmental impact remain pertinent for politicians, scientists, filmmakers, preservationists, and researchers. Today, the strategies for addressing the global environmental crisis, particularly concerning climate change, encompass a wide spectrum, ranging from educational initiatives to more radical forms of protest. The legacy of Walter Ulbricht and Erich Honecker's avoidance of problems and their commerce-driven, environment-damaging approach continues to cast a shadow over European populations. Cleaning up the (photo)chemical industry pollution around Bitterfeld-Wolfen, often likened to Monika Maron's "town called B.," remains an ongoing endeavor. Responsible implementation of conditions for a sustainable film industry and a healthy environment also continues to preoccupy us in the digital age.

Josephine Diecke is assistant professor of film studies at the University of Zurich. She has conducted extensive research on the history of film color technologies, moving image preservation, and digital methods for computer-assisted text and video analysis. Her expertise stems from her work as a research associate on the SNSF project "Film Colors: Technologies, Cultures, Institutions" (University of Zurich), as the academic coordinator of the Volkswagen Foundation project "Digital Cinema-Hub" (Philipps-Universität Marburg), and as a film lab technician for various service providers. She holds a PhD from the University of Zurich with a thesis on Agfacolor and Orwocolor, and is coeditor of the *Open Media Studies* blog.

Notes

1. *Flugasche* was first published in West Germany by S. Fischer Verlag in 1981. It was translated into English by David Newton Marinelli in 1986, and Maron fled to West Germany in 1988.
2. Maron, *Flight of Ashes*, 8.
3. Agfa initially launched Agfacolor, the first German subtractive color process that registered color in one single multilayer filmstrip: as of 1936, in the form of a reversal film and, as of 1939, also as a negative-positive process. Unlike Kodak's Kodachrome process, used as of 1935, the color components were integrated into the Agfacolor film strip, not added during the developing process. The term "chromogenic" is derived from the chromogenic reaction that occurs between dye couplers (that form cyan, magenta, and yellow each in individual layers) and the oxidized color developer in the developing bath. While Kodachrome was mainly marketed for amateur use and Technicolor No. IV for professional use, Agfacolor could be used for both as well as for color photography and thus covered both photographic and cinematographic needs. On the chromogenic processes used by Agfacolor and Kodachrome, see Flueckiger, "Agfacolor" and "Kodachrome."
4. Lovejoy, "Celluloid Geopolitics"; and Diecke, "Agfacolor in (Inter)National Competition."
5. In 1964, Wolfen Filmfabrik changed the trademark of their photochemical products from Agfa to ORWO—an acronym for "Original Wolfen." As a result, color film materials were also renamed from Agfacolor to Orwocolor; see Präsidium des Ministerrates, "Bericht Sicherung des Exportes," 25–29.
6. Silke Fengler thoroughly examined this case; see Fengler, *Entwickelt und fixiert*, 190–91.
7. Abt. Grundstoffindustrie, "Information für das Sekretariat des Zentralkomitees der SED," 157–61.
8. Assofoto, or Assoziation Fotochemische Industrie der sozialistischen Länder (Association of Photochemical Industry of Socialist Countries), was an economic organization that facilitated cooperation and coordination among the photochemical industries of socialist countries during the Cold War era. Established in 1973, It aimed to foster collaboration in the production of photographic materials, chemicals, and related technologies among Eastern Bloc countries; see Fengler, *Entwickelt und fixiert*, 200.
9. See Abt. Grundstoffindustrie, "Information für das Sekretariat des Zentralkomitees der SED," 159.
10. See Abt. Grundstoffindustrie, "Information für das Sekretariat des Zentralkomitees der SED," 159.
11. The "Timeline of GDR Environmentalism," created by the DEFA Film Library to accompany Jörg Foth's film *Biology!* chronicles significant moments in East Germany's environmental history, including both government initiatives and the increasing environmental consciousness among its people; see DEFA Film Library, "Timeline of GDR Environmentalism."
12. In the GDR, VEB stood for Volkseigenene Betrieb, a company owned by the state (i.e., the people).
13. Particularly within currents of media ecology, experimental media archaeology, and new materialism, the sustainable handling and processing of photochemical film processes in (post)production is an important topic. For current initiatives by filmmakers, researchers, and conservators to emphasize the material aspects of film production,

see Catanese and Parikka, "Handmade Films and Artist-Run Labs"; Knowles, *Experimental Film*; and Cherchi Usai, "'I Can't Breathe.'"
14. See Pinch and Bijker, "Social Construction," 44.
15. The methodological approach described here is based on considerations that Noemi Daugaard and I developed in the framework of the *Filmcolors: Technologies, Cultures, Institutions* research project at the University of Zurich, which was funded by the Swiss National Science Foundation and supervised by Prof. Dr. Barbara Flueckiger. See Daugaard and Diecke, "Farbfilmverfahren."
16. Huff, "Über die Umweltpolitik"; Obertreis, "Von der Naturbeherrschung."
17. Huff, "Über die Umweltpolitik," 538; Obertreis, "Von der Naturbeherrschung," 118.
18. Obertreis, "Von der Naturbeherrschung," 118.
19. See DEFA Film Library, "Timeline of GDR Environmentalism."
20. See Huff, "Über die Umweltpolitik," 536–45.
21. See Obertreis, "Ökozid," 118; Huff, *Natur und Industrie im Sozialismus*, 412–13.
22. Schulze, "Flow Vis for the Space Race," in this volume.
23. Brune, "Das Filmkopierwerk"; Jander, "Die Weiterentwicklung"; Jander, "Die Entwicklung."
24. Processing, or developing, chromogenic negative film consists of several steps: (1) a *color developer* reduces the silver halide crystals that have been exposed to light so that a metallic silver image appears; the oxidized developer reacts with dye couplers on unexposed silver halide, forming visible color dyes in each emulsion layer; (2) a *bleach* converts the developed metallic silver into silver halide; (3) a *fixer* removes the remaining silver halide, which can then be recovered from the fixer bath; (4) *washing* the film with water removes fixer bath residues and complex silver compounds; (5) a *stabilizing* bath contains a wetting agent to prevent dry spots on the surface of the processed film and formaldehyde to improve the stability of the dyes. See Anklin-Mühlemann and Gschwind, *Script zur Vorlesung*, 84–98.
25. Mansfeld, "Sulfate. Sulfite. Sulfide."
26. Wilkening, "Neue Entwicklungs-Verfahren"; Robert Hardonyi, "Über die Sprühentwicklung"; Keiler et al., "Die Pastenentwicklung."
27. Meyer, "Über Phenidon-Entwickler." Keiler, "Über Möglichkeiten"; Dittrich, "Messung und Registrierung"; Meyer, "Verfahren."
28. Gräfe and von Engelhardt, "Effective Films," in this volume.
29. Sujew, "Konstrukteur und Wissenschaftler," 90.
30. See Huff, "Umweltpolitik," 536–45.
31. This and the following translations are my own.
32. Bezirksverwaltung Halle, "Information," 420.
33. Richter, "'Silbersee' in Bitterfeld."
34. Bezirksverwaltung Halle, "Information," 420–21.
35. See Huff, *Natur und Industrie*, 413; and Obertreis, "Ökozid," 121.
36. Pollakowski and Boetzel, "Technik und Technologie."
37. *Beratung der Kopierwerkstechniker sozialistischer Staaten*.
38. Keiler, "Beratung der Kopierwerkstechniker."
39. Keiler, "2. Beratung der Kopierwerkstechniker," 51.
40. Keiler, "3. Beratung der Kopierwerkstechniker," 187; Keiler, "4. Beratung der Kopierwerkstechniker," 318.
41. Pollakowski et al., "Bemerkungen zur Stabilität"; Pollakowski and Keiler, "Bleichbäder und Umweltschutz"; Keiler and Pollakowski, "Persulfat/Chinon-Bleichbäder."

42. Keiler and Pollakowski, "Persulfat/Chinon-Bleichbäder."
43. Dittrich, "Messung und Bekämpfung von Chemikalienstäuben."
44. Pollakowski: "Kopierwerksabwässer und Umweltschutz (I)"; "Kopierwerksabwässer und Umweltschutz (II)"; "Kopierwerksabwässer und Umweltschutz (III)."
45. Pollakowski, "Kopierwerksabwässer und Umweltschutz (I)," 139.
46. Pollakowski, "Kopierwerksabwässer und Umweltschutz (II)," 182.
47. Pollakowski, "Kopierwerksabwässer und Umweltschutz (II)," 182.
48. See Gräf and von Engelhardt's chapter in this volume.
49. Huff characterizes the discrepancy between the SED's reaction to environmental and economic issues from the 1970s onward as a mixture of secrecy, denial, ignorance, and suppression, which discredited official environmental policies. See Huff, "Uber die Umweltpolitik," 549–54.
50. Pietsch and Epperlein, "Hat die Silberhalogenidfotografie eine Zukunft?"; Anon., "Wieviel Silber?"
51. Meyer, "Die Fotografie und das Silber," 312.
52. Epperlein and Spigiel, "Das Silber und die Fotografie," 23.
53. A.B., "Die Rückgewinnung."
54. Grupinski and Schreiber, "Die Bedeutung"; Mansfeld's three-part series: "Silberrückgewinnung"; "Silberrückgewinnung (1. Fortsetzung)"; "Silberrückgewinnung (2. Fortsetzung)."
55. Methods for recovering silver from developed film and fixer baths have been known for several decades. The development of silver recovery techniques began in the early to mid-twentieth century, as the photographic industry sought ways to recycle and reuse silver, a valuable resource, from spent photographic materials. See, for instance, Stevens, Pirnazari, and Saavedra, "Removal of Silver from Photographic Solutions"; Sathaiyan, Adaikkalam, Kader, and Visvanathan, "Silver Recovery from Photographic Process Wastes."
56. Von Prittwitz, "Katastrophenparadox und Handlungskapazität," 350.
57. Von Prittwitz, "Katastrophenparadox und Handlungskapazität."
58. Anon., "Das Silberproblem."
59. Anon., "Das Silberproblem."
60. Epperlein and Spigiel, "Das Silber und die Fotografie," 19.
61. Dittrich and Grüßner, "Zum gegenwärtigen Stand," 357.
62. Dittrich and Grüßner, "Untersuchungen der Gerätetechnik" and "Zum Einfluß der Zusammensetzung."
63. Götz Pollakowski, "Zur Silberrückgewinnung," 124.
64. Epperlein and Liebscher, "Fotografie und Silber," 177.
65. Klaue, "Jeder Film ein Abenteuer," 296.
66. Anon., "Wieviel Silber."
67. Epperlein and Spigiel, "Das Silber und die Fotografie," 19.
68. Pietsch and Epperlein, "Hat die Silberhalogenidfotografie," 41.
69. Meyer, "Die Fotografie und das Silber," 312–13; Anon., "Fotos ohne Silbersalz"; Epperlein, "Modifizierte Verarbeitungsverfahren"; Epperlein, "Über einige neuere Entwicklungen."
70. Anon., "Informationen: Neuer Kodak-Film"; Anon., "Informationen: Silberarmer Film."
71. Huff, *Natur und Industrie*, 412.
72. Huff, *Natur und Industrie*, 413.
73. Huff, "Über die Umweltpolitik," 523.

Bibliography

A.B. "Die Rückgewinnung des Silbers aus Fixierbädern." *Bild und Ton* 2, *Beilage Fototechnik* (1949): 372.

Allen-Robertson, James. "The Materiality of Digital Media: The Hard Disk Drive, Phonograph, Magnetic Tape and Optical Media in Technical Close-Up." *New Media & Society* 19(3) (2017): 455–70.

Anklin-Mühlemann, Regula, and Rudolf Gschwind. "Script zur Vorlesung Farbphotographie." Unpublished dissertation, Universität Basel, 2001.

Anonymous. "Das Silberproblem in der Fotografie." *Bild und Ton* 22(8) (1969): 254.

———. "Fotos ohne Silbersalz." *Bild und Ton* 25(2) (1972): 49.

———. "Wieviel Silber gibt es noch?" *Bild und Ton* 29(2) (1976): 76.

———. "Informationen: Neuer Kodak-Film ohne Silber." *Bild und Ton* 31(2) (1978): 36.

———. "Informationen: Silberarmer Film für die Polygrafie." *Bild und Ton* 38(6) (1986): 163.

Bezirksverwaltung Halle. "Information über die Situation im Chemiefaserbereich des VEB FCK Wolfen bezogen auf Umweltbelastungen sowie die Anlagen- und Produktionssicherheit, vom 26.9.1988." *Dokumente zur neueren Geschichte des VEB Filmfrabrik Wolfen: Ausgewählte Dokumente und Quellen zur Filmherstellung 1964–1990*, edited by Jochen Steinecke, 420–22. Magdeburg: Verlag Steinecke, 2014.

Brune, Wolfgang. "Das Filmkopierwerk." *Bild und Ton* 10(10) (1957): 265–67.

Catanese, Rossella, and Jussi Parikka. "Handmade Films and Artist-Run Labs: The Chemical Sites of Film's Counterculture." *NECSUS: European Journal of Media Studies* 23 (2018).

Cherchi Usai, Paolo. "'I Can't Breathe': Extinction Rebellion to Film Preservation." *Journal of Film Preservation* 103(10) (2020): 9–13.

Daugaard, Noemi, and Josephine Diecke. "Farbfilmverfahren und Historiografie(n): Ein interdisziplinärer Ansatz." *ffk Journal* 5 (2020): 153–70.

DEFA Film Library. "Timeline of GDR Environmentalism." *Biology!* (DVD), 2022. Retrieved 5 September 2023 from https://www.umass.edu/defa/sites/default/files/Timeline per cent20of percent20GDR percent20Environmentalism_0.pdf.

Diecke, Josephine. "Agfacolor in (Inter)National Competition." *Color Mania. The Material of Color in Photography and Film*, ed. Barbara Flueckiger, Eva Hielscher, and Nadine Wietlisbach, 211–21. Zurich: Lars Müller / Fotomuseum Winterthur, 2020.

Dittrich, Klaus. "Messung und Registrierung der Entwicklungszeit als Mittel der Betriebs- und Bearbeitungskontrolle." *Bild und Ton* 29(8) (1976): 241–44.

———. "Messung und Bekämpfung von Chemikalienstäuben in Bäderansatzräumen von Kopierwerken." *Bild und Ton* 31(11) (1978): 333–39.

Dittrich, Klaus, and Theodor Grüßner. "Zum gegenwärtigen Stand der Silberrückgewinnung in Kopierwerken." *Bild und Ton* 33(12) (1980): 357–65.

———. "Untersuchungen der Gerätetechnik zur Entsilberung fotografischer Fixierlösungen." *Bild und Ton* 34(3) (1981): 69–77.

———. "Zum Einfluß der Zusammensetzung fotografischer Fixierbäder auf die Silberrückgewinnung." *Bild und Ton* 34(12) (1981): 357–62.

Epperlein, Joachim. "Modifizierte Verarbeitungsverfahren für Silberhalogenid-Medien mit verringertem Silbergehalt." *Bild und Ton* 36(6) (1983): 173–77.

———. "Über einige neuere Entwicklungen auf dem Diazo-Gebiet." *Bild und Ton* 36(1) (1983): 13–19.

Epperlein, Joachim, and Siegfried Liebscher. "Fotografie und Silber im Jahre 1988." *Bild und Ton* 41(6) (1988): 177–79.

Epperlein, Joachim, and Susanne Spigiel. "Das Silber und die Fotografie." *Bild und Ton* 31(1) (1978): 19–23.
Fengler, Silke. *Entwickelt und fixiert: Zur Unternehmens- und Technikgeschichte der deutschen Fotoindustrie, dargestellt am Beispiel der Agfa AG Leverkusen und des VEB Filmfabrik Wolfen (1945–1995)*. Vol. 18, Bochumer Schriften zur Unternehmens- und Industriegeschichte. Essen: Klartext-Verlag, 2009.
Flueckiger, Barbara. "Agfacolor Neu/Agfacolor." *Timeline of Historical Film Colors*. Retrieved 5 September 2023 from https://filmcolors.org/timeline-entry/1276/.
——. "Kodachrome." *Timeline of Historical Film Colors*. Retrieved 5 September 2023 from https://filmcolors.org/timeline-entry/1277/.
Fossati, Giovanna. *From Grain to Pixel: The Archival Life of Film in Transition*. Amsterdam: Amsterdam University Press, 2009.
——. "Obsolescence and Film Restoration: The Case of Colored Silent Films." *Technè: Conserver l'art contemporain à l'ère de l'obsolescence technologique*, edited by Hélène Breuil and Cécile Dazord, 102–6. Paris: Centre de recherche et de restauration des musées de France, 2013.
Grupinski, L., and G. Schreiber. "Die Bedeutung des Silbers in der Fotografie." *Bild und Ton* 4(4) (1951): 118–21.
Hardonyi, Robert. "Über die Sprühentwicklung." *Bild und Ton* 16(3) (1963): 72–74.
Huff, Tobias. *Natur und Industrie im Sozialismus: Eine Umweltgeschichte der DDR*. Göttingen: V&R Academic, 2015.
——. "Über die Umweltpolitik der DDR: Konzepte, Strukturen, Versagen." *Geschichte und Gesellschaft* 40 (2014): 523–54.
Jander, Alexander. "Die Weiterentwicklung der DEFA-Kopierwerke." *Bild und Ton* 19(5) (1966): 148.
——. "Die Entwicklung des VEB DEFA-Kopierwerke." *Bild und Ton* 22 (10) (1969): 292–96.
Keiler, Johann-Albrecht. "Über Möglichkeiten zur Leistungssteigerung bei der fotochemischen Kinefilmbearbeitung." *Bild und Ton* 23(9) (1970): 267–69.
——. "Beratung der Kopierwerkstechniker sozialistischer Staaten." *Bild und Ton* 30(3) (1977): 91.
——. "2. Beratung der Kopierwerkstechniker sozialistischer Staaten." *Bild und Ton* 31(8) (1978): 50–51.
——. "3. Beratung der Kopierwerkstechniker sozialistischer Staaten." *Bild und Ton* 33(6) (1980): 187.
——. "4. Beratung der Kopierwerkstechniker sozialistischer Staaten." *Bild und Ton* 34(10) (1981): 318.
Keiler, Johann-Albrecht, Reinhard Lorenz, Theodor Grüßner, and Götz Pollakowski. "Die Pastenentwicklung von Kinefilm." *Bild und Ton* 25(2) (1972): 37–42.
Keiler, Johann-Albrecht, and Götz Pollakowski. "Persulfat/Chion-Bleichbäder: Ökologische und ökonomische Aspekte." *Bild und Ton* 38(1) (1985): 23–25.
Klaue, Wolfgang. "Jeder Film ein Abenteuer: Vor fünfzig Jahren wurde das Staatliche Filmarchiv gegründet." *Apropos Film: Das 6. Jahrbuch der DEFA-Stiftung*, edited by Ralf Schenk, Erika Richter, and Claus Löser, 292–300. Berlin: Bertz + Fischer, 2005.
Knowles, Kim. *Experimental Film and Photochemical Practices*. London: Palgrave Macmillan, 2020.
Lovejoy, Alice. "Celluloid Geopolitics: Film Stock and the War Economy, 1939–47." *Screen* 60(2) (2019): 224–41.

Mansfeld, Sigurd. "Silberrückgewinnung aus verbrauchten Fixierbädern." *Bild und Ton* 10(2) (1957): 40.
———. "Silberrückgewinnung aus verbrauchten Fixierbädern (1. Fortsetzung)." *Bild und Ton* 10(3) (1957): 70.
———. "Silberrückgewinnung aus verbrauchten Fixierbädern (2. Fortsetzung)." *Bild und Ton* 10(3) (1957): 101–6.
———. "Sulfate. Sulfite. Sulfide." *Bild und Ton* 10(11) (1957): 295–99.
Maron, Monika. *Flugasche*. Frankfurt am Main: S. Fischer, 1981.
———. *Flight of Ashes*. Trans. David Newton Marinelli. London: Readers International, 1986.
Meyer, Kurt. "Die Fotografie und das Silber." *Bild und Ton* 22(10) (1969): 311–13.
———. "Über Phenidon-Entwickler." *Bild und Ton* 23(3) (1970): 85–87.
———. "Verfahren der Kinefilmentwicklung." *Bild und Ton* 23(11) (1970): 343–46.
Obertreis, Julia. "Von der Naturbeherrschung zum Ökozid? Aktuelle Fragen einer Umweltzeitgeschichte Ost- und Ostmitteleuropas." *Zeithistorische Forschung / Studies in Contemporary History* 9(1) (2012): 115–22.
Pietsch, Herward, and Joachim Epperlein. "Hat die Silberhalogenidfotografie eine Zukunft?" *Bild und Ton* 32(2) (1979): 37–41.
Pinch, Trevor J., and Wiebe E. Bijker. "The Social Construction of Facts and Artifacts: Or How the Sociology of Science and the Sociology of Technology Might Benefit Each Other." *The Social Construction of Technological Systems: New Directions in the Sociology and History of Technology*, edited by Wiebe E. Bijker, Thomas P. Hughes, Trevor J. Pinch, and Deborah G. Douglas, 17–50. Cambridge, MA: MIT Press, 1987.
Pollakowski, Götz. "Kopierwerksabwässer und Umweltschutz (I)." *Bild und Ton* 36(5) (1983): 139–42.
———. "Kopierwerksabwässer und Umweltschutz (II)." *Bild und Ton* 36(6) (1983): 182–84.
———. "Kopierwerksabwässer und Umweltschutz (III)." *Bild und Ton* 36(7) (1983): 207–10.
———. "Zur Silberrückgewinnung aus fotografischen Behandlungslösungen." *Bild und Ton* 40(4) (1987): 124–25.
Pollakowski, Götz, and Doris Boetzel. "Technik und Technologie der Kopierung und Bearbeitung von Kinefilm." *Bild und Ton* 28(9) (1975): 279–85.
Pollakowski, Götz, and Johann-Albrecht Keiler. "Bleichbäder und Umweltschutz." *Bild und Ton* 35(8) (1982): 250–52.
Pollakowski, Götz, Johann-Albrecht Keiler, Christoph Gerisch, Gerhard Fiedler, Theodor Grüßner, and Klaus Dittrich. "Bemerkungen zur Stabilität von Hexacyanoferrat-Bleichbädern." *Bild und Ton* 33(3) (1980): 87–88.
Präsidium des Ministerrates. "Bericht über die Sicherung des Exportes der fotochemischen Erzeugnisse des VEB Filmfabrik Wolfen." *Dokumente zur neueren Geschichte des VEB Filmfrabrik Wolfen: Ausgewählte Dokumente und Quellen zur Filmherstellung 1964–1990*, edited by Jochen Steinecke, 25–29. Magdeburg: Verlag Steinecke, 2014.
Richter, Christoph. "'Silbersee' in Bitterfeld: Sinnbild für die Ökokatastrophe." *Deutschlandfunk*, 8 November 2019. Retrieved 19 January 2021 from https://www.deutschlandfunk.de/silbersee-in-bitterfeld-sinnbild-fuer-die-oekokatastrophe.697.de.html?dram:article_id=462984.
Sathaiyan, N., P. Adaikkalam, J. A. M. A. Kader, and S. Visvanathan. "Silver Recovery from Photographic Process Wastes." *Encyclopedia of Materials: Science and Technology*, 8621–23. Amsterdam: Elsevier, 2001.
Stevens, M. R., M. Pirnazari, and F. Saavedra. "Removal of Silver from Photographic Solutions." *Studies in Environmental Science*, 23:525–34. Amsterdam: Elsevier, 1984.

Sujew, W. "Konstrukteur und Wissenschaftler im Bunde gegen Umweltverschmutzung." *Bild und Ton* 27(3) (1974): 74, 90.

Von Prittwitz, Volker. "Katastrophenparadox und Handlungskapazität: Theoretische Orientierungen der Politikanalyse." *Policy-Analyse*, edited by Adrienne Héritier, 328–55. Wiesbaden: Springer Fachmedien Wiesbaden, 1993.

Wilkening, Albert. "Neue Filmentwicklungs-Verfahren." *Bild und Ton* 14(30) (1961): 69.

PART II

IMAGINARIES OF SELF AND OTHER

CHAPTER 4

Between Cooperation and Competition

Cold War Imaginaries and Representations of US-Soviet Encounters in Space

Darina Volf

Introduction

In the early 1960s, US astronauts began training for the first human flight into outer space. Although in the West, the general public knew little about their Soviet rivals, Soviet preparations for manned spaceflight were going full speed ahead as well. But even before Yuri Gagarin left the Earth for his historic flight in April 1961, the space race had begun on movie theater screens with the premiere of the Soviet science fiction film *Nebo Zovyot* (The sky beckons, 1959, dirs. Mikhail Karyukov and Aleksandr Kozyr, USSR).

Even though science fiction films envisioning space travel have a much longer history, before the late 1950s film productions in this genre rarely focused on national categories or international interactions. Rather, very early space films tried to imagine encounters of Earthlings with the inhabitants of other celestial bodies, most often Mars. Examples include: *Himmelskibet* (*A Trip to Mars*, 1918, dir. Holger-Madsen, Denmark), *Aelita* (1924, dir. Yakov Protazanov, USSR),[1] and the US serial *Flash Gordon's Trip to Mars* (1938, dirs. Ford Beebe and Robert F. Hill). *Le Voyage dans la Lune* (*A Trip to the Moon*, 1902, dir. Georges Méliès, France) envisioned beings on the Moon. Even in the early 1950s, after the Cold War had begun, space films did not focus on conflict between different nations or a competition to reach outer space but rather dealt with more fundamental questions, such as the rationale for space travel, and challenges related to the human advance into space.[2] When a race into space was mentioned at all, it was in passing, at best—as, for example in *Destination Moon* (1950, dir. Irving Pichel, USA) or *Project Moonbase* (1953, dir. Richard Talmadge, USA).

In the mid-1950s, with the Cold War simultaneously moving into a mode of competitive coexistence and extending into the newly evolving field of space technology,[3] the idea of outer space as an arena for national, or rather systemic competition started to appear more frequently in filmic representations as well. This new focus is hardly surprising, since the films were both products and reflections of this period shaped by an emerging space rivalry. But was there any room left for visions of space cooperation?

This chapter reveals that Cold War cultural and media imaginaries and representations of international encounters in space oscillated between cooperative visions and competitive displays of national achievements. Focusing on feature films and the press, powerful mass media of the time, it is based on the assumption that these media played an important role in co-constructing public discourses about space travel and exerted significant impact on space policies and technologies. Visions of space cooperation generated and disseminated in public discourses helped to pave the way for the first international space missions and thus fostered the development of technologies for cooperative projects, such as compatible docking mechanisms. This impact is best understood by conceptualizing space technology as a "public technology"; the notion of technologies as "publicly shaped entities" draws attention to the role played by cultural and societal influences in the creation and transformation of technology.[4]

The coexistence of cooperation and competition in Cold War outer space imaginaries can be explained by considering the specific nature of space exploration. Achievements in space were perfectly suited to showcasing national prowess and strength, but the advance into outer space was also "a project that transcends national claims and appeals to the global."[5] The 1959 Soviet film *Nebo Zovyot* offers a good example of the coexistence of competition and cooperation with respect to cultural representations of space exploration. According to the East German newspaper *Neues Deutschland*, the film focuses on the choice between the "peaceful joining of all creative forces of the world for the purpose of fast space exploration oriented to the common-good" and the "unreasonable rivalry for prestige and supremacy at any cost."[6]

While the practice of space exploration was more in accordance with the latter statement, the idea of joint space exploration entered public discourse through media and popular culture imaginaries. The first part of this chapter analyzes this discourse. To show that similar ideas evolved at about the same time on both sides of the Iron Curtain, I will use two films that follow a very similar plot: *Nebo Zovyot* and an episode from the US television series *Men into Space*, produced by Ziv Television Programs in 1959–60. The East German–Polish coproduction *Der schweigende Stern* (*The Silent Star*, 1960, dir. Kurt Maetzig, GDR/Poland) moreover proves that visions of cooperative endeavors in outer space were also developed in non-space-faring countries, which also supports the argument that interest in space-related topics was widespread at the time and not limited to the United States and the USSR.[7]

The second part of this chapter explores the transformative power of such imaginaries. Drawing on the concept of sociotechnical imaginaries, it engages in particular with the idea of space cooperation, as imagined and disseminated to the public, and examines how such public discourses then reflexively shaped the practice of space exploration. Science studies scholar Sheila Jasanoff defines sociotechnical imaginaries as "collectively held, institutionally stabilized, and publicly performed visions of desirable futures, animated by shared understandings of forms of social life and social order attainable through, and supportive of, advances in science and technology."[8] According to Jasanoff, science fiction serves as one repository of sociotechnical imaginaries.[9] By presenting space cooperation as a desirable future, science fiction films helped promote the idea of space cooperation. There is an anecdote suggesting that the 1969 Hollywood film *Marooned* (dir. John Sturges, USA) inspired the 1975 Apollo Soyuz Test Project (ASTP), the first joint US-Soviet manned space mission.[10] At the same time, the shift to cooperative projects and practices did not completely wipe out the sense of competition among the partners. On the contrary, archival sources presented in this part of the chapter suggest that, behind the scenes, both ASTP cooperation partners were still driven by competition, and that this motivation affected not only the public presentation of the project but also specific managerial and technical choices and solutions within the ASTP.

When the relationship between the United States and the USSR again deteriorated into rivalry in the late 1970s, space competition came to the fore, culminating in President Ronald Reagan's announcement of the Strategic Defense Initiative in 1983. A brief look at the 1980s, in the third section of this chapter, however, shows that despite the renewed dominance of the competitive paradigm in the relations of the superpowers and their abandonment of cooperative endeavors, the idea of cooperation remained present and shaped the manner in which the superpowers' competition was carried out. Furthermore, despite weak prospects for a new US-Soviet space mission, the idea of their cooperation was maintained through media and cultural imaginaries until the chances for a new rapprochement improved in the mid-1980s.

Divided We Fall: Imaginaries of Space Cooperation at the Peak of the Space Race

In the late 1950s, the USSR and the United States were each working on manned orbital capsules to send the first humans into space. Even though Leonid Sedov—the head of the Soviet delegation of scientists to a meeting of the American Rocket Society in November 1959—denied that there were Soviet plans to put a man in space, it was evident that the American public and experts expected the first Soviet manned spaceflight in the very near future.[11] Several US newspaper articles posed the question of whether the US would be able to surpass the Soviet Union, thereby implying the existence of a competition.[12] On the other side of the Iron Curtain, all the information related to the Soviet space program was kept secret, but there was great interest in information about the US space program. Historian Asif Siddiqi has demonstrated that Soviet scientists and engineers involved in the space program followed the coverage from the United States very carefully. Articles about the advancement of the American Mercury program had a direct impact on Soviet plans for the manned spaceflight, and the Soviet decision to launch the first *Sputnik* satellite was fostered by media reports of US preparations for the launch of a satellite.[13] Valentina Ponomaryova, a member of the first Soviet group of female cosmonauts, also confirmed in an interview the significance of US plans for Soviet space program decisions.[14] Space endeavors were thus considered by both sides to be a race between the superpowers, with media coverage and public debates playing a key role in the development and perpetuation of this competitive way of thinking.

Nebo Zovyot addressed the idea of a space competition as a problem, as its actual message was the promotion of space cooperation. The film depicts the efforts of a Soviet team, led by space scientist Evgeny Kornev (Ivan Pereverzev), to land on Mars and a similar US mission, led by astronaut Robert Clark (Konstantin Bartashevich), which is accompanied by the sensation seeking journalist Erwin Verst (Gurgen Tonunts). The choice to forefront a Mars mission is not surprising; even though the destination for the real space race would eventually become the Moon, Mars had captured the fantasy of Soviet space enthusiasts early on.[15] Moreover, the description of Mars as the "red planet" had clear ideological undertones, and the film's impressive images of a red Mars rising might be interpreted as a symbol of the rise of "red communism."

When the US crew must make a stopover at a Soviet space station, Kornev proposes to Clark that they join forces in working toward their common goal, to reach Mars. Clark reacts with surprise to his rival's offer, but Kornev explains that space exploration is an endeavor that concerns all people and should be conducted for the benefit of all mankind. Though Clark does not seem averse to Kornev's proposal, he is unable to accept because of his superiors in the United States.[16] Even though the joint flight to Mars does not materialize, the rivals are ultimately brought together again, because the US spaceship runs into trouble on its flight to Mars. Instead of a triumphant landing on the red planet, the Soviets rescue the US spaceship *Typhoon*, carrying Clark and Verst, and the film ends with a joint return to Earth. Because of the rescue mission, the Soviet spaceship *Rodina* (Motherland) is not able to set out for Mars; nevertheless, the *Rodina* crew is greeted by cheering crowds celebrating the safe return of both crews to Earth. The film's central message thus appears to be that cooperation and mutual help are both necessary and desirable—even if they entail sacrifice—and far superior to competition and rivalry. This message is underlined by Kornev, who ultimately expresses his hope that the incident has become a lesson about pointless rivalries in space.

Despite this explicit message—promoting space cooperation and condemning rivalry for space supremacy—competition remains subtly present in *Nebo Zovyot*. It manifests itself in the clearly defined criteria that are used to assess the performance of each competitor in their effort to conquer Mars. These criteria involve the commitment to universal, humanistic norms and values, such as altruism, mutual aid, and cooperation regardless of national borders. In this respect, the film presents its audience with an unequivocal winner: it is the Soviet, Kornev, who takes the high road, first proposing a joint flight and later interrupting his attempt to reach Mars without hesitation; and it is Kornev who extols the benefit for humanity of helping each other and renders space rivalry meaningless. In the process, *Nebo Zovyot* additionally depicts a Soviet cosmonaut sacrificing himself to save lives. In contrast, the American space explorers appear reckless, selfish, and more interested in sensationalism than scientific progress and the well-being of all. Under pressure from their profit-seeking superiors, the US team declines the Soviet offer to combine forces and sets out for Mars alone, despite unfavorable conditions; the overhasty launch puts the life of a Soviet cosmonaut working on the launch pad at risk.[17]

Nebo Zovyot was not only distributed and screened in the socialist bloc but also re-edited and adapted for the US market under the title *Battle Beyond the Sun* (1962, dir. Francis Ford Coppola, USA). In order to obfuscate its Soviet origins and remove all pro-Soviet bias, *Battle Beyond the Sun* starts by presenting US developments in space technology, and all the characters are given English-sounding names. Unlike the original, in *Battle Beyond the Sun* the competitors in the space race are not Americans and Soviets but two fictional empires called North and South Hemis. According to the narrator, these empires arose in the wake of the Great Atomic War, which is represented in a scene displaying a mushroom cloud that was absent in the original film.[18] Ultimately, the end of the American version differs significantly from the Soviet original. While *Nebo Zovyot* concludes by promoting personal friendship among the members of the Soviet and American crews, without raising any hopes of rapprochement between the two countries, *Battle Beyond the Sun* presents the rescue in space as a first step to ending the rivalry on Earth. In the words of the narrator in the US version: "The way was now open for both sides to begin working together in peace and, out of their combined knowledge and effort, all men would come to know a better world in which to live."[19]

At about the same time, a similar idea of transnational space cooperation appeared in the East German–Polish coproduction, which was based on Stanisław Lem's novel *Astronauts* and titled *The Silent Star*. In contrast to *Nebo Zovyot*, *The Silent Star* presents us with an international and racially diverse crew on a joint flight to Venus. The crew consists of scientists from eight different countries, including the Soviet Union, the United States, Japan, and an unspecified country in Africa. There are no direct references to the space race, and the main adversary of the Earthlings is the population of Venus, which—the scientists later learn—had planned to exterminate humankind but instead perished in a self-induced nuclear disaster. Nevertheless, here too Cold War ideological rivalries find their way into the film.[20] For instance, *The Silent Star*, contains references to selfish American capitalists who try to prevent the American professor Hawling from participating in the Soviet-led mission, as well as references to the nuclear destruction of Hiroshima. *The Silent Star* became one of the most expensive but also most successful productions of the GDR state-owned DEFA Feature Film Studio.[21] As in the case with *Nebo Zovyot*, a reedited, shortened, and adapted US version of the East German–Polish coproduction entered American and British markets under the title *First Spaceship on Venus* in 1962. This second version was "cleaned up," with no ideologically pro-socialist elements or references to the bombing of Hiroshima.

It is no surprise that parallel ideas—of mutual assistance in outer space and of space cooperation transcending national borders and ideological boundaries—also appeared on the western side of the Iron Curtain. The US television series *Men into Space* (1959–60) presents future space travel as an international enterprise.[22] The episode titled "Mission to Mars" (May 1960, dir. William Conrad, USA), for example, bears a startling similarity to *Nebo Zovyot*, with a significant difference: this time, it is the Americans who interrupt their mission to Mars to help the Soviet crew, who are obsessed with winning the space race. Gary Westfahl argues that, despite its opening scene being an actual race in space, "Mission to Mars" is the "episode [of the show] that works hardest to counter feelings of competitiveness in space."[23] As we saw in *Nebo Zovyot*, however, competition between the film's ideological rivals is not erased from the film's narrative but rather transformed and refocused. Unsurprisingly, the American space program is presented as superior to the Soviet program: the US crew members carefully plan the launch, show respect for human life, and are prepared to help each other in their time of need. These values are especially foregrounded at the end of the episode, when the US president welcomes astronaut Col. Edward McCauley (William Lundigan) and praises his rescue mission: "By abandoning your mission to save human lives, you have at one stroke done more to make the American position clear than anything that has been done in the last fifty years."[24]

A few years later, shortly after the US landing on the Moon in July 1969, the film *Marooned* entered US and European cinemas. The film, based on Martin Caidin's eponymous 1964 novel, is about a US spacecraft having difficulties returning to Earth. Unlike the films discussed above, however, *Marooned* presents no clear dichotomy: not between good and evil, superior and inferior space programs, or helpful and helpless nations. When the US crew encounters technical problems, the Soviet spaceship unexpectedly appears to help, but it is too small to bring them all back to Earth, and the American rescue mission must be postponed because of unfavorable weather conditions. Neither of the two space powers is able to save the lives of the stranded astronauts on its own, and the rescue mission's ultimate success depends on joint Soviet-US efforts.

Despite the generally favorable depiction of both superpowers in the film, there are some references to anxieties on the part of the Americans regarding the Soviets' motivations and actions. When the Soviet spacecraft approaches the marooned Americans, ground control suggests that US crew members observe them closely, since no one knows what they intend to do.[25] That Soviet help shows up is also unexpected for the American crew, as earlier in the film NASA administrator Charles Keith (Gregory Peck) inquired whether the Soviets could help and was told the trajectory of the Soviet spacecraft would make it impossible.[26] In the media coverage context of the time, this scene might have been interpreted as highlighting the openness of the American space program, in contrast to the Soviets' secrecy—which the US often identified in the media as the main obstacle for closer American-Soviet space cooperation.[27] Nevertheless, the depiction of the Soviets in *Marooned* is more positive than that in the "Mission to Mars" TV episode analyzed above, indicating that the general perception of space rivalry between the United States and the USSR had shifted in the decade between the end of the 1950s and the late 1960s.

In their sociopolitical contexts, these science fiction productions served as a rich source for the development of what Sheila Jasanoff has called "sociotechnical imaginaries."[28] By envisioning technological innovations—such as orbital space stations or spacecraft capable of landing on distant planets—science fiction films and TV shows presented forward-looking fantasies of what was technically attainable. At the same time, they portrayed visions of future social orders and disseminated normative views about what was socially desirable. They stressed the advantages of joining forces in space and underlined appeals for cooperation by pointing out dystopic futures, represented by images of nuclear destruction in both East and West.

In the second half of the 1960s, such individual visions of space cooperation became what Jasanoff describes as "collectively held reference points."[29] The idea of space cooperation became more powerful and started overtaking the previously dominant perception of space as an area for national and systemic competition. Changing media coverage also contributed to the further spread of imaginaries originating in the science fiction genre while reinforcing cooperative visions of space explorations. As a result, by the end of the 1960s, public discourse, as reflected in the press, articulated space cooperation as a desirable norm, and public expectations toward space programs changed accordingly.

US media's emphasis on the desirability of, as well as US interest in, space cooperation was not entirely new in the second half the 1960s.[30] Addressing Soviet doubts about the seriousness of the recent US interest in cooperation, the *New York Times* admitted in 1965 that there was reason to believe that previous US proposals for cooperation had been motivated by the fact that the country was lagging behind the USSR in the past; however, after the success of the US Gemini 5 mission, the newspaper continued, "there should be no illusions in Moscow now about the genuineness of the President's new offer of partnership."[31] In contrast to this positive image of US interest in cooperation, both the *New York Times* and the *Washington Post* depicted the Soviets as responsible for the current stalemate in developing joint space efforts.[32] Analogous to filmic representations, the media thus pushed the idea of space cooperation, even as it fostered a certain competition by evaluating the cooperative performance of the Cold War rival. According to the *Washington Post*, for example, there was a clear winner of the competition for cooperation: "While the United States has been a tireless suitor for Soviet cooperation, its overtures have been either ignored or met so half-heartedly as to make the resulting projects almost meaningless."[33]

The Soviet Union also tried to present its space program as peaceful and cooperative. Articles about space exploration in Soviet government-controlled media such as the major newspapers *Izvestia* and *Pravda* often concluded with a phrase affirming the USSR's interest in developing international space cooperation.[34] To demonstrate cooperativeness, the party leaders and space program authorities referred to examples of cooperation with other socialist countries.[35] At the same time, Soviet leadership claimed that the Soviet Union was generally open to cooperation with all nations as a means to promote progress and peace, both on Earth and in outer space. These connections in official Soviet media discourse—between promoting peace on Earth, friendship among nations, and cooperation in space—gave the USSR a justification for its reluctance to establish cooperation with the United States usually depicted as a threat to global peace. Accordingly, a declaration published in *Pravda* after the first spacewalk mission in 1965 stressed the USSR's interest in "space cooperation with all nations"; but the very next paragraph condemned the aggressive imperialism of the United States, especially in Vietnam, indicating that prospects for US-Soviet space cooperation were rather weak.[36] In 1970, however, president of the USSR Academy of Sciences Mstislav Keldysh responded positively to a letter from NASA administrator Thomas Paine, which proposed an expansion of the limited space cooperation between the United States and the USSR. It was this exchange of letters between Keldysh and Paine that paved the way for the first joint manned space mission—the ASTP.

The ASTP: Cooperation before the World's Eyes, Competition behind the Scenes

To understand this Soviet shift toward cooperation with the United States in 1970, it is useful to revisit the idea of the public co-construction of technology and examine the influence that public perceptions, opinions, and expectations voiced in the media had on space programs and, thus, on technologies developed to achieve program goals. As we saw above, the authorized NASA history of the ASTP emphasizes the impact of the film *Marooned* on Philip Handler, the president of the National Academy of Sciences. Handler said that he mentioned the film to Soviet officials during his visit in Moscow in May 1970 and that the Soviet side was impressed, in particular by the fact that a Soviet cosmonaut was a hero in an American film.[37]

In fact, the story was not entirely new to the Soviets, as Caidin's novel had been released in the Soviet Union in 1967 under the title *V plenu orbity* (Trapped in orbit).[38] Nevertheless, references to the film became a popular narrative when telling the history of the ASTP. One week prior to the launch of the American Apollo and Soviet Soyuz spacecrafts, for example, the British *Daily Telegraph* introduced an article with a reference to the film and continued: "Next week Life follows Art as the Americans and Russians attempt the first-ever link-up of their spacecraft in the Apollo-Soyuz Test Project."[39] On the other side of the Iron Curtain, the film was mentioned in the Czechoslovak book *Sojuz volá Apollo* (Soyuz calls Apollo), published in 1976.[40] It is unsurprising that the idea that a film inspired the beginning of the superpowers' cooperation in the manned spaceflight was so appealing to different actors and adopted for various audiences. It was able to both explain the seemingly sudden shift from fierce rivalry to friendly cooperation and hide the real motivations and interests behind the shift. In addition, it incorporated an attractive narrative about an enlightening experience that changed everything.

In actuality, the space authorities' interest in cooperation resulted from competitive rather than cooperative considerations, however. While developing compatible rescue systems to enable emergency assistance became an important and repeated public justification for the ASTP,[41] behind the scenes both the American and Soviet authorities hoped to maintain or improve the competitiveness of their respective space programs through cooperation.[42] The "impact" of *Marooned* on superpowers' cooperation should also be understood in this way. It is hardly plausible that a film, alone, would explain the sudden Soviet acceptance of the US offer of cooperation. Still, I argue that there was an impact—but one that was indirect, mediated, and only to be understood in the context of other films and media coverage of the time. Together, these contributed to the spread of positive ideas about cooperation in public discourse, and the discursive demand for increased space cooperation tied in with other motives that space actors had to cooperate.

One of these motives was NASA's interest in discovering more about the Soviet space program: how it functioned and, in particular, what were its future plans. Keeping their plans secret was an advantage to Soviet space officials, not least because it enabled them to retroactively deny the existence of goals they did not achieve, such as a manned lunar landing. The secrecy around Soviet endeavors in space was a disadvantage to NASA in the competition for public recognition and prestige, also, because NASA constantly feared Soviet "surprises" similar to the launch of *Sputnik* in October 1957. Cooperation was thus seen as a means of getting more direct access to the Soviet space program, as NASA deputy administrator George Low explained during a hearing before the Senate Committee on Aeronautical and Space Sciences in 1972.[43] During his visit to the United States, Academy of Sciences president Keldysh was reportedly told that one reason behind the US interest in cooperation was fear of renewed Soviet surprises.[44]

Unlike NASA, which had little information about the Soviet space program, Soviet space authorities could benefit from the openness of information about the American space program and adjust their plans and schedules accordingly. The Soviet side was interested in cooperation with NASA for different reasons, albeit reasons still rooted in competitive rather than cooperative thinking. Already in the early days of space exploration, Soviet space scientists had felt pressure to demonstrate their interest in cooperation to the international community and present themselves as reliable partners for joint space projects. When, in 1960, the Soviet Union refused to participate in the International Academy of Astronautics, as the Soviet government considered the arguments against cooperation weightier,[45] the leading scientists of the

Academy of Sciences of the USSR were aware that it could be used to accuse them of being uncooperative.[46]

As NASA's international space activities increased, the Soviet Union came under increasing pressure to cooperate.[47] In the early 1970s, this pressure was reinforced by European as well as US media, as they reported more and more frequently on NASA's attempts to initiate cooperation, emphasizing the ignorance and reluctance of the Soviets. This intensified media coverage in favor of space cooperation can be explained first and foremost by NASA's public relations work. In his public statements, for example, NASA administrator Paine spoke of a number of letters sent to the Soviets and complained about not receiving any answer.[48] The idea of space cooperation was also taken up and disseminated in European public discourse, and the media in France, Germany, and Great Britain reported on US attempts to approach the Soviets and initiate cooperation.[49] In light of the Cold War rivalry, the Soviet Union could therefore not afford to stand aside and allow NASA to claim unchallenged leadership in international space activities.

The relaxation of tensions in several other areas of the East-West conflict was another factor in the increased references to international space cooperation in early-1970s media coverage. First, the détente promoted by general secretary of the Communist Party of the USSR Leonid Brezhnev and US president Richard Nixon from 1969 to 1974 somewhat neutralized the fears that had previously defined their relations, creating better conditions for political rapprochement between the superpowers. Second, the extension of contacts and cooperation between the United States and the Soviet Union in the less politicized and less prestige-laden domains of science, technology, and trade demonstrated that broad cooperation was possible. Third, famous photographs of the Earth taken by Apollo astronauts in space—including *Earthrise* and *Blue Marble*—raised awareness of the Earth's vulnerability and the common destiny of humanity, whose interests were better served by international cooperation than conflict.[50] Fourth, the public was captivated by images of "space brotherhood," as historian Thomas Ellis has demonstrated in his article focusing on public representations of interactions between cosmonauts and astronauts.[51] Finally, as mentioned above, a normative shift generated by the dissemination of sociotechnical imaginaries via popular culture, particularly science fiction films, influenced public perception of the feasibility and desirability of cooperation in outer space.

In the United States, public interest in the space race with the Soviet Union decreased, while support for cooperation between the superpowers increased. While rejection and skepticism seem to have prevailed in the early 1960s, with most US citizens finding it fairly important to be ahead of the Soviets in space exploration, by the mid-1960s the situation was changing.[52] In 1965, more than 50 percent of those polled said that it did not matter at all whether or not the Soviet space program was ahead.[53] Moreover, public support for space spending was decreasing.[54] By 1969, when the Moon race ended, 60 percent of US citizens favored joint US-Soviet exploration of outer space.[55] Though we do not have similar opinion polls from the USSR, some articles published in the Soviet press in the late 1960s and early 1970s attempted to justify space spending, thereby indicating that public support for space exploration was also declining in the Soviet Union after the initial enthusiasm about *Sputnik*, Gagarin, and other space achievements had dissipated.[56]

Due to hitherto limited access to Russian archives, it is difficult to reconstruct how and to what extent global public opinion and reporting on space cooperation affected Soviet space program decisions. Nevertheless, those involved in Soviet space activities closely monitored US newspaper articles about US cooperation proposals and Soviet noncooperation and often translated them into Russian.[57] Moreover, a report prepared in the early 1970s reflected on the possibility of extending international space cooperation to offer a counterweight to the image of the United States as the leading power in cooperative endeavors in outer space.[58] The close monitoring of newspaper articles and musings regarding potential cooperation indicate that Soviet officials were concerned about how the Soviet lack of cooperation looked, especially in comparison to the United States. Given these pressures, an ambitious cooperative space project may have been seen as a perfect opportunity to dispel doubts in the media about the Soviet willingness to cooperate.

An additional motive for Soviet space officials to cooperate with NASA arose with the triumphant US lunar landing in 1969. Since US astronauts seemed to have overtaken the Soviets in the space competition, Soviet space program authorities took up the challenge to prove the superiority, or at least parity of its space program. This was all the more important as Brezhnev was less enthusiastic about the space program than his predecessor had been and preferred space missions that promised political benefits.[59] A cooperative Soviet-US endeavor where the United States and the USSR would meet at eye level would signal US recognition of the Soviets as an equal partner in space, which carried political implications as well. The US participants in the ASTP were well aware of this Soviet motive, and NASA's public affairs office recommended declining a Soviet proposal to make a joint movie about the ASTP, with the argument that the film would help the Soviets in their ambition to demonstrate space parity.[60]

Though there was no joint film project, the American ASTP team could not prevent the joint mission from being presented to the global public as a project conducted on an equal footing, because the demonstration of space parity was important to retain Soviet cooperation. The demonstration of parity had a profound effect on all project decisions that required the consent of both parties, including the ASTP's public presentation, management, and working language, as well as the technology developed. Since the use of existing docking technology would require that one spacecraft play the active and one the passive role during the docking, US and Soviet ASTP engineers agreed to develop an androgynous docking mechanism.[61] This allowed both spacecrafts to fulfill either the passive or the active role, so that neither would appear inferior. Because US and Soviet space technology had developed along different lines in the years before the ASTP, the shift to cooperation required further adjustments and technological innovation. For instance, an airlock module was needed to resolve the problem presented by the different atmospheres in the Apollo and Soyuz spacecraft. The ASTP can thus be seen as a clear example of the "public co-construction of space technology."[62] The developed docking mechanism and airlock module were concrete results of the impact of demands for cooperative space exploration in public discourse, as well as the decisions of space officials in light of their public image regarding the development of space technologies and specific technical choices.

The demonstration of parity was also important for decisions made about the public representation of the project, at least for those made jointly by US and Soviet ASTP participants. For example, at the Paris Air Show in 1973, a model of the docked Apollo and Soyuz spacecraft was exhibited in a space equidistant from the US and Soviet pavilions.[63] Public representations of the ASTP that were not coordinated and mutually agreed upon, however, enabled the rivals to present their partisan views on the ASTP. For instance, each side released its own publication describing the mission.[64] In 1976, the book *Soyuz i Apollon* (*Soyuz and Apollo*), written by Soviet ASTP engineers and scientists, was released by the publishing house Politizdat; in the US in 1978, a book on the ASTP titled *The Partnership: A History of the Apollo-Soyuz Test Project*, written by historians Linda and Edward Ezell, was published by NASA. Both books praise the idea of space cooperation and emphasize the benefit of the project for progress and humanity;[65] but apart from these cooperative aspects, each publication also includes oblique references that tout national superiority and achievements in space. *Soyuz i Apollon* begins with the launch of Sputnik and other—mostly Soviet—achievements in space but does not mention the lunar landing.[66] In contrast, *The Partnership* discusses the landing on the Moon in detail but mentions *Sputnik* and Yuri Gagarin only in passing. Furthermore, while the Soviet account traces the idea of space cooperation back to the Soviet space pioneers Konstantin Tsiolkovsky and Sergei Korolev and emphasizes Soviet efforts to establish international space cooperation, the US account presents the United States as the driving force behind international space activities and the initiative of NASA officials as the birth of the ASTP.[67]

Such details demonstrate that competition was present even within the joint US-Soviet ASTP space mission, which had a commitment to cooperation at its core. As public attention changed emphasis—shifting from being the first to being the more cooperative superpower—both space programs had to adjust to the challenge and show off their commitment to cooperating in the service of humanity. In public, US and Soviet ASTP participants presented the joint mission as the first step in establishing long-term space cooperation. However, a closer look at the motives, practices, and presentation of the ASTP to the outside world reveals that the ASTP did not entail an abandonment of space competition but rather its continuation by other means.

After ASTP: The Odyssey between Cooperation and Competition Continues

After the high point of the ASTP mission—the famous "handshake in space" in July 1975—the global public eagerly awaited news about the next steps in US-Soviet cooperation, and media coverage raised expectations of a prompt resumption of joint space activities. In October 1975, the *New York Times* reported that the crews of the joint Apollo-Soyuz flight favored new shared missions, quoting Soviet cosmonaut Alexei Leonov as saying that "it would be a shame for the United States and Soviet Union to maintain parallel space programs without cooperative ventures."[68] *Pravda*, in turn, quoted US astronauts Thomas Stafford and Donald Slayton, who both believed in the continuation and expansion of Soviet-US space cooperation.[69]

As before, Soviet space program authorities in the Academy of Sciences closely monitored and collected articles from the Western press, which mainly presented the successful joint flight in a positive light and reflected on the possibility and timing of the next US-Soviet mission.[70] While the space program managers under Keldysh's leadership at the Academy of Sciences had been reserved about discussing further cooperative projects until the ASTP had been successfully completed, after the mission they pushed for a prompt decision on the next joint flight. However, NASA now chose a more cautious approach and postponed planned talks several times. Then, with the end of détente and the deterioration of US-Soviet relations in the late 1970s, the prospects for another cooperative venture in space dwindled even further, and, at the beginning of the next decade, the relationship between the two space programs was once again characterized by intensifying competition.

Nevertheless, in the previous years the idea of international space cooperation had become embedded in public debates in East and West and did not disappear from the civilian space program. US and Soviet officials continued efforts to present their space activities as peaceful, cooperative, and oriented toward the common good—with the sole difference being that they no longer cooperated with each other. On both sides, cooperation in space now took place within the ideological alliances of the Cold War.[71] The Soviet Union invited cosmonauts from socialist countries and countries sympathetic to socialism to participate in its manned spaceflight, loading the flights with political symbolism.[72] The US authorities undertook cooperation within the Spacelab program, which developed a science laboratory in concert with Western European scientists and offered Western European astronauts the opportunity to travel into space from 1983 to 1998. By disseminating such cooperative images of their space programs, the Soviet Union and the United States found themselves competing at cooperation.[73]

Though the prospects for joint US-Soviet space missions decreased in the early 1980s, the idea of superpowers' cooperation in space exploration was kept alive in cultural and media representations. For example, both Arthur C. Clarke's 1982 novel *2010: Odyssey Two* and its film adaptation, *2010: The Year We Make Contact* (1984, dir. Peter Hyams, USA) demonstrated the benefit and necessity of cooperation between the two superpowers. To learn the cause of the failure of the *Discovery One* mission to Jupiter—the theme of Clarke's first novel, *2001: A Space Odyssey* and its film adaptation (1968, dir. Stanley Kubrick, UK/USA)—the United States and the Soviet Union depend on each other, as neither nation is able to fulfill the task on its own. In *Odyssey Two*, despite worsening tensions on Earth and instructions to return to Earth separately, the US and Soviet crews must join forces to confront worrisome incidents in space, which eventually impel US and Soviet political leaders to seek peace on Earth. The novel and film further reminded audiences of the real joint US-Soviet space venture of the past, as the Soviet spaceship in the film is named after cosmonaut Alexei Leonov, the commander of the Soyuz spaceship in the ASTP.

With manned spaceflight to Jupiter, *Odyssey Two* imagined a technically possible future. More importantly, however, it drew a line between desirable and undesirable futures and presented the socially desirable one, making the film a fitting example of Jasanoff's concept of sociotechnical imaginaries. In an interview, both director Hyams and author Clarke clarified that they had intentionally promoted the idea of space cooperation at a time when the political climate for joint US-Soviet space ventures was unfavorable. Moreover, Clarke cited the narrative of *Marooned* as an inspiration for the ASTP, thereby implying that he felt his novel and the film could possibly be the impetus for a renewed attempt at superpower space cooperation.[74] This confident statement demonstrates that the creators of science fiction were aware of their power to shape public discourses by presenting their vision of a desirable society and exercised it accordingly. Amid Cold War tensions, *2010: The Year We Make Contact* thus issued a warning by portraying a world on the verge of nuclear war but concluded with an optimistic vision of peace on Earth.

A *New York Times* article underlined Clarke and Hyams's call for cooperation and peace by quoting politicians and scientists who were afraid of the militarization of space triggered by Reagan's Strategic Defense Initiative attempt to create a US missile-defense system. In the same article, Clarke and Hyams demanded the pooling of resources and efforts, arguing that, without cooperation in a task as monumental as space exploration, civilization was doomed.[75] On 13 September 1984, a videotape of Clarke pleading for the renewal of Soviet-American space cooperation, in which he mentions superpower space cooperation as the main idea of *Odyssey Two*, was shown by political proponents of the renewal of space cooperation with the Soviet Union to the US Senate Foreign Relations Committee to bolster their argument.[76] That October, a joint resolution passed by both bodies of the US Congress impelled President Reagan to seek space cooperation with the Soviet Union.

This series of events clearly demonstrates how the perpetuation of the idea of space cooperation in public discourse, even at the peak of Cold War tensions, helped pave the way for the resumption of talks on space cooperation. In April 1987, the US and Soviet governments signed a new agreement to cooperate on space exploration, in which they proposed establishing joint working groups and agreed to a number of cooperative space projects.[77]

Conclusion

From the dawn of the space age in the 1950s, expectations, norms, and concerns articulated in public debates significantly shaped the development of space technology. In the heated atmosphere of the escalating Cold War, public discourse and media coverage charged the advancement of space technology with various meanings. Because the relationship between the Cold War rivals was marked by enmity, anxiety, and mistrust, each superpower interpreted any progress in space technology on the part of its rival as a potential threat to national security, which generated public pressure on decision-makers to promote and finance the development of technologies capable of responding to a security challenge. At the same time and before the eyes of the global public, the peculiarities of the Cold War made outer space a battlefield in the rivalry between systems and achievements in space a symbol of ideological superiority. Finally, the fact that there were only two nations initially able to travel into space fostered a perception of the space program as both a matter of national pride and respective Soviet and US exceptionalism.[78] All this resulted in an early understanding of space exploration in terms of a competition and placed the focus on the so-called "space firsts."

However, as this chapter has demonstrated, a shift in public expectations and norms in the second half of the 1960s—fostered first by the dissemination of cooperative visions in the science fiction genre and then in the print media—transformed public perception of the most desirable mode of space exploration from competitive to cooperative. Willingness to cooperate, broad participation, and the scientific value of space exploration became the new criteria for the public evaluation of superpower performance. The superpowers, in turn, had to adjust their space travel plans to ensure public support for their programs, though they kept the competitive aspect in mind even during their joint ASTP mission. The pivotal role in this shift in perception was played by the sociotechnical imaginaries of space films that envisioned human space ventures as an area of cooperation that could even overcome differences and rivalries on Earth. In the worrisome atmosphere of the Cold War, outer space, with its unknown challenges and potential dangers, and space exploration as a transnational task were perfectly suited to the projection of desirable sociotechnical futures of peace and cooperation on Earth. As these visions became embedded in public discourse, they eventually came to shape the development of space technology itself and contributed to the recalibrated directions of the space programs on both sides.

Darina Volf is a postdoctoral researcher at both the Collegium Carolinum Research Institute for the History of the Czech Lands and Slovakia and the University of Munich. Her research project is titled "The Apollo-Soyuz Test Project (ASTP): Cooperation Despite Systemic Rivalry" and focuses on the history of Soviet-US cooperation in outer space in the mid-1970s. Her further research interests include cultural history, the history of science and technology, the history of socialism, and memory studies. She is the author of *Über Riesen und Zwerge: Tschechoslowakische Amerika- und Sowjetunionbilder 1948–1989* [On giants and dwarfs: Czechoslovak images of America and the Soviet Union from 1948 to 1989]. Her most recent articles have been published in journals such as *Minerva* and *East Central Europe*.

Notes

1. For more on *Aelita*, see the chapter by Doru Aurel Pop in this volume.
2. Westfahl, *The Spacesuit Film*, 20.
3. On the role of the space race in the systemic Cold War competition between the United States and USSR, also see the chapters by Doru Aurel Pop and Ettore Costa in this volume.
4. Trischler and Bud, "Public Technology."

5. Siddiqi, "Competing Technologies," 425.
6. "Der Himmel ruft," *Neues Deutschland*.
7. The chapters by Ettore Costa and Doru Aurel Pop also substantiate this point.
8. Jasanoff, "Future Imperfect: Science," 4.
9. Jasanoff, "Imagined and Invented Worlds," 337.
10. The anecdote is mentioned in the official NASA history of ASTP. See Ezell and Ezell, *The Partnership*, 10.
11. At the time, Leonid Sedov was believed to be the father of *Sputnik*. The name of the "real father" of *Sputnik*, Sergei Korolev, became publicly known only after his death in 1966. For reports about the Soviet delegation to the United States, see Finney, "U.S.-Soviet Effort in Space Studied," *New York Times*; "Sedov Disavows Spaceman Plans," *Washington Post*.
12. Gamarekian, "We Hope to Orbit a Man by '61"; "Russians Seen Ready to Put Men in Orbit," *Washington Post*.
13. Siddiqi, *Challenge to Apollo*, 193 and 250.
14. Gerovitch, *Voices of the Soviet Space Program*, 219.
15. Examples are the silent science-fiction film *Aelita* (mentioned above), which is based on a novel by Aleksei Tolstoy, and Alexander Dovzhenko's unrealized screenplay *V glubinakh kosmosa* (In the depths of the cosmos, 1954). Siddiqi also points out (*Challenge to Apollo*, 5) that Mars was popular among Soviet engineers in the GIRD group (Group for the Study of Reactive Motion).
16. *Nebo Zovyot*, TC: 23:00–23:40.
17. *Nebo Zovyot*, TC: 25:00–26:50.
18. *Battle beyond the Sun* (1962, dir. Francis Ford Coppola, USA), 00:01:58–00:02:20. Retrieved 25 February 2021 from https://www.youtube.com/watch?v=R7Zp0BNQ3IU.
19. *Battle beyond the Sun*, TC: 01:06:00–01:06:30.
20. For more information on the difficulties that arose during the production of the film, see Soldovieri, "Socialists in Outer Space." Also see Heiduschke, "The Birth of DEFA Genre Cinema."
21. DEFA stands for Deutsche Film Aktiengesellschaft (German Film Company); the East German studios and production facilities were in operation from 1946 to 1992.
22. Westfahl, *The Spacesuit Film*, 54.
23. Westfahl, *The Spacesuit Film*, 67.
24. "Mission to Mars," TC: 23:30–24:00. (May 1960 episode of *Men into Space*, dir. William Conrad, ZIV Television Programs USA.) Retrieved 25 February 2021 from https://www.dailymotion.com/video/x6glvgu.
25. *Marooned*, TC: 01:55:00–01:55:50.
26. *Marooned*, TC: 52:30–52:50.
27. On Soviet secrecy, see Siddiqi, "Cosmic Contradictions."
28. Jasanoff, "Future Imperfect," 4.
29. Jasanoff, "Future Imperfect," 28.
30. See, for example, Schwartz, "Why Not a Soviet-American Gemini?"; "U.S. and Soviet Are Urged to Pool Space Efforts," *New York Times*; "It Takes Two," *Washington Post*.
31. "Cooperation in Space," *New York Times*.
32. See, for example, Simons, "Soviet Rejects Negotiated"; Finney, "Joint Space Test Stalled in Soviet."
33. Reistrup, "Two Space Giants Lack Cooperation."
34. "S vozvrashcheniem, slavnye syny otchizny" [Welcome back, glorious sons of fatherland], *Izvestia*.

35. "Rech tovarishcha B. N. Petrova" [Speech by comrade B. N. Petrov], *Pravda*.
36. "K kommunisticheskoi partii i vsemu sovetskomu narodu" [To the Communist Party and the whole Soviet nation], *Pravda*.
37. Ezell and Ezell, *The Partnership*, 9–10; Slayton, *Deke!* 277.
38. Martin Keydin [Caidin], *V plenu orbity*.
39. Gatland, "Space . . . Proving Ground for Détente."
40. Pacner, *Sojuz volá Apollo*, 7–8.
41. For example, the jointly prepared ASTP press kit identifies the development of compatible rendezvous and docking systems as the principal objective of the mission and emphasizes its significance in improving the safety of the spaceflight. See "Apollo Soyuz Test Project: Press Kit," 1975. National Archives at Fort Worth, USA. RG 255-E267 (ASTP Transmittals), box 5, 9.
42. The following is only a brief outline of the most important motives. For more about the evolution of ASTP, see Volf, "The Evolution."
43. "Hearing before the Senate Committee on Aeronautical and Space Sciences, material requested for the record on pages 54–5 by Senator Weicker," 23 June 1972. National Archives at Atlanta, USA. RG 255–77A0090, ASTP Files, box 2.
44. "Stenogramma zasedania Prezidiuma Akademii nauk SSSR" [Stenographic record of a session of the Presidium of the Academy of Sciences of the USSR], 14 December 1972. Archive of the Russian Academy of Sciences (ARAN), Moscow, Russia. Fond 579, opis' 13, delo 199.
45. The International Academy of Astronautics, a nongovernmental organization for encouraging international cooperation in space activities, was established in Stockholm in 1960.
46. "Ob uchrezhdenii Mezhdunarodnoi akademii astronavtiki v ramkach mezhdunarodnoi astronavticheskoi federatsii" [On founding the International Academy of Astronautics in the framework of the International Astronautical Federation], 13 May 1960. Russian State Archive of Contemporary History (RGANI), Russia, Moscow. Fond 5, opis' 35, delo 133, listy 66–68.
47. On NASA's international activities, see Logsdon, *After Apollo*, 109–10; Krige, "Introduction and Historical Overview," 14–16; Sebesta, "The Politics of Technological Cooperation."
48. Lyons, "NASA Seeks Links."
49. See, for example: "Zusammenarbeit in der Raumfahrt gefordert," *Frankfurter Allgemeine Zeitung*; "Vorschlag: Sowjetischer Kosmonaut," *Frankfurter Allgemeine Zeitung*; "Head of NASA Urges Close Cooperation," *Daily Telegraph*; *Air & Cosmos*, 7 February 1970, 15.
50. Geppert, "The Post-Apollo Paradox," 9–10.
51. Ellis, "'Howdy partner!'"
52. Karash, *The Superpower Odyssey*, 37.
53. Bainbridge, "The Impact of Space Exploration," 12.
54. Roy, Gresham, and Christensen, "The Complex Fabric," 670.
55. Bainbridge, "The Impact of Space Exploration," 51–52.
56. For example: Chembrovsky, "Zemlja—Kosmos—Zemlja"; Konovalov, "Iz pokoleniia tridcatiletnikh."
57. See "Sovetsko-amerikanskoe sotrudnichestvo v razlichnykh oblastiakh issledovania kosmicheskogo prostranstva—spravka" [Soviet-American cooperation in various fields of space research—information], 1971. ARAN. Fond 1678, opis' 1, delo 110; some translated articles are in Fond 1678, opis' 1, delo 93 and delo 142.

58. "Otchet po teme 'DAL'-4': Kniga 1: Opredelenie perspektiv razvitia kosmicheskikh issledovanii po nauchnym programmam mezhdunarodnogo sotrudnichestva" [Report on the topic DAL'-4: Book 1: The definition of perspectives for the development of space exploration in international space programs], 15 July 1974. ARAN. Fond 1862, opis' 2, delo 10, listy 7–9, 78, and 81.
59. Siddiqi, *Challenge to Apollo*, 552.
60. "Memo from Assistant Administrator for Public Affairs to Program Director ASTP, Apollo/Soyuz Film," 20 September 1973. National Archives at Fort Worth, USA. RG 255–64, box 7.
61. See also Jenks, *Collaboration in Space*, 88–91.
62. Trischler and Bud, "Public Technology," 189.
63. "Otchet o rezul'tatakh peregovorov delegacii Akademii nauk SSSR po organizacii sovmestnoi ekspozitsii 'Soyuz'-'Apollon' na 30-om Parizhskom salone 23 maia—3 iiunia 1973 goda" [Report on the results of talks of a delegation of the USSR Academy of Sciences about the organization of a joint exhibition Soyuz-Apollo on the 30th Paris Air Show from 23 May to 3 June 1973], 23 March 1973. ARAN. Fond 1678, opis'1, delo 294, listy 112–15.
64. Ezell and Ezell, *Partnership*; Bushuev, *"Soyuz" i "Apollon": Rasskazyvaiut sovetskie uchenye, inzhenery i kosmonavty—uchastniki sovmestnykh rabot s amerikanskimi spetsialistami* ["Soyuz" and "Apollo": Soviet scientists, engineers, and astronauts—participants of the joint work with American specialists narrate].
65. See, for example, Bushuev, *Soyuz i Apollon*, 7–8; Ezell and Ezell, *Partnership*, 1, 6, and 356–57.
66. Bushuev, *Soyuz i Apollon*, 9–14.
67. Bushuev, *Soyuz i Apollon*, 8; Ezell and Ezell, *Partnership*, 1–14.
68. Schmeck, "Soyuz and Apollo Crews."
69. "Orbita druzhby i sotrudnichestva," *Pravda*; Strel'nikov, "Dobro pozhalovat."
70. "Opredelenie osnovnykh napravlenii razvitiia kosmicheskikh issledovanii po nauchnym programmam mezhdunarodnogo sotrudnichestva, chast' 2" [The definition of main directions in the development of international cooperation in space research, part 2], 30 December 1975. ARAN. Fond 1862, opis' 2, delo 32.
71. An exception was the French space program, which maintained relations with both the US and Soviet space programs.
72. Sheehan, *The International Politics of Space*, 60.
73. Shaw, "Competing at Cooperation."
74. Broad, "Science Facts."
75. Broad, "Science Facts."
76. "East-West Cooperation in Outer Space."
77. "Agreement Between."
78. Siddiqi, "Competing Technologies," 430.

Bibliography

"Agreement Between the United States of America and the Union of Soviet Socialist Republics Concerning Cooperation in the Exploration and Use of Outer Space for Peaceful

Purposes." Retrieved 25 February 2021 from https://www.archives.gov/files/presidential-libraries/events/centennials/nixon/images/exhibit/agreement-of-cooperation.pdf.

Air & Cosmos, 7 February 1970, 15.

Bainbridge, William Sims. "The Impact of Space Exploration on Public Opinions, Attitudes, and Beliefs." In *Historical Studies in the Societal Impact of Spaceflight*, edited by Steven J. Dick, 1–74. Washington, DC: National Aeronautics and Space Administration, 2015.

Broad, William J. "Science Facts Help Propel Science Fiction in the Film '2010.'" *New York Times*, 2 December 1984, H1.

Bushuev, Konstantin. *Soyuz i Apollon: Rasskazyvaiut sovetskie uchenye, inzhenery i kosmonavty—uchastniki sovmestnykh rabot s amerikanskimi spetsialistami* [Soyuz and Apollo: Soviet scientists, engineers, and astronauts—participants in the joint work with American specialists narrate]. Moscow: Politizdat, 1976.

Chembrovsky, Oleg. "Zemlja—Kosmos—Zemlja" [Earth—Space—Earth]. *Izvestia*, 22 May 1969, 3.

"Der Himmel ruft" [The sky calls]. *Neues Deutschland*, 7 August 1960, 6.

"East-West Cooperation in Outer Space: Hearing before the Committee on Foreign Relations." United States Senate, Ninety-eighth Congress, Second Session, on S.J. Res. 236, 13 September 1984. Washington, DC: US Government Printing Office, 1984.

Ellis, Thomas. "'Howdy Partner!': Space Brotherhood, Detente and the Symbolism of the 1975 Apollo-Soyuz Test Project." *Journal of American Studies* 53(3) (2019): 744–69.

Ezell, Edward C., and Linda N. Ezell. *The Partnership: A History of the Apollo-Soyuz Test Project*. Washington, DC: National Aeronautics and Space Administration, 1978.

Finney, John W. "U.S.-Soviet Effort in Space Studied: Preliminary Talks Are Held by Scientists of 2 Nations—Glennan Is Hopeful." *New York Times*, 20 November 1959, 1.

———. "Joint Space Test Stalled in Soviet: U.S. Going Ahead with Echo Shot as Moscow Wavers." *New York Times*, 19 January 1964, 3.

Ford, Gerald R. "Remarks on the Launch of the Soyuz Spacecraft, 15 July 1975." The American Presidency Project. Retrieved 26 February 2021 from https://www.presidency.ucsb.edu/documents/remarks-the-launch-the-soyuz-spacecraft.

Gamarekian, Edward. "We Hope to Orbit a Man by '61." *Washington Post*, 28 June 1959, L14.

Gatland, Kenneth. "Space . . . Proving Ground for Détente." *Daily Telegraph*, 11 July 1975, 12–13.

Geppert, Alexander C. T. "The Post-Apollo Paradox: Envisioning Limits during the Planetized 1970s." In *Limiting Outer Space: Astroculture after Apollo*, edited by Alexander C. T. Geppert, 3–26. New York: Palgrave Macmillan, 2018.

Gerovitch, Slava. *Voices of the Soviet Space Program: Cosmonauts, Soldiers and Engineers Who Took the USSR into Space*. New York: Palgrave Macmillan, 2014.

"Head of NASA Urges Close Cooperation." *Daily Telegraph*, 14 July 1969, 24.

Heiduschke, Sebastian. "The Birth of DEFA Genre Cinema, East German Sci-fi Films, New Technologies, and Coproduction with Eastern Europe: *Der schweigende Stern* (*The Silent Star*, Kurt Maetzig, 1960)." In *East German Cinema: DEFA and Film History*, ed. Sebastian Heiduschke, 69–75. New York: Palgrave Macmillan, 2013.

"It Takes Two." *Washington Post*, 15 October 1967, B6.

Jasanoff, Sheila. "Future Imperfect: Science, Technology, and the Imaginations of Modernity." In *Dreamscapes of Modernity: Sociotechnical Imaginaries and the Fabrication of Power*, edited by Sheila Jasanoff and Sang-Hyun Kim, 1–33. Chicago: The University of Chicago Press, 2015.

———. "Imagined and Invented Worlds." In *Dreamscapes of Modernity: Sociotechnical Imaginaries and the Fabrication of Power*, edited by Sheila Jasanoff and Sang-Hyun Kim, 321–41. Chicago: The University of Chicago Press, 2015.

Jenks, Andrew. *Collaboration in Space and the Search for Peace on Earth*. London: Anthem Press, 2022.

Karash, Yuri. *The Superpower Odyssey: A Russian Perspective on Space Cooperation*. Reston, VA: American Institute of Aeronautics and Astronautics, 1999.

Keydin [Caidin], Martin. *V plenu orbity* [Trapped in orbit]. Moscow: Mir, 1967.

"K kommunisticheskoi partii i vsemu sovetskomu narodu" [To the Communist Party and the whole Soviet nation]. *Pravda*, 20 March 1965, 1.

Konovalov, Boris. "Iz pokoleniia tridcatiletnikh" [From the generation of thirty-year-olds]. *Izvestia*, 2 June 1970, 3.

Krige, John. "Introduction and Historical Overview: NASA's International Relations in Space." In *NASA in the World: Fifty Years of International Collaboration in Space*, edited by John Krige, Angelina Long Callahan, and Ashok Maharaj, 3–22. New York: Palgrave Macmillan, 2013.

Logsdon, John. *After Apollo? Richard Nixon and the American Space Program*. New York: Palgrave Macmillan, 2015.

Lyons, Richard D. "NASA Seeks Links to Soviet Science but Moscow Response to 7 Letters Has Been Tepid." *New York Times*, 12 March 1970, 9.

"Orbita druzhby i sotrudnichestva" [Orbits of friendship and cooperation]. *Pravda*, 5 October 1975, 6.

Pacner, Karel. *Sojuz volá Apollo* [Soyuz calls Apollo]. Prague: Albatros, 1976.

"Rech tovarishcha B. N. Petrova" [Speech by comrade B. N. Petrov]. *Pravda*, 23 October 1969, 2–3.

Reistrup, John V. "Two Space Giants Lack Cooperation." *Washington Post*, 19 November 1967, B6.

Roy, Stephanie A., Elaine C. Gresham, and Carissa Bryce Christensen. "The Complex Fabric of Public Opinion on Space." *Acta Astronautica* 47 (2000): 665–75.

"Russians Seen Ready to Put Men in Orbit." *Washington Post*, 19 May 1960, A23.

Schmeck, Harold M. "Soyuz and Apollo Crews Urge New Joint Missions." *New York Times*, 14 October 1975, 26.

Schwartz, Harry. "Why Not a Soviet-American Gemini?" *New York Times*, 14 June 1965, 32.

Sebesta, Lorenza. "The Politics of Technological Cooperation in Space: US-European Negotiations on the Post-Apollo Programme." *History and Technology* 11(2) (1994): 317–41.

"Sedov Disavows Spaceman Plans." *Washington Post*, 1 December 1959, A6.

Shaw, Anthony J. "Competing at Cooperation: Technological Design and Propaganda Representations of Soviet and American Space Programs, 1981–1986." *Quest* 20 (2013): 41–49.

Sheehan, Michael J. *The International Politics of Space*. London: Routledge, 2007.

Siddiqi, Asif A. *Challenge to Apollo: The Soviet Union and the Space Race, 1945–1974*. Washington, DC: NASA History Division, 2000.

———. "Sputnik 50 Years Later: New Evidence on Its Origins." *Acta Astronautica* 63(1–4) (2008): 529–39.

———. "Competing Technologies, National(ist) Narratives, and Universal Claims: Toward a Global History of Space Exploration." *Technology and Culture* 51(2) (2010): 425–43.

———. "Cosmic Contradictions: Popular Enthusiasm and Secrecy in the Soviet Space Program." In *Into the Cosmos: Space Exploration and Soviet Culture*, edited by James T. Andrews and Asif A. Siddiqi, 47–76. Pittsburgh: University of Pittsburgh Press, 2013.
Simons, Howard. "Soviet Rejects Negotiated Accord with U.S. on Space Data Sharing." *Washington Post*, 6 October 1964, A1.
Slayton, Donald K. *Deke! U.S. Manned Space: From Mercury to the Shuttle*. New York: Forge Books, 1994.
Soldovieri, Stefan. "Socialists in Outer Space: East German Film's Venusian Adventure." *Film History* 10 (1998): 382–98.
Strel'nikov, B. "Dobro pozhalovat,' ekipazh 'Apollona" [Welcome, Apollo crew]. *Pravda*, 21 September 1975, 6.
"S vozvrashcheniem, slavnye syny otchizny" [Welcome back, glorious sons of fatherland]. *Izvestia*, 19 March 1965, 2.
Trischler, Helmuth, and Robert Bud. "Public Technology: Nuclear Energy in Europe." *History and Technology* 34(3/4) (2018): 187–212.
"U.S. and Soviets Are Urged to Pool Space Efforts." *New York Times*, 6 June 1965, 70.
Volf, Darina. "Evolution of the Apollo-Soyuz Test Project: The Effects of the 'Third' on the Interplay between Cooperation and Competition." *Minerva* 59(3) (2021): 399–418.
Westfahl, Gary. *The Spacesuit Film: A History, 1918–1969*. Jefferson, NC: McFarland, 2012.
"Zusammenarbeit in der Raumfahrt gefordert" [Space cooperation demanded]. *Frankfurter Allgemeine Zeitung*, 8 September 1969, 5.

CHAPTER 5

Postcolonial Science Heroes in East German Science-Fiction Films

Evan Torner

Introduction

The notion of a *Held der Technik* (hero of technology) was prevalent in the former German Democratic Republic (GDR). Literary scholar Hans-Edwin Friedrich studies the history of this type of character, which he calls the "serial heroic engineer," in early science-fiction (sci-fi) literature and explores how they transitioned into positive archetypes for socialist projections of power and influence in the twentieth century.[1] In the West, the widely accepted term for this character type has been "science hero," which can be understood as a hero who uses science and technology in a positive sense to reach beyond the bounds of human possibility.

The science hero has been a pivotal type in white European and North American genre fiction and has secured two ideological tenets. The first is the notion of "science" as a quasi-magical, ambiguous force that helps humans overcome any obstacle; this stands in sharp distinction from a definition of science as a process of sustained inquiry with caveats and limitations, including human biases. The second ideological notion is that certain humans, by right of their scientific and technological development, have a moral obligation to intervene in the affairs of others, to redistribute resources and colonize, as they push the boundaries of their own knowledge. These characters are cisgender, white men, who are modeled on the majority of US and/or (Western) European scientists—like Robert Oppenheimer and Wernher von

Braun—or inspired by white artist-geniuses, such as Leonardo da Vinci. Scientist and industrialist Bruce Wayne, alias Batman—the iconic sci-fi character developed by US artists Bob Kane and Bill Finger—can thus be considered a science hero, as he often uses chemistry to solve crimes and deploys his own invented technologies to bring villains to justice. Another science hero would be Edward Stratemeyer's early-twentieth century character Tom Swift, modeled on Henry Ford, Thomas Edison, Glenn Curtiss, and others; in most of these stories, he invents a new gadget that then, conveniently, helps him resolve the plotline. The figure of the science hero thus has served as a powerful validation of scientific inquiry and engineering in popular culture.[2]

This chapter examines depictions of science heroes in several East German coproductions and film cooperations with Eastern European countries of the 1960s and 1970s. While a coproduced film involves a transnational, state-level cooperation, a film cooperation referred to an exchange—such as gaining access to shooting locations or employing local actors as extras—within the Eastern Bloc. Both sci-fi coproductions and film cooperations delivered socially progressive interpretations of the science hero figure. The 1960 GDR-Polish coproduction *Der schweigende Stern* (*The Silent Star*, dir. Kurt Maetzig) was the first sci-fi film in the world to introduce a multiracial, multicultural space crew.[3] Other Eastern Bloc sci-fi films made with East Germany's DEFA film studios featured mixed-gender space crews and/or scientists and crew members who are Black, Indigenous, or People of Color (BIPOC) and supposedly drawn from the global Socialist Bloc. These included *Signale—Ein Weltraumabenteuer* (*Signals: A Space Adventure*, 1970, dir. Gottfried Kolditz, Poland/GDR); *Eolomea* (1972, dir. Herrmann Zschoche, 1972), mostly shot in Bulgaria; and *Im Staub der Sterne* (*In the Dust of the Stars*, 1976, dir. Gottfried Kolditz, GDR), mostly shot in Romania.

These productions formed an active postcolonial counterpoint to the ostensibly still-colonialist Western sci-fi tradition.[4] By this I mean that a serious reckoning with the aftermath of Western colonialism is noticeable in the production of each of these films.[5] Already in 1916, the Russian communist leader Vladimir Lenin had written an essay titled "Imperialism, the Highest Stage of Capitalism," which described how capitalists needed colonies in order to continue the growth of capital. These points were deepened in *The Wretched of the Earth*, published by philosopher Frantz Fanon in 1961, which demonstrates how the subjugation of BIPOC in the Global South produces permanent mental dysfunction that must be overcome with violence.[6] Given that the socialist critique of capitalism was thus clearly linked to both colonialism and racism, it is no surprise that a communist response to postcolonial conditions should promote the agency and integration of BIPOC into their media representations and regime, in addition to addressing worker impoverishment and white-imposed racial hierarchies. If the workers of the world were intended to laud the scientists leading them, the questions of *who* gets to be a science hero in these sci-fi films and *how* these science heroes act beg to be answered.

Unfortunately, the four sci-fi films in question do not, at the end, measure up to their socially progressive aspirations. The creation of a "new" socialist type of science hero in these films was intentional and implemented in all sincerity. The specters of tokenism and colonial appropriation of nonwhite bodies and voices cannot be easily exorcized from these films, however. The filmmakers behind *The Silent Star* and *Signals* insisted on having prominent BIPOC figures on their space crews; in *Eolomea*, however, most BIPOC are confined to Earth, and all the narratively decisive roles are held by white Europeans; and the conclusion of DEFA's "space" sci-fi cycle, *In the Dust of the Stars*, puts Romanian actors in brownface to play an oppressed class that is saved by a group of all-white science heroes. Examining *how* these films fail in their postcolonialist endeavors is productive both in understanding the historical dynamics at play and in measuring the aspirations and achievements of our own contemporary moment. The postcolonial circumstances of the Cold War and anti-racist activism of the present thus prompt us to continue interrogating the multicultural whiteness of past representations of science meritocracy.[7]

This chapter first outlines the colonial and racial heritage of the science hero archetype. It then illustrates several attempts in socialist countries, such as the Soviet Union and the GDR, to decouple the figure of the science hero from white supremacist, capitalist thinking and remold it in the context of a global, anti-racist socialism. As noted above, the weight of the earlier, white-male archetype can still be felt in these multicultural socialist productions, however. Furthermore, they illustrate that science *itself* is neither neutral nor free of racism. As social theorist David Theo Goldberg argues, white-dominant "anti-racist" societies benefit by labeling all racism "irrational" as opposed to systemic.[8] White anti-racist societies, such as the GDR, tended to locate racism in outmoded folk superstition, disavowing the outsized role that both science and modern systems of exploitation have played in securing the fiction of "race" for the long term. The chapter thus uses BIPOC critiques of race to examine the inability of white socialist sci-fi productions to grant agency and scientific authority to their BIPOC characters.

An East German Intervention

In 1958, director Kurt Maetzig's explicit goal for *The Silent Star* was to break the chain of derogatory representational tropes and to include, for once, the nonwhite world in what ought to be a globally shared future modernity. In a publicity interview with the West German magazine *Der Spiegel*, he stated that "a film should be created that can be compared in its scientific-utopian logic to Fritz Lang's 1929 film *Woman in the Moon* and has a universally-understood statement binding people together, similar to the 1932 Ufa film *F.P.1 Doesn't Answer*."[9] Here, Maetzig both calls upon the inspirational and utopian qualities of earlier German sci-fi films and conceives of *The Silent Star* as a kind of restitution for past fascist portrayals of the future. In the opening credits, Maetzig describes "people from all over the world" as being depicted on screen, and he ends his film with a multiracial clasping of hands similar to the logo Pablo Picasso designed for Joris Ivens's global agitprop documentary *Lied der Ströme* (*Song of the Rivers*, 1954, GDR).

But it was precisely because of such ambitions that the mixed-race space crews in *The Silent Star* and *Signals* appear tokenist to audiences, both past and present. A second-order awareness of multiculturalism as being *itself* ideological would have been operant, revealing both the private entrenchment of a notion of white "normalcy" within the target GDR audience and the association of a shared, world modernity with GDR hegemonic power. The audience knew it was being asked to consent to a future, seamless, lived reality of solidarity inscribed within these images; but the images of global solidarity were perhaps received as even more speculative than the technologies the films otherwise portray with sets and props. Thus Maetzig, on one hand, expanded the range of science heroes in the socialist film imaginary, while, on the other hand, he also revealed the absolute limits of tokenist inclusion to an audience already inclined to see diversity as imposed from the outside. This was not simply due to everyday racist prejudice of the average white European viewer; it also had deep-seated roots in expectations developed within the tradition of German-language sci-fi films and literature.

White Germanness and Science Hero Origins

In the late nineteenth and early twentieth centuries, as literary scholar John Rieder argues, European colonial empires waned, and a new moral center was sought in science, representing progress for all mankind.[10] This relatively sudden turn away from active colonial subjugation in favor of a gesture of humanistic progress meant steering the very notions of science and technology away from white European superiority and power and toward resources shared by all of humanity. The result was that science heroes defaulted to patriarchal, white supremacist templates. Heroes such as Tom Swift are often linked not only with real scientists, such as Edison and Ford, but also with a time when ostensibly square-jawed white male adventurer-scientists used the results of their experimentation and engineering to enhance and expand the limits of where humans could go and what they could do.

Science itself is also represented differently in fiction at different historical moments. Scientists at the origins of sci-fi—such as Mary Shelley's Victor Frankenstein in the early nineteenth century—are notably *not* science heroes. Rather, they are frequently tormented "geniuses," whose creations turn against them and who gave rise to the "mad scientist" trope over two centuries. Science heroes, by contrast, emerged as optimistic male figures. Arne Saknussemm, for example, in *Journey to the Center of the Earth*—the 1864 novel by Jules Verne—is a white European male visionary whose scientific knowledge allows him to go where seemingly none have gone before and who leaves clues for future explorers so they might follow in his path. Like Captain Nemo and other Verne protagonists, Saknussemm was actually inspired by Raphael Semmes, whose ingenuity and adaptability in naval warfare was popularized in mid-1800s nonfiction. (As a Confederate officer in the Civil War, Semmes served the attempt to keep millions of Black people in bondage.) These characters illustrate how the trope that science reflects an obsessive psychological state gave way to the idea that science steels a certain kind of "new man" for adventure within imperial designs.

In the German context, the two main authors of early science fiction, Kurd Laßwitz and Hans Dominik, both relied on science heroes in their major works. In Laßwitz's novel *Auf zwei Planeten* (*Two Planets*, 1897), Grunthe, Saltner, and Torm form a grim, male polar expedition that then reveals evidence of a "benevolent" invasion by Mars. Under the circumstances, these "explorers" and other scientists like them are handed full agency over the fate of the entire globe—despite (or because of) their lack of diversity in background or knowledge. Laßwitz admittedly represents diversity and situated knowledge *elsewhere*—namely, through the figure of Friedrich Ell, a human-Martian hybrid who helps avert war between the Earth and Mars. Laßwitz's novel ultimately advocates for cultural diversity and understanding within the scientific community. The science heroes developed later by Hans Dominik have no "Ell"-style hybrid character to counterbalance his proud, white, German men dedicated to scientific advancement (ethics and nuance be damned).[11] Laßwitz and Dominik typify the two main approaches to science heroes in the years leading up to 1945: they are either bland and impersonal or feverishly engaged in scientific and exploratory work on behalf of a colonizing national power. Their literary purpose is to serve as powerful avatars for imagined white, male, German readers, who can then project their own decision-making capacities into the "what if" space of sci-fi rather than grapple with the inherent ambiguity of human experience.

The literary tradition that preceded East German sci-fi had been intended for engineers, white-collar workers, and students aspiring to be scientists.[12] Its roots were to be found not only in the literary experiments of Laßwitz and Dominik but also in Weimar-era German communist sci-fi. This tradition largely trended toward the paranoid intelligentsia of the early German Communist Party, and the communist sci-fi of the 1920s was quite pessimistic. Konrad Loele's bitter *Züllinger und seine Zucht* (Züllinger and his kin, 1920) depicts a future proletariat defanged by the aristocracy, military, and church. Arthur Zapp's *Revanche für Versailles!* (Revenge for Versailles! 1924) sees the future threat of gathering nationalist youth groups undermining the potential for France and Germany to resolve their differences.[13]

In the GDR after 1949, despite both the atomic bomb and the Holocaust, this pessimism had transformed into relative scientific optimism. The young textile engineer Lothar Weise, for example, wrote seven sci-fi books in his short career, each concerning some kind of crisis or mystery in outer space. One of them, *Das Geheimnis des Transpluto* (The mystery of Transpluto, 1963), starts paradigmatically when a crew member at the moon station, Kolya Stepanov, is alerted to meteors headed toward his position. Meanwhile, Dr. Helford Runig and his assistant, Sil Goder, immediately decipher that, no, these are not meteors but rather visitors from another planet. None of these three characters are particularly well-developed as protagonists, serving instead to focus readers' attention outward onto scientific mysteries and dilemmas—e.g., on questions such as: What is that object? Where is it from? What is its apparent objective? They invite the reader into the pleasures of discovery and introduce the possibility of scientists as part of cosmic-level political discussions.

Part of most science heroes, however, is a worldview that subjugates nonwhite people to white (Christian) European needs and desires—a bitter pill to swallow along with the thrill of adventure to be found in these fictions. Science heroes invite identification with the colonizer by default, as the alternative would be to identify with a dim, poorly developed, and equally fictional depiction of the colonized. The role of science and technology in colonialism has also been well-documented, however, whereby Indigenous knowledge is supplanted by Western "science," which often turns out to have worse social outcomes for the colonized, in the form of exacerbated oppression by the colonizer.[14] To have "adventurer" scientists "discover" lands (or planets) is to tell the tale of foreign interlopers who claim moral superiority upon entering into territories not their own and, in so doing, unfortunately equate the "natives" with the territory rather than dignifying the well-grounded, centuries-old Indigenous knowledge of the land. Although some in the GDR attempted to push against this trend, the strain against the country's white colonial impulses is also apparent.

Eastern Bloc Science Heroes

Cold War–era socialists were ostensibly against racism and imperial capitalism. As we have seen, in the early 1960s they also started creating their own sci-fi film science heroes, with some attention paid to representing an anti-racist, postcolonial future. By the 1970s, when sci-fi films were being churned out at full speed throughout the Socialist Bloc, however, Soviet authorities had advanced a populist campaign in support of the sciences and space travel. The beacon of the space race in socialist countries had been ignited by *Sputnik* in 1957,[15] which had galvanized international scientific cooperation and genuine feelings of solidarity in socialist sci-fi media.

By the mid-1960s, the kernel of colonialism existing within the sci-fi science hero was becoming increasingly apparent, however. In a 1965 commission for the communist Tudeh Party of Iran, for example, Azerbaijani carpet artist Jafar Mujiri immortalized Soviet-era space explorers; in the center of the carpet, below a typical depiction of Lenin, is a depiction of a "New Soviet Man" rocketing into space from Moscow. According to Leon Trotsky in a 1924 essay, the New Soviet Man would become a new force in the speculative reshaping of the world: "Through the machine, man in Socialist society will command nature in its entirety, with its grouse and its sturgeons. He will point out places for mountains and for passes. He will change the course of the rivers, and he will lay down rules for the oceans."[16] The New Soviet Man in Mujiri's artwork represents a figure that exudes confidence in building and confronts each challenge with determination: the resolute figure's left hand reaches to the stars, the right hand holds a blazing sun. Rocket ships with the Soviet Union's hammer-and-sickle insignia either follow the figure's trajectory or arc around the Earth, giving the impression of outer space saturated with Soviet spaceships. The striking red border of the carpet contains cosmonauts, the real-life Soviet science heroes of the mid-1960s: Gherman Titov, Valery Bykovsky, Valentina Tereshkova, and Yuri Gagarin, among others. Mujiri immortalizes what, in retrospect, is a highly important moment in Soviet space history, which encapsulates global pride and solidarity among communist artists and scientists as well as Moscow's own space travel ambitions, which would diminish only after the *Soyuz 11* disaster in 1971.[17]

The view of the future implicit in the Marxism-Leninism of the 1960s and 1970s conflated space travel, scientific inquiry, communist inevitability, and the socialist personality that would meet all challenges.[18] In this context, science fiction and global communist science heroes were not necessarily top-down impositions from Soviet-Russian leadership in Moscow but often emerged bottom-up from a populace optimistic about and eager for a shared global futurity.[19] In the wave of popular enthusiasm after *Sputnik*'s success in 1957—when the USSR launched an orbiting satellite into Earth's atmosphere before the United States—this was not a mere propaganda point; Soviet achievements seemed to stand in for the socialist world's. And there were authors and readers, mostly male, ready to write and consume fiction about it all.

In retrospect, it is fascinating that the meritocratic vanguard of Marxism-Leninism opened the possibility of multiracial science heroes. Sara Lennox has argued that East Germany was "better on race" than West Germany.[20] This relative positive also extended to fellow socialist countries, such as Yugoslavia, as demonstrated in the work of Sunnie Rucker-Chang.[21] Being "better on race" was based on three ideological pillars. The first pillar was the open *acknowledgment* of the postcolonial condition: that the world we know is primarily shaped through the forces of colonialism and colonial capitalism, and that structural inequality plays a larger role in immiserating nonwhite people than any solitary, personal act of racism ever will. The second pillar was the unconditional *support* of liberation movements around the world. This included the US civil rights movement of the 1960s; the fact that this movement was erroneously framed in terms of communist revolution rather than what would later be called "identity politics" does not diminish the impact of a global power such as the GDR lending legitimacy to its goals.[22] The third pillar was the *aspiration* toward a future in which racism is abolished by directly addressing structural issues. Equal treatment under the law and economy was seen as only possible after the Communist Party–led working class overthrew their bourgeois-capitalist oppressors worldwide. This future would include Africans, Asians, and Latin Americans, all able to take on the science hero roles formerly occupied by white colonizers.

Nevertheless, as Mujiri's carpet illustrates, Soviet space accomplishments took center stage, with all others relegated to the sidelines. Indeed, Russian whiteness and its centrality in the multiethnic Soviet empire has recently re-entered scholarly discussion. Eugene Avrutin makes a compelling case for a Russian people genuinely culpable for their history of (antisemitic) ethnic cleansing and, at the same time, genuinely open to Black communists from around the world.[23] Interracial marriage was not regulated in the USSR and Eastern European socialist countries, the way it was in the US, and Soviet domestic media portrayed anti-Black systems in the West more accurately than media in the "free democracies" of the United States, UK, West Germany, and France. And yet, international science and technology spheres generally favored the contributions of (white) Russians and East Germans. Marxism-Leninism may thus have offered glimpses of a science-driven, race-blind meritocracy, but innovation and science were still associated with the scientific and technical elite of Moscow and its close allies.[24]

In the GDR, there was an acute emotional investment in both the scientific basis of the Soviet-guided government and the systemic rewards accruing to white powers in what we today call the Global North. Genuine anti-racist bonds of solidarity were formed; but neither popular biases nor the idea of science promoted during the space race vanished. BIPOC scholars with backgrounds in the GDR, including Peggy Piesche, Patrice G. Poutrus, and Angelika Nguyen, describe a miserable status quo with respect to BIPOC treatment in GDR society.

Representation of BIPOC in GDR media, including DEFA films, was tokenist at best, and openly colonialist at worst. Ten years ago I coined the term "race-time continuum" to describe how the GDR aspired to a meritocratic, multiracial communist future while also preserving harmful white ideology against BIPOC agency and achievements.[25] My work is inspired by Priscilla Layne, who writes of the processes through which white artists in West Germany appropriated Black culture for their own claims.[26] It also builds on the work of André Carrington, who writes that "to comprehend the significance of Blackness in speculative fiction and media, sci-fi critics must . . . put into practice sophisticated approaches to understanding race that are relevant to the field of cultural production writ large as well as its social surroundings."[27] Blackness may be, as anthropologist and journalist Damani Partridge posits, a "universal claim" far beyond that of the Enlightenment that asserts the power and agency of those who resist and strive for freedom in a society, rather than those who affirm a (white) normative state, even in the name of science and progress.[28]

The white Eastern Bloc sci-fi coproductions and film cooperations with diverse casts have, indeed, a distinct history that inform their approach to race. To return to the three pillars, it is clear that the attempted anti-racist dimension of GDR cultural production at least *acknowledged* the global postcolonial condition, unlike comparable science fiction in West Germany or the United States. While Cold War US sci-fi, such as the original *Star Trek* television series, typically extended a multicultural, US-style empire into the cosmos, the GDR sought to depict humble scientists and brave socialists whose mission, above all, was the advancement of science for the improvement of human conditions and knowledge worldwide. The bright, multiracial future of the United States had to be staged, often with great tension, against extant systems of the time. Carrington writes that, although for *Star Trek* "a diverse cast seemed to embody the future on its set," the reality for Nichelle Nichols—who played the Black woman chief communications officer—was that she had no lasting guarantee that her already tokenist role on the show would be renewed.[29] Her lines were repeatedly cut or shortened. Her later work in diversity and inclusion at NASA turned out to be the life's legacy of which she was proudest. American race relations manifest in the tension between the representations on screen and in reality, just as was the case for socialist race relations represented in GDR media.

The GDR was not exactly what political scientist Cedric Robinson would call a "racial regime," in which all mechanisms of society were directed toward white-supremacist ideology.[30] But the dominant narrative of anti-fascism in the GDR also brought a common blindness on the part of its white populace toward its own ongoing prejudices. Hans Frey has recently argued that, in fact, the purpose of GDR sci-fi was to fill the "unspoken need . . . for an anti-fascist [Hans] Dominik."[31] In hindsight, the anti-fascist approach counters the "reverse colonization" dynamic that literary scholar David Higgins observes in some US Cold War sci-fi (e.g., *Dune*, *Star Wars*), in which white, reactionary impulses are conquered or dominated by superior, alien forces that require messianic interventions by white people to reach a solution. This appropriation of victim identity has faint traces in the DEFA coproductions *The Silent Star* and *In the Dust of the Stars*—insofar as both contain alien races—but is otherwise largely absent. Instead, these white-centered sci-fi stories tell shared "[celebrations of] anticolonial forces struggling against domination and oppression."

In other words, the GDR and its Eastern Bloc allies do not offer a radical departure from colonial dynamics but rather affirm the role of the more-scientifically-informed, paternalistic white scientists who will "lift up" those in the Global South. The dynamics of the World Council in *Eolomea* also conform to this trope: Professor Oli Tal concludes the film by telling the multiracial council that his extralegal course of science, which led to his interstellar travel, was somehow necessary and that he was in no need of oversight by the international political body. Similarly, *In the Dust of the Stars* presents a white savior narrative, albeit through the lens of anti-fascist, socialist solidarity.

Four Sci-Fi Films

The Silent Star (1960) was based on Polish author Stanisław Lem's first popular sci-fi novel, *Astronauci* (*The Astronauts,* 1951), a hard sci-fi epic in which five male cosmonauts, sent on a fact-finding mission to Venus, discover a civilization that annihilated itself in an atomic war. In contrast to the novel, the film adaptation introduces an eight-person space crew on spaceship *Cosmocrator*, which includes the Soviet captain Arsenyev (Mikhail N. Postnikov), US nuclear physicist Professor Hawling (Oldrich Lukes), Japanese physician Sumiko (Yoko Tani), East German pilot Brinkmann (Günther Simon), Indian mathematician Sikarna (Kurt Rackelmann), Chinese biologist and linguist Chen-Yu (Tang Hua-Ta), Polish engineer Soltyk (Ignacy Machovski), and African communications specialist Talua (Julius Ongewe). In investigating mysterious transmissions from Venus, they discover an alien civilization that has already destroyed itself. It is clearly a cautionary tale against nuclear war. Brinkmann, Chen-Yu, and Talua all die helping the spaceship leave Venus and return to Earth. *Signals: A Space Adventure* (1970) was the Eastern Bloc's response to Stanley Kubrick's *2001: A Space Odyssey* (1968, USA). In it, a spaceship named *Ikaros*, largely manned by nonwhite crewmembers, suddenly has a technical malfunction and is adrift in space, and a crack crew of socialist cosmonauts, largely white, is sent to rescue them. Hermann Zschoche's *Eolomea*, produced later in 1972, concerns the research of Dr. Maria Scholl (Cox Habbema) into Professor Oli Tal's (Rolf Hoppe) secret mission to the planet Eolomea, which is more than one hundred light years away. Dan Lagny (Ivan Andonov), a shiftless ex-boyfriend of Scholl's, is stationed on an asteroid near the location where eight ships have vanished . . . and eventually becomes the pilot of one of them. Finally, *In the Dust of the Stars* (1976) depicts a *Star Trek*–esque all-white, polyamorous space crew, consisting of four women and two men, that is responding to an emergency distress signal from the planet TEM-4. Once there, they discover that the inhabitants of TEM-4 have enslaved another planet's race, the Turi, played by Romanian extras in brownface. The cosmonauts must decide what is more important: going to get help from Earth, or staying behind to help the local resistance.

The white, Eastern European filmmakers involved in all four productions sincerely intended their sci-fi films to portray a future in which racism and patriarchy had been abolished. Genre film director Gottfried Kolditz—who directed both *Signals* and *In the Dust of the Stars*—is proudly described by film historian Ralf Schenk as having cast "Serbians, Czechs, Poles, Russians, Romanians, Mongols, [and] Egyptians" in his productions.[32] A full ten years earlier than Kolditz's films, Kurt Maetzig's *The Silent Star* underwent a tortured script-rewriting process, documented by historian Burghard Ciesla, which revolved around the specific nationalities, genders, and races to be represented on the spaceship.[33] The script had long since departed from Lem's novel and, significantly, from Lem's protagonist, Robert Smith, a Black American science hero raised in Russia. *Eolomea*'s multinational and multiracial Space Council was originally intended to include "9 young people, quiet and self-assured, with one African and two Asians among them."[34] In the final film, however, there are over a dozen BIPOC and South Asians in the Space Council sequence. GDR sci-fi films are thus great examples of what historian Quinn Slobodian has described as "socialist chromatism," or the supposed rejection of both scientific and cultural racism through the tactical inclusion of nonwhite representation in art. "In the postwar years," he writes, "East German scholars followed the UNESCO statement in discarding racial hierarchy while preserving the kernel of hereditary racial difference."[35] What remains in this "kernel" of racial hierarchy is of interest to us here.

To evaluate the science heroes in East Germany's four sci-fi films, one need only ask: Who is working which jobs? *The Silent Star* clearly situates race as an obstacle that has already been overcome in professional spheres, although racism still exists in spaces related to the United States.[36] To illustrate this idea, let us turn to one of the opening scenes, in which an international translation committee is trying to decode an unknown signal from Venus.

The committee primarily consists of scientists from India, who are standing around looking concerned, while the camera tracks in on Chinese linguist Chen Yu as he actively writes in a notebook. Having surveyed this consensual South-South collaboration, with a German voice-over expounding how Chen Yu single-handedly solved the problem of world hunger, the camera then tracks over to the (supposedly) Indian mathematician Sikarna, played by a German actor Kurt Rackelmann in brownface. The camera move depicts overwhelming organization and consensus among the diverse experts; but it also asks the viewer to consent to Rackelmann's brownface, in what Katrin Sieg has called "ethnic drag," against the backdrop of Indians and others who are simply performing their physiognomy.[37] Another problematic depiction arises in the interviews of the spaceship crew on the launch platform. The composition is arranged as follows: as various workers and reporters of color circulate in the background, two BIPOC maintenance workers stare, backs to the camera, at two white reporters interviewing US professor Hawling—who was inspired by Wernher von Braun, a former Nazi scientist at the forefront of US aerospace endeavors during the Cold War.[38] Attention and authority is thus visually conferred onto the white media event at the center of the predominantly nonwhite maintenance workers laboring on behalf of the mission. This vision of somatic consent is then immediately contrasted with the servile "N—— waiter" (as the screenplay puts it) in the New York City sequence meant to convey existing US racial hierarchies.[39]

To the credit of *The Silent Star*'s filmmakers, however, they put effort into keeping the original German voice of African communications officer Talua, albeit post-dubbed by himself later on. Actors of Color in later DEFA sci-fi films—with the exception of African American expatriate Aubrey Pankey in *Signals*—were not given this level of presence; their voices are all post-dubbed by white German voices, leaving only their bodies and gestures part of the narrative. Julius Ongewe, who was studying medicine in Leipzig when he was cast as Talua in *The Silent Star*, nevertheless experienced wage discrimination; as he was cast on the basis of his skin color, not his acting talents, he was paid less than his fellow actors.[40] As a communications officer, his character is also given the lowest rank of authority, and his noble sacrifice at the end appears only dubiously justified.

A decade later, *Signals: A Space Adventure* would pick up the same sci-fi universe as *The Silent Star*. It was to be a "generally humanistic film," as the GDR Cultural Committee described it, that affirmed the present socialist vision of modernity and challenged the "cultural pessimism" of Western productions.[41] This film was in dialogue with Kubrick's *2001* and Andrei Tarkovsky's *Solaris* (1972), both films that also overwhelmingly cast white actors and centered on white people's concerns. The DEFA film studio consciously responded by maintaining a multiracial space crew, as well as the trope of an international and seemingly diverse World Council that oversaw scientific advancement. The commitment to BIPOC in this film, however, was half-hearted. In the script, the African crew member, played by Aubrey Pankey, does not even have a name; instead, he is referred to as an "old gray-haired African."[42] In the story, his spaceship is the one to be rescued. In the film, Gojko Mitić—a Serbian-born DEFA star known for his main roles as a Native American in GDR westerns—plays the token "ethnic" hero, alongside white, Polish crew member Veikko (Piotr Pawłowski), as the co-travelers spends most of their time trying to save spaceship *Ikaros* from disaster. The film then ends in a highly implausible beach frolic involving all space crew members. In terms of intellectual and physical competence, jobs, and race, the film evolves a clear dynamic between the feminized, nonwhite, disempowered *Ikaros* crew in need of help and the masculine, white, empowered science-hero rescuers.

Eolomea, a film about a secret mission to colonize a distant planet, gives the tokenist space crew a brief cameo at the end. The film features only one Black crew member, in a distant background, while focusing on an "international council" at Earth Central, which considers grounding all flights until several lost ships are recovered. Earth Central is represented by a conference room containing a richly dressed mix of people from all nations of the world. This council, which in fact displays very few white people, is nevertheless chaired by a white German man (Wolfgang Greese). Like Rackelmann in *The Silent Star*, he has a quiet, earnest-faced BIPOC man beside him, as a figure of consent. When an Ethiopian woman in a pink sari, filmed in extreme long shot, stands up to speak, the chairman specifically informs her that she is not a member of the council; "but we'd love to hear your opinion," he adds. Professor Maria Scholl then immediately disregards the Ethiopian woman's reasonable request to have them send out rescue squads. This tightly controlled vision of "openness" parallels the GDR's own early-1970s entry into the UN. In a fantasy of being a greater-than-equal partner on the world stage, it mirrors Higgins's "fiction of benevolence" in a socialist form. While the gathering of this council is a welcome nod toward an idea of global decision-making among predominantly white and nonwhite nations, it also undercuts the agency of the latter group.

By the time director Kolditz filmed *In the Dust of the Stars* in 1976, little enthusiasm remained for the trope of the multiracial crew. Here, the space crew was represented by six white men and women, with female Captain Akkala (Jana Brejchová) in charge. In other words, women's emancipation in science was now positioned above racial representation among the crew. Their enemies are the Temers of TEM-4, evil white overlords who have enslaved the Indigenous Turi people and make them work underground. The story strongly resembles a colonial-era rescue fantasy, as the crew commits its superior technology to fight for a better material existence for premodern humanity. The plot seems like a re-import of the science-hero race politics of *The Time Machine* (1895) by H. G. Wells—with its Morlock and Eloi ciphers for nonwhites and whites respectively—instead of representing continuity with the internationalist ambitions of *The Silent Star*. By the mid-1970s, it thus seems, the "postcolonial" ideals of Eastern European socialist science fiction had faded, leaving behind the notion of a multiracial coalition of science heroes out to save all of humankind.

In *In the Dust of the Stars* the scientific method is also represented differently. It largely manifests through characters who talk on-screen with an artificial intelligence, which then vocalizes data back to them. In *The Silent Star*, it was quite clear what kind of scientist the characters were—be it biologist, linguist, or physician—and what their country or continent of origin was. They seemed to apply their scientific knowledge to the mystery at the center of the film almost mechanically. In *In the Dust of the Stars*, in contrast, the captain and the largely interchangeable all-white crew have jobs that are unclear, and their research methodology mainly involves formulating questions for the artificial intelligence. Despite the fact that these crew members then take up a valuable struggle to save the Turi from their enslaved condition toward the end of the film, it is, in fact, the crew's reliance on the computer and lack of curiosity about their situation that lands them in their predicament in the first place. The crew arrives on planet TEM-4 as a result of following a distress signal communicated to them via the computer, which Temer leader Ronk (Milan Beli) then dismisses as fake. Ronk and his chief (Ekkehard Schall) use mind control to erase the concerns of five of the crew members; but the skeptical Suko (Alfred Struwe) traces the distress signal to the caverns below and discovers the exploitation of the Turi by the Temer. The film is thus a Cold War metaphor of two white, technologically superior powers—one socialist and one capitalist—struggling over the bodies and minds of a technologically inferior populace representing the Global South. Rather than welcoming the Global South into the scientific community of a space-faring world—as both *The Silent Star* and *Eolomea* do in their limited ways—*In the Dust of the Stars* thus reverts to earlier, colonial notions of white saviors as vehicles of scientific and social progress.

Conclusion

In conclusion, it should be reasserted that most East German citizens were convinced that their state represented an internationally inclusive, postcolonial power. Sci-fi films, the imaginary playground of East German technicians and engineers, could correspondingly project a cinematic future that allegedly critiqued and progressed away from colonial logics. Nevertheless, these GDR films still display persistent colonial relationships and ideas—through an audiovisual regime of white prominence and nonwhite tokenization, white savior narratives, and science heroes as anti-fascist, white European figures. In *The Silent Star*, these figures are Brinkmann, Arsenyev, and Soltyk, experts in piloting, astronomy, and robotics, leaving the feminized sciences—such as medicine and biology—to the Asian and African crew members. Even Professor Hawling—the white US scientist who slowly comes around to communism—gets a lion's share of screen time, whereas Talua is sacrificed to the perils of Venus in order to save the crew. This implicit dynamic evolves, over time, to the blatant use, in *In the Dust of the Stars*, of Romanian actors in brownface who must be saved by an all-white communist space crew.

Representations such as these matter. Science-fiction films form a basis of learning through imagination, an experimental field in which audiovisual propositions are offered to the broad audiences that are willing to entertain them. If the GDR cosmonaut Sigmund Jähn (1937–2019) was championed as a pioneer for socialism, could not have Matthew Henson (1866–1955), the Black-American explorer who reached the North Pole, also conceivably had a place in the US's science-hero pantheon?

Damani Partridge's concept of "Blackness as a universal claim" may help us situate these tokenizing DEFA sci-fi films in relation to more contemporary work on race, sci-fi, and the former GDR.[43] After all, sci-fi signifies not just "science fiction," but also "speculative fiction."[44] This involves the freedom to ask "What if?" without having to justify one's own social position in terms of a "science" that has a poor track record of including anyone other than white men. Multiculturalism need not come at the expense of speculation about the sciences. Socialist sci-fi speculated about how society might transform in response to its political system and space travel. With BIPOC experiences at the center, a new space age can still be imagined that truly embraces *all* of humanity.

Evan Torner is an associate professor of German studies and film and media studies at the University of Cincinnati, where he serves as the undergraduate director of German studies and the director of the UC Game Lab. He coedited the volume *Immersive Gameplay: Essays on Role-Playing and Participatory Media* and is one of the founding editors of the *Analog Game Studies* journal. His doctoral work, focusing on questions of race and East German genre cinema, has led to numerous chapters and articles on East German science-fiction films, westerns, musicals, and war films. His book manuscript, "Race and State Capitalism in German Science-Fiction Cinema," is currently under review.

Notes

1. Friedrich, *Science Fiction*, 181.
2. Also see, in this volume, the chapters by Darina Volf and Etorre Costa.
3. Torner, "Casting for a Socialist Earth," 130.
4. Please note that in film history, socialist sci-fi films—including GDR coproductions and cooperations with other socialist partners—are still considered separately from "German" sci-fi films. In the book accompanying the 2017 Berlin International Film Festival's sci-fi film retrospective, *Future Imperfect* (Rother and Schaefer 2017), Matthias Schwartz's chapter "Archaeologies of a Past Future" is unable to comment on the fact that the DEFA productions made with socialist partners were far more racially inclusive and globally minded than anything made in West Germany prior to the 1980s.
5. Postcolonial theory draws from a wide range of thinkers who should be mentioned in brief. The work of Edward Said, Gayatri Spivak, and Hamid Naficy all stand out in this respect.
6. Peruvian sociologist Anibal Quijano later described a "coloniality of power" that links the very *notion* of race to white capitalism in the West. See Quijano, "Coloniality of Power."
7. Hudson, "Vampires of Color"—cited in Torner "Casting for a Socialist Earth."
8. Goldberg, *The Threat of Race*, 36.
9. "Happy-End auf Venus." *Der Spiegel*, 8 January 1958. Translated by the author.
10. Rieder, *Colonialism and the Emergence*.
11. See Fischer, *Fantasy and Politics*; Fritzsche, *Science Fiction Literature*.
12. Fritzsche, *Science Fiction Literature*, 115.
13. Fisher, *Fantasy and Politics*, 158.
14. See Arnold, *Science, Technology and Medicine*; Roy, "Science Still Bears the Fingerprints."
15. Also see, in this volume, the chapters by Mario Schulze, Darina Volf, and Ettore Costa.
16. Trotsky, "Revolutionary and Socialist Art."
17. *Soyuz 11* was the only crewed mission to board the world's first space station, *Salyut 1*. After the space capsule depressurized on reentry into the Earth's atmosphere, killing all three crewmembers, there was a notable (though not commented) decrease in both popular support and funding for Soviet space travel.
18. Schaber, *Das rote Weltall*.
19. On this point also see, in this volume, the chapters by Doru Aurel Pop and Etorre Costa.
20. Lennox, "Reading Transnationally."
21. Rucker-Chang, "The Uses of Blackness in Yugoslavia."
22. The term "identity politics" came out of the queer Black scholar group called the Combahee River Collective in 1977.
23. Avrutin, *Racism in Modern Russia*.
24. For more on this topic, see the chapter by Darina Volf in this volume.
25. Torner, "The Race-Time Continuum."
26. Layne, *White Rebels in Black*.
27. Carrington, *Speculative Blackness*, 3.
28. Partridge, *Blackness as a Universal Claim*.
29. Carrington, *Speculative Blackness*, 68–81 and 68.

30. Robinson, xii.
31. Frey, *Vision und Verfall*, 82.
32. Schenk, "Ein Mann für alle Filme," 109.
33. Ciesla, "Droht der Menschheit Vernichtung?"
34. BArch DR 117 / 30551. "Dispos."
35. Slobodian, "Socialist Chromatism," 27.
36. Ciesla, "Droht der Menschheit Vernichtung?"
37. Sieg, *Ethnic Drag*, 2.
38. Ivanova, *Cinema of Collaboration*, 136–37.
39. BArch—DR117 / 11. Drehbuch Endfassung, 20 January 1959.
40. Torner, "Casting for a Socialist Earth."
41. BArch—DR1-Z / 297. "Stellungnahme zum Film Signale—Ein Weltraumabenteuer," 25 August 1970.
42. BArch—DR 117 / 20525 / Dispositionen.
43. Partridge, *Blackness as a Universal Claim.*
44. In this connection, also see in Doru Aurel Pop's chapter, in this volume, on the concept of anticipation.

Bibliography

Arnold, David. *Science, Technology and Medicine in Colonial India*. Cambridge: Cambridge University Press, 2000.

Avrutin, Eugene M. *Racism in Modern Russia: From the Romanovs to Putin*. London: Bloomsbury Academic, 2022.

Banerjee, Anindita. "T/Racing Revolution between Red October and the Black Atlantic." In *Science Fiction Circuits of the South and East*, edited by Anindita Banerjee and Sonja Fritzsche, 23–50. New York: Peter Lang, 2018.

Carrington, André M. *Speculative Blackness: The Future of Race in Science Fiction*. Minneapolis: University of Minnesota Press, 2016.

Ciesla, Burghard. "Droht der Menschheit Vernichtung?" In *apropos: Film 2002*, 121–36. Berlin: Das Neue Berlin, 2002.

Cornils, Ingo. *Beyond Tomorrow: German Science Fiction and Utopian Thought in the 20th and 21st Centuries*. Rochester, NY: Camden House, 2020.

Fischer, William B. *The Empire Strikes Out: Kurd Lasswitz, Hans Dominik, and the Development of German Science Fiction*. Bowling Green, OH: Bowling Green University Popular Press, 1984.

Fisher, Peter S. *Fantasy and Politics*. Madison: University of Wisconsin Press, 1991.

Frey, Hans. *Vision und Verfall: Deutsche Science Fiction in der DDR*. Berlin: Memoranda Verlag, 2023.

Friedrich, Hans-Edwin. *Science Fiction in der deutschsprachigen Literatur: Ein Referat zur Forschung bis 1993*. Tübingen: Max Niemeyer, 1995.

Fritzsche, Sonja. *Science Fiction Literature of East Germany*. Bern: Peter Lang, 2006.

Goldberg, David Theo. *The Threat of Race*. Malden, MA: Blackwell, 2009.

"Happy-End auf Venus." *Der Spiegel*, 8 January 1958.

Haupts, Tobias. "The Empty Sky: A Brief History of German Science Fiction Film." In *Fu-*

ture Imperfect: Science. Fiction. Film, edited by Connie Betz, Rainer Rother, and Annika Schaefer, 64–81. Berlin: Stiftung Deutsche Kinemathek, 2017.

Higgins, David M. *Reverse Colonization: Science Fiction, Imperial Fantasy, and Alt-Victimhood*. Iowa City: University of Iowa Press, 2021.

Hudson, Dale. "Vampires of Color and the Performance of Multicultural Whiteness." In *The Persistence of Whiteness: Race and Contemporary Hollywood Cinema*, edited by Daniel Bernardi, 199–240. New York: Routledge, 2008.

Ivanova, Mariana. *Cinema of Collaboration: DEFA Coproductions and International Exchange in Cold War Europe*. New York: Berghahn Books, 2019.

Layne, Priscilla. *White Rebels in Black: German Appropriation of Black Popular Culture*. Ann Arbor: University of Michigan Press, 2018.

———. "Space Is the Place: Afrofuturism in Olivia Wenzel's *Mais in Deutschland und anderen Galaxien* (2015)." *German Life and Letters* 71(4) (2018): 511–28.

Lennox, Sara. "Reading Transnationally: The GDR and American Black Writers." In *Art Outside the Lines: New Perspectives on GDR Art Culture*, edited by Elaine Kelley and Amy Wlodarski, 111–29. London: Brill, 2011.

Martín-Albo, Ángel Mateos-Aparicio. "The Frontier Myth and Racial Politics." In *The Postnational Fantasy: Essays on Postcolonialism, Cosmopolitics and Science Fiction*, edited by Masood Ashraf Raja, Jason W. Ellis, and Swaralipi Nandi, 100–24. Jefferson, NC: McFarland, 2011.

Mazierska, Ewa, and Alfredo Suppia. "Marxism and Science Fiction Cinema." In *Red Alert: Marxist Approaches to Science Fiction Cinema*, edited by Ewa Mazierska and Alfredo Suppia, 1–24. Detroit: Wayne State University Press, 2016.

Nguyen, Angelika. "Die ostdeutsche Geschichte ist offiziel weiß." 2022. Retrieved 7 November 2023 from https://www.youtube.com/watch?v=clhUefkGvqQ

Partridge, Damani. *Blackness as a Universal Claim: Holocaust Heritage, Noncitizen Futures, and Black Power in Berlin*. Berkeley: University of California Press, 2022.

Piesche, Peggy. "Black and German? East German Adolescents before 1989: A Retrospective View of a 'Non-existent Issue' in the GDR." *AICGS Humanities* 13 (2002): 37–59.

Poutrus, Patrice G. "Aufnahme in die 'geschlossene Gesellschaft': Remigranten, Übersiedler, ausländische Studierende und Arbeitsmigranten in der DDR." In *Handbuch Staat und Migration in Deutschland seit dem 17. Jahrhundert*, edited by Jochen Oltmer, 853–94. Oldenburg: De Gruyter, 2015.

Power, Aidan. "Modern Inclinations: Locating European Science Fiction Cinema of the 1960s and 1970s." In *Future Imperfect: Science. Fiction. Film*, edited by Connie Betz, Rainer Rother, and Annika Schaefer, 82–95. Berlin: Stiftung Deutsche Kinemathek, 2017.

Quijano, Anibal. "Coloniality of Power, Eurocentrism, and Latin America." *Nepantla* 1(3) (2000): 533–80.

Rieder, John. *Colonialism and the Emergence of Science Fiction*. Middletown: Wesleyan University Press, 2014.

Robinson, Cedric. *Forgeries of Memory and Meaning: Blacks and the Regimes of American Theatre and Film before World War II*. Chapel Hill: University of North Carolina Press, 2007.

Rother, Rainer, and Annika Schaefer, eds. *Future Imperfect: Science. Fiction. Film*. Berlin: Stiftung Deutsche Kinemathek, 2017.

Roy, Rohan Deb. "Science Still Bears the Fingerprints of Colonialism." *Smithsonian Magazine*, 9 April 2018. Retrieved 7 November 2023 from https://www.smithsonianmag.com/science-nature/science-bears-fingerprints-colonialism-180968709/

Rucker-Chang, Sunnie. "The Uses of Blackness in Yugoslavia: Dimensions and Legacies of an Idea." Talk delivered at the University of Kentucky, 1 October 2019.

Schaber, Elisabeth. *Das rote Weltall: Bildnarrative der Raumfahrt in der visuellen Kultur in der DDR*. Weimar: Böhlau Verlag, 2021.

Schenk, Ralf. "Ein Mann für alle Filme: Gottfried Kolditz bei der DEFA 1955–1982." In *Publikumspiraten: Das Genrekino der DEFA und seine Regisseure (1946–1990)*, ed. Stefanie Mathilde Frank and Ralf Schenk, 83–109. Berlin: Schriftenreihe der DEFA Stiftung, 2022.

Schwartz, Matthias. "Archaeologies of a Past Future: Science Fiction Films from Communist Eastern Europe." In *Future Imperfect: Science. Fiction. Film*, edited by Rainer Rother and Annika Schaefer, 96–117. Berlin: Stiftung Deutsche Kinemathek, 2017.

Sieg, Katrin. *Ethnic Drag: Performing Race, Naton, Sexuality in West Germany*. Ann Arbor: University of Michigan Press, 2002.

Slobodian, Quinn. "Socialist Chromatism: Race, Racism and the Racial Rainbow in East Germany." In *Comrades of Color: East Germany in the Cold War World*, edited by Quinn Slobodian, 23–40. New York: Berghahn Books, 2015.

Torner, Evan. "The Race-Time Continuum: Race Project in DEFA Genre Film." Dissertation, University of Massachusetts Amherst, 2013.

———. "Casting for a Socialist Earth: Multicultural Whiteness in the East German/Polish Science-Fiction Film *Silent Star*." In *Liverpool Companion to World Science Fiction Film*, edited by Sonja Fritzsche, 130–49. Liverpool: Liverpool University Press, 2014.

Trotsky, Leon. "Revolutionary and Socialist Art." In *Literature and Revolution*, 1924. Retrieved 8 November 2023 from https://www.marxists.org/archive/trotsky/1924/lit_revo/ch08.htm.

Uhl, Karsten. *Technology in Modern German History: 1800 to the Present*. New York: Bloomsbury, 2022.

Weise, Lothar. *Das Geheimnis des Transpluto*. (East) Berlin: Buchgemeinschaft der Freien Deutschen Jugend im Verlag Neues Leben, 1963.

CHAPTER 6

The Culture of Anticipation and Scientific Fiction in Socialist Romania

Doru Aurel Pop

Introduction

Socialist Romania developed a thriving science fiction (sci-fi) culture between 1948 and 1989. Initially nurtured under Soviet influence, Romanian sci-fi also incorporated its own elements into films, literature, comic books, and other popular culture productions. Part of an international phenomenon, sci-fi was embedded in the competition between communism and capitalism and amplified by the Cold War space race. The emergence of a uniquely Romanian science-fiction culture and its contributions to global science fiction have received little attention from scholars, who have tended to focus on productions made in the USSR and more prominent Central European countries.[1]

This chapter focuses on the creation of what were called "anticipation films" in Romania (*filme de anticipație*). Here, the idea of "anticipation" was borrowed from the French literary concept of scientific anticipation literature (*littérature d'anticipation scientifique*) and then developed in a local context. To analyze the Romanian anticipation film genre, the chapter undertakes a broader interpretation of the cultural processes through which anticipation almanacs, popular books, novels, and comics created under the communist regime influenced social interactions. This remarkable cultural trend emerged when the Communist Youth Union's Central Committee began publishing some of Romania's first science-fiction stories in its scientific magazine, *Science and Technique* (*Ştiinţă şi tehnică*), in October 1955.

The enduring impact of Romanian anticipation films is examined in a case study that looks at the "fantastic-scientific" films of Ion Popescu-Gopo, one of Romania's most important film directors of the time. The first national science-fiction films were produced by the state-owned Buftea Film Studios, founded in 1950. These films were extremely popular, with millions of spectators each, and were modeled after Soviet fantastic-scientific (*nauchnaia fantastika*) films. This genre was practiced across the Socialist Bloc, including by East German authors and their Polish and Hungarian counterparts, among others.[2] Inspired by Konstantin Tsiolkovsky's popular works, the fantastic-scientific genre featured a stronger scientific dimension than Western science fiction, which was primarily fantasy driven.

To properly understand the culture of anticipation in Romania, we must adopt an interdisciplinary approach and situate film analysis in its specific circumstances. Many science-fiction media productions were available in Romania during the Cold War, from children's magazines such as *Cutezătorii* (The Bold) to various forms of science promotion, including journals and booklets, that are important in understanding specific cultural practices of science fiction in the socialist countries.[3] At the same time, concrete social conditions in communist societies were often difficult, creating a paradoxical interplay between futuristic ideals and actual social conditions.

Battle of the Genres: *Science Fiction* versus the Fantastic-Scientific

While the science-fiction genre had a fantastic and sometimes gratuitously spectacular dimension in Anglo-American and Western culture, an alternative notion predominated in the Soviet Union and its satellite countries, with different practices visible in literature and cinema. Following the French literary tradition of integrating scientific achievements into fictional narratives, Cold War Romanian writers and critics consciously avoided the term "science fiction" and instead used the terms "anticipation" (*anticipație*) or "scientific fiction" (the Romanian term for which is *ficțiune științifică*).

Ion Hobana, a science-fiction novelist and literary critic and one of the pioneers of the genre in Romania, explained this distinction in introducing readers to one of the first sci-fi anthologies in Romanian, a collection of mostly French authors, in 1965.[4] Hobana suggested that the proper term was "fantastic-scientific" or «scientific anticipation" rather than "science fiction." Later, in one of the first comprehensive histories of the sci-fi phenomenon in Romanian literary criticism, Florin Manolescu also distinguished between what he called SF (science fiction) and FS (fantastic-scientific).[5] In this theoretical landmark, Manolescu established an opposition between the Soviet concept of the fantastic-scientific, which he considered to be preoccupied with social progress and the scientific evolution of society, and science fiction, which he characterized "amazing stories" and sensational narratives that were less scientific.

This conceptual distinction was an important element in the culture wars between the Soviets and their Western adversaries. While both SF and FS are oriented toward the future, the former presents a future in which adventure takes place in space, a future used as a "background" for fantasy; in contrast, the latter addresses profound social issues and is presented as charting the progress of humankind.[6] This explains, in part, why many Western science-fiction productions were often treated with disdain by Eastern European authors; the Polish writer and futurologist Stanisław Lem, for example, regarded the "poverty" of these works, which he considered to be less scientific and more fictional, with contempt.[7]

In comparing typical Western sci-fi films, such as *E.T. the Extraterrestrial* (1982, dir. Steven Spielberg, USA), with Yakov Protazanov's *Aelita* (1924, USSR) or Andrei Tarkovsky's *Solaris* (1972, USSR), it is evident that the former barely focuses on science at all: science and scientists are barely present in the American film, and it lacks any purposeful scientific exploration or discoveries. The only common ground with the Soviet films is a belief in the existence of intelligent life in the cosmos. In contrast, fantastic-scientific and even fictional scientific depictions in scientific fiction (the Russian term for which was *nauchnaia fiktsiya*) have a profound anticipative dimension. The notion of scientific anticipation (the French term for which is *anticipation scientifique*), originally used to describe the literature of Jules Verne, involves the ability to make scientific and technical predictions as well as to anticipate solutions for social and political issues.

Often presented in the Socialist Bloc as an alternative to American sci-fi, such "anticipation" also formed the basis for ideological education in Soviet society, since the function of fantastic-scientific works was to "instruct" readers and spectators as much as to amuse them. In the Soviet Union, literature and cinema were at the forefront of the larger educational effort, as they were thought to reach the masses better than academic studies; major fantastic-scientific works had been published and even adapted into influential films since the 1920s. An example is *Aelita*, which remains one of the major Soviet fictional works combining science and imaginary worlds. The eponymous novel was written by Aleksey Tolstoy in 1923, and Protazanov's screen adaptation premiered in 1924. In the film adaptation, a Soviet engineer manages to create a rocket propelled by an innovative pulse engine and arrives on Mars with Gusev, a former Red Army soldier. Here they discover a society ruled by social engineers, called the Elders, who control and exploit the Martian inhabitants. The Soviet visitors instigate a communist revolution on Mars, which leads to the proclamation of a "Union of Soviet Socialist Republics of Mars." The film thus establishes a transparent parable promoting the possibility of exporting revolution even to other planets, since the ultimate goal of the communist revolution was its expansion to the entire known world.

Despite this propagandistic message, *Aelita* remains a remarkable cinematic work, bringing its own revolution in filmmaking. *Aelita* displayed surrealist costumes and avant-garde visuals, which made the fantastic-scientific an extremely popular genre, and Dmitri Shostakovich is said to have played music for the film live in select movie theaters. The Soviet masterpiece was also internationally influential. Made at least three years before Fritz Lang's better-known *Metropolis* (1927, Germany), it introduced robots into global cinema, and its robotic army of faceless slaves serving the ruling class on Mars clearly inspired Lang's vision of robotic laborers mechanically marching.

After World War II, the output of science fiction increased in all socialist countries, especially after the credo of socialist realism was rejected following Stalin's death in 1953. In addition to numerous literary and cinematic productions, magazines and fanzines appeared. Some of the most widely read authors were Stanisław Lem (Poland) and Arkady and Boris Strugatsky (USSR), whose works were later adapted by Andrei Tarkovsky in two Soviet-era masterpieces: *Solaris* and *Stalker* (1979, USSR), which was adapted from the Strugatsky brothers' 1972 novel, *Piknik na obochine* (*Roadside Picnic*). Not only was the Soviet sci-fi genre was not only extremely popular, but it also pervaded almost all aspects of cultural life. Just as it had taken Soviet society "by storm,"[8] the fantastic-scientific imaginary reached and gained momentum in Romania. Under socialism, a thriving science-fiction culture evolved that manifested in various media, from anticipation literature and cinema to anticipation cartoons, comics, and other publications designed for mass consumption.

Space: The Final Ideological Frontier

Science-fiction narratives existed long before the competition between the two superpowers during the Cold War. The historical evolution of science fiction—from the fantastic travel stories of antiquity to the classical utopias of Swift and the technological fantasies of Jules Verne—reveals several themes within the genre, including various forms of social critique, providing a "diagnosis" of society, serving as an instrument to understand the present, and "mapping" possible alternative futures.[9] Science-fictive imagination was integral to both contemplating possibilities for the future and constructing what seemed like impossible worlds, while fantasy narratives offered readers the tools to deal with events and situations otherwise incomprehensible.

The popularity of the sci-fi genre manifest in popular culture and entertainment today can make it hard to understand the importance of utopian imagination in the former Soviet Union and its satellite countries. Early Soviet poets like Vladimir Mayakovsky and visual artists like Kazimir Malevich and Vladimir Tatlin, however, understood communism as an ideology oriented toward the future. Soviet writers like Alexey Tolstoy, who authored *Aelita*, used the genre to promote revolutionary ideals. A driving slogan of the Soviet Union, as featured in a famous poster created by the Russian constructivist Aleksander Rodchenko, was "The Future Is Our Only Goal." The first communist society in the world was construed as an "industrial-futuristic" world, with its internal dialectical engine based on the deep contradiction between the praxis of the present and the dream of a desired future.

As the competition between communism and capitalism grew more intense, the science-fiction genre was integrated into the ideological conflict. The zenith of this conflict was reached during the Cold War, especially as the space race accelerated during the 1950s and 1960s. Space exploration became an integral part of the propaganda of both superpowers, with each trying to position its political system at the forefront of the global imagination. In the early 1950s, American science and technology seemed to be one step behind their Soviet competitors. The US Congress reacted to the remarkable achievements of the Soviets by creating the National Aeronautics and Space Administration (NASA) in 1958. In an effort to counteract the new image of the Soviet Union on the world stage, the United States made sustained attempts to achieve a favorable position, yet the Americans were in second place when it came to technological advancement.

By 1958, the Soviet space program managed to place the first mammal in orbit, with the dog Laika circumnavigating the Earth aboard *Sputnik 2*. In 1959, *Lunik I* orbited the Moon, with the following missions successfully landing and transmitting images from "the dark side" of this celestial body. By 1961 the Soviets were launching space probes to Venus, and in April 1961 cosmonaut Yuri Gagarin became the first man in space aboard the *Vostok 1*, followed three weeks later by Alan Shepard, who completed a suborbital flight in May 1961. At the 22nd Congress of the Communist Party of the Soviet Union in October 1961, Chairman Nikita Khrushchev claimed his nation's outperformance of the United States in the space race, boasting that the Soviet people were now firmly advancing on their "road to communism." The "Khrushchev Report," which was immediately translated and published in many other languages, including Romanian, promised a "new era" for the development of scientific research and predicted that the Soviet Union would "leave behind" the United States by 1980.[10] Relevantly, this "Report" opened with remarks about the leadership of the Soviet Union, supported by its successful conquest of space. This pushed President Kennedy to present his famous "Moon speech" when addressing the students of Rice University in September 1962, in which he vowed that his space program could fulfill its potential only "if we in this Nation are first [. . .] we intend to be first."[11] Putting a man on the moon became the new challenge for the two superpowers.

It would be interesting to further analyze the impact of all these scientific and political events. When the cost of the space race skyrocketed, however, it ultimately bankrupted the Soviet Union. Some estimates show that in the United States the cost of the space race ballooned from $58 million in 1958, to $11.5 billion (including civilian and military spending) in 1982. While the Soviets launched the Mir space station in 1986, the failure of the Phobos missions to Mars attested that the Soviet Union was no longer ahead.[12]

Popularizing the Conquest of Space

In the 1960s and 1970s, the "conquest of space" was presented as the main scientific achievement of the Soviet Union by party propaganda, and literature, cinema, and many other cultural products were influenced by this rhetoric. Placing Earth's first artificial satellite into orbit in 1957 became a major scoring point for the Socialist Bloc, causing the so-called "*Sputnik* shock" in the West and a "cosmos frenzy" in the East. A fierce cultural as well as economic and political competition between the two superpowers ensued, as space conquest and exploration had an enormous impact on ideological and cultural output on both sides of the Iron Curtain. As the conquest of space became an integral part of the public imagination in East and West through various media, one contributing factor was the rapid development of the science-fiction genre in literature and film.

In the East, the Soviet space triumphs were considered proof that socialism was superior to capitalism, and, as a result, in Eastern Europe science and space-related communications began to emphasize "cosmic" education in the early 1960s. The Romanian Communist Party used many popular books dealing with "cosmic" problems in its ideological confrontation with the West. For instance, Romanian authors Dumitru Andreescu and Ion Pascaru also discussed "the conquest of Cosmos" in a popularizing booklet about the exploration of cosmic bodies.[13] Published by the Romanian Society for Spreading Science and Culture in 1962, this booklet was based on Konstantin Tsiolkovsky's pioneering works and anticipated the possibility of creating interplanetary stations.

Such forms of scientific popularization were designed to convince readers that socialism was a launching pad to the cosmos, by foregrounding the great achievements of the Soviet space program while denouncing capitalism as backward. Most of the great scientific achievements of the Soviet Union, dubbed "the homeland of cosmonautics," were credited to its Leninist vision, which announced that only socialism could "liberate science from the shackles of capitalism"—a claim also echoed by Western communist scientists.[14] The Romanian Council for Spreading Scientific-Cultural Knowledge published such works systematically; in one of them, Dumitru Andreescu compared the Soviet and the American space programs and concluded that US achievements were merely "modest" and "ill prepared."[15] During the 1960s and 1970s, Romanian space technology experts thus announced the "unquestionable superiority" of socialist science and technology. As illustrated by encyclopedias for children, such as *Asaltul cosmosului* (The siege of the cosmos) and *Omul în spațiul cosmic* (Man in cosmic space), readers were encouraged to believe in the possibility of conquering space and colonizing Mars.[16]

Cultural Star Wars

The technological and political developments triggered by the competition for space conquest also had extremely important cultural, literary, and cinematographic consequences. In fact, the birth of Romanian science-fiction cinema can be linked to the popularization of space travel. Ion Popescu-Gopo's *Pași spre lună* (*Steps to the Moon*, 1963, Romania) is arguably the first true science-fiction production of the country's socialist film industry. In a compressed historical exploration of the human desire to fly into space, this film perfectly adapted to the characteristics of the fantastic-scientific genre blooming in the Soviet Bloc. Literally placing the scientific ahead of the fantastic, the Romanian director crafts a true anticipatory narrative, in which the main protagonist, a cosmonaut from the year 2000, suffers an accident and slides into a time-travel fantasy, meeting Prometheus, Galileo Galilei, the Baron of Münchausen, and even H. G. Wells. The director took a similar approach in his most-awarded short animation film, *Scurtă istorie* (*A Brief History*, 1956, Romania), also a fast-tracked account of the evolution of mankind from the Stone Age to present day. The theme of the amateur cosmonaut would become a recurrent topic in the films of Popescu-Gopo. In *Footsteps to the Moon*, however, the cosmonaut who travels through time meets historical figures and, more importantly, encounters several scientists and astronomers who were impeded in or persecuted for their exploration of space and the universe.

In this context it is relevant to note that Popescu-Gopo was also one of the most important officials in the Romanian film industry at the time. Not only the first Romanian filmmaker to be awarded a Palme d'Or at Cannes (1957), he was also president of the Romanian Association of Cinematographers. His involvement can be seen as indicative of the broader predisposition of

Romanian cinema to science fiction. Emblematic of the socialist search for alternatives to typical US formulas, Popescu-Gopo was less preoccupied with spectacular storytelling and focused on issues such as humanity, solidarity, and humanism.

This is explicit in another Popescu-Gopo production that thematizes space and time travel. *Comedie fantastică* (*Fantastic Comedy*, 1975, Romania) is yet another film with a similar subject. A humanoid robot from the imaginary planet Neptons arrives on Earth in search of a new source of energy. Just as humanity was confronted with a global oil crisis in the 1970s, the entire fictive universe lacks fuel. Consequently, the robot investigates the invention of a homemade rocket designed by a schoolteacher and propelled by a new energy source. The protagonist is indeed a simple schoolteacher who is fascinated by space travel and manages to create a rocket in his backyard. Unlike the "mad scientists" of Hollywood productions—such as Dr. No in the James Bond franchise—this is a banal science teacher who takes schoolchildren to museums, shows them how science works, and presents the moon landings. He is a man of the people, with no hidden agenda. Science and scientific achievements are not spectacular. Instead, they are made available for everybody by a schoolteacher who gives his young students lectures on complex mathematical concepts. This absent-minded scientist, an idealist who wants to enlighten his fellow human beings, is confronted with the lack of fantasy on the part of his individualistic countrymen, however. They are represented by "the Tapir," the scientist's enemy and the leader of a vulgar and ignominious gang. Jealously believing that the scientist used the spacecraft to steal the heart of his love interest, a history teacher named Miss Christina, the Tapir destroys the rocket.

In the second part of the film, Popescu-Gopo expands on a captivating idea that he had originally used in his earlier film *Faust XX* (1966, Romania). As the schoolteacher prepares to launch his space vehicle, he also manages to recreate the mechanisms of evolution. In a dreamlike sequence, the rocket travels for two million years with an onboard android watching over an evolutionary tank filled with seawater. This female robot ensures the welfare of an embryo and brings a human child to life. The boy is raised by this android and educated with the help of homemade movies created by his father, the professor. The child downloads educational files from a computer, listening to a collection of information from the twentieth century. In order to protect his son from learning about the awful side of humanity, however, the teacher has removed all the negative chapters from history, even editing *Hamlet*'s tragic parts.

The creation of a "perfect" man, scientifically engineered and grown by science but lacking all human contact, becomes a critical metaphor denouncing the dangers of gratuitous scientific improvement. For this perfect envoy of all humanity, a young man who has been educated for twenty-one years by the best in human sciences, remains unhappy. When the child becomes an adult, he finally returns to Earth in a holodeck-like invention and enjoys real life. It is worth mentioning in passing that *Star Trek* did not feature a holodeck until 1988! In fact, this film anticipates several important elements that would become trademarks of later sci-fi productions. When the young man in the rocket asks his monitoring android if it is true that the robots do not dream, for instance, it clearly anticipates Philip K. Dick's 1968 novel *Do Androids Dream of Electric Sheep?* and its film adaptation *Blade Runner* (1982, dir. Ridley Scott, USA).

Fantastic Comedy is distinguished by its profoundly humanistic message, which emphasizes an optimistic view of science and the importance of learning and education. By the end of its investigation, the extraterrestrial robot searching for a new energy source concludes that fantasy is the most important resource of planet Earth. This particular combination of the fantastic-scientific with the philosophical was a distinctive marker of the 1970s Romanian approach to science-fiction films, literature, and popular culture productions.

How We Learned to Fear the Atom Bomb

Although there were major differences between Western science-fiction and Eastern European fantastic-scientific works, the "atomic scare" was a recurring theme on both sides of the Iron Curtain. Public imagination was caught up in the US/USSR nuclear arms race, and writers and filmmakers responded to the atomic scare in their artistic works. After the Soviet Union tested its first atomic bomb in 1949, the nuclear competition escalated, with tests of the hydrogen bomb in 1953 and the creation in 1961 of the Czar Bomba, a fifty-megaton behemoth that generated an unprecedented thermonuclear blast.

David Seed outlines how, in Western storytelling, science-fiction literature and film were impacted by the threat of a worldwide nuclear conflagration, which led to the creation of many remarkable works of fiction.[17] Joyce A. Evans presents the repertoire of related themes and topics in Hollywood film narratives from 1946 to 1964, noticing an increase following accelerated use of atomic technology, with some films taking a turn for the catastrophic.[18] Similar Cold War anxieties also emerged in Eastern European fantastic-scientific cinema, with many films addressing fears related to atomic technology and the testing and uses of nuclear power.

During this troubled period, Ion Popescu-Gopo wrote and directed his first full-length film. *S-a furat o bombă* (*A Bomb Was Stolen*, 1961, Romania) was shot in black and white and featured surrealist overtones and no dialogue. Romania's first fantastic-scientific film, it shows that the anxieties of artists living in the Eastern Bloc did not differ greatly from those in the West. The production was shot in Buftea's newly built "Red studios," developed by the Stalinist leader Gheorghe Gheorghiu-Dej for his daughter, who dreamed of movie stardom. The film's transparently allegorical plot follows a young man, who has accidentally found a stolen nuclear bomb, in his attempts to get rid of it. Finally, through the power of love, he transforms the bomb into a creative weapon.

The film opens with a campestral scene in which the young man, who remains nameless, holds a flower in his hands (similar to the hero of Popescu-Gopo's animation film *A Short History*), only to be surrounded by heavily armed men dressed in parodic hazmat suits. They are employed by another group of men, "capitalists" dressed in typical "Western" attire, who are watching the detonation of a nuclear bomb. Several years before Stanley Kubrick's exceptional *Dr. Strangelove or: How I Learned to Stop Worrying and Love the Bomb* (1964, USA), Popescu-Gopo ridiculed the military-industrial complex and its fascination with destructive technologies. His comedic representation of the escalating nuclear armament race includes American gangsters (one of the gang members even wears a cowboy hat), the uncoordinated businessmen and futuristic buildings of the XOX company, and the mindless military.

As its title indicates, the film is about the nuclear bomb; but it is not simply antinuclear in its attitude. Unlike Kubrick, the Romanian director uses comedy to anticipate a positive future. In the film's final scene, in which people from all social conditions and ethnic backgrounds come together, the director presents his viewers with an explicit proposal for the peaceful use of nuclear energy. This production was clearly attuned to the politics of the Romanian Communist Party (at the time, the Romanian Workers Party), as it uses allegory and parody to criticize Western society and its entertainment industry, replete with capitalist consumerism and exploitation. (The film shockingly included scenes of a strip tease, a first for socialist cinema!) In the society of the future, where the young man meets a young blond woman wearing angel wings, it anticipates that all energy will be used only for the happiness of people of all races, ages, and genders, who will be joined in a common destiny. The story's melodramatic conclusion shows the young man and his angelical partner walking in a field of flowers sprouting from fragments of the bomb. As noted by Romanian film critics, this "anticipation fantasy" proved to be one of the first socialist blockbusters, drawing in more than two million spectators.[19]

The fantastic-scientific approach was continued by other antimilitaristic films, such as Alexandru Tatos's *Secretul armei secrete* (*The Secret of the Secret Weapon*, 1986, Romania), in which fantasy characters denounce the dangers of war. Once more, cinema promoted party politics. Declaring 1986 the "International Year of Peace," President Nicolae Ceaușescu embarked on a project to receive the Nobel Peace Prize. An editorial in the 1986 issue of

Cinema magazine, the most important publication on cinematography in Romania at the time, argued that filmmakers and artists had a responsibility to contribute to peace. Another article explained that documentary filmmakers had been enlisted to make productions, such as: *Si vis pacem, para pacem* (If you want peace, prepare for war, dir. Titus Mesaroș, Romania), *România, pentru o lume fără arme* (Romania for a world without weapons, dir. Ion Moscu, Romania), and *Solia de pace a poporului român, a președintelui României* (The peace envoys of the Romanian people, dir. Pantelie Țuțuleasa, Romania).[20]

The Fantastic-Scientific as a Form of Social Criticism

The most important work of Romanian anticipation cinema is *Faust XX* (1966, Romania) also written and directed by Ion Popescu-Gopo. In a fantastic-philosophical black-and-white production, the filmmaker creates an elaborate story that explores the complex issues of the "human condition." The story refers directly to the classic German legend, recounting the troubles of a self-absorbed scientist who meets a man promising him eternal youth. The film's opening scene shows the terminally ill university professor, who is fascinated by the myth, leaving a theatrical performance of *Faust*. There he is approached and tempted by a Mephistopheles figure, a strange man presenting himself as a medical doctor. Later appearing in the company of a couple of comedic and stupid "demons," the man is an emissary from Hell who claims to have discovered a scientific solution to the problem of mortality. At first the professor refuses the monstrous proposal, which he describes as a "crime against humanity," only to finally succumb to the promise of youth. After accepting the transfer of his mind to a younger vessel, he discovers that, to his dismay, the only body available is that of his young assistant, who is romantically involved with his daughter. Ignoring any psychosexual consequences, the professor instead worries about the quality of the vessel, since he has a tense relationship with the young man. The assistant has "grandiose" ideas about intergalactic travel and spaceships carrying embryos, envisioning a child of humanity traveling for two million years—the theme Popescu-Gopo pursued a decade later in *Fantastic Comedy*.

While *Faust XX* can be compared to René Clair's tragicomic *La Beauté du diable* (improperly translated as *Beauty and the Devil*, 1950, France)—as both stories address the relationship between science, invention, and demonic intervention—Popescu-Gopo's film has an added philosophical dimension, which is best explained in reference to the work of the Hungarian Marxist thinker György Lukács. In his critical interpretation of Goethe's drama *Faust*, which had been translated into Romanian by Pavel Apostol in 1958, Lukács argues that the myth is a manifestation of bourgeois ideology, an expression of the contradictions residing deep within the structure of modern capitalism.

The "Faustian man," Lukács notes, is dominated by an internal contradiction, simultaneously believing in the progress of humanity and depending upon illusions and fictions.[21] In the Marxist thinker's analysis of Goethe's fictional character, this contradiction is deeply rooted in capitalist civilization, and the Faustian myth exposes the historical roots of evil in human society. Mephistopheles is a manifestation of an intellectual "parasite" disrupting reason and hindering the development of mankind in late capitalist culture. Even as the bourgeoisie dreams of progress, it accepts that this progress comes at the high price of tragedy. Observing a connection between *Faust* and Hegel's *The Phenomenology of Spirit*, Lukács interprets the legend as illustrative of the Weltanschauung of the German Enlightenment.[22] The belief that human perfection and progress was achieved by breaking the shackles of the Middle Ages obscures the fact that the bourgeois mind is trapped in the same contradictions and unable to be free. Lukács claims that socialism is the only "progressive force" capable of leading humanity toward a positive future. He promotes a triumphalist view of a future socialist society without succumbing to fear or hatred, without truth-distorting myths, and without evading reality.[23] The earthly paradise promised by communism is not a utopian future, it is a society liberated from the entrapment of capitalist and bourgeois illusions.

Popescu-Gopo's *Faust XX* follows Lukács's subtle line of criticism, denouncing reactionary and irrational influences that promise a better life through scientific discoveries and a better world in which science would help mankind defeat nature by giving into "the Mephistophelian principle." The modern Mephistopheles of the film is a scientist, a medical doctor using scientific discourse and scientific practices to put his promises of immortality into practice, which Satan himself (parodically called "The Boss") is reluctant to accept. Mephistopheles mocks the Devil's antiquated strategies, transferring the electromagnetic impulses of one brain into a recipient brain via a magnetophone recording. After transferring the consciousness of the elderly professor into the body of his philandering assistant, he convinces the dean of faculty to go through the same procedure. By offering the professor and other intellectuals the ability to choose a young and healthy body—in a comedic scene, young people are practically fished out of a swimming pool—Mephistopheles tries to rule the world as he pleases with the help of his scientific minions.

The film's social critique targets a mode of understanding history and the diabolic nature of capitalism itself. As Lukács points out, the "phantasmagoric" illusions of the Faustian mind, which have not lost the inclination toward irrationalism and mysticism, ultimately lead to fascism. The film thus finally denounces the ideal of creating a "perfected human being"; in the words of the professor, nobody has the right to decide the lives of other human beings. The story openly addresses anti-capitalist themes. The amusing ending sees Mephistopheles himself lose his mind, and the archdemon "inspector" commands him back to Hell with the formula: "Believe and do not verify!" Expressing his admiration for humanity, Mephistopheles decides instead to stay on Earth as a "futureless devil" and empty-minded creature who, in a funny turn of events, brings to life the unborn child of Faust.

The Popularization of Science through Culture

Attention to scientific and technological developments was fundamental in the Soviet Union and the Socialist Bloc, and the scientific process was promoted as the means to achieving communist ideals. Nikolai Krementsov documents the enormous scientific establishment that was constructed under Stalin and the institutional effort that included thousands of institutes and hundreds of thousands of scientists employed by the state.[24] After the Bolshevik revolution of 1918, there was also an explosion of "popular science" outlets in the Soviet Union, which were later exported to the satellite socialist countries. The total number of popular science publications grew exponentially, representing more than a fifth of total publications; according to the 1930 Gosplan analysis of publications printed in the Soviet Union, out of the 17,000 magazines in existence, 32.7 percent were dedicated to scientific topics.[25]

Major institutions for scientific propaganda abroad and at home included the All-Union (Soviet) Society for the Dissemination of Political and Scientific Knowledge, which was created in 1947 and overtook many popularization activities, and the Znaniye Society, which controlled the Znaniye Publishing House and other institutions involved in so-called mass enlightenment. According to some estimates, the circulation of Znaniye publications exceeded 60 million copies in the Soviet Union.[26]

The most important popular science publication in the Soviet Union, *Science and Life* (*Nauka i Zhizn'*), was also the oldest Russian scientific magazine in print, selling 3.5 million copies during the 1960s and 1970s. Hundreds of magazines, such as the popular *Knowledge Is Power* (*Znaniye-Sila*, the Russian translation of the famous Baconian expression), and numerous television programs glorified the progress made by Soviet research and science.[27] After *Science and Life* was introduced in 1934, many competing publications began contributing to the Soviet science culture; among many others, these included *Technology for Youth* (*Tekhnika Molodezhi*), *Young Technician* (*Yunyj Tekhnik*, as of 1959), *Herald of Knowledge* (*Vestnik Znanie*), and *Sparks of Science* (*Iskry Nauki*). These media outlets provided a vast array of information on fields from astronomy and geography to physics and medicine. They ranged from the practical, as with *I Want to Know Everything* (*Hochu Vsye Znat'*), a monthly magazine featuring "do it yourself" advice, to *The Scientific Worker* (*Nauchnii Sotrudnik*), a specialized magazine for the thousands of workers at Soviet research institutions.

After World War II, all socialist countries had their own versions of such journals and magazines, including East Germany's *Youth and Technology* (*Jugend und Technik*), Poland's *Young Technician* (*Młody technik*), and Hungary's *Popular Technology* (*Népszerű Technika*). Like the Soviet science magazines, they provided not only propaganda about the latest scientific achievements of socialist research but also information about the state of international science. Science education was concurrently made available to hundreds of millions of citizens. Romanian citizens had access to Soviet popular-science literature. Starting with the popular "Sciences for Entertainment" series, created by Yakov Perelman, many Soviet authors were made easily available translated into Romanian, which provided a way for Soviet scientific and educational propaganda to be disseminated in this series and other works.

The Emergence of the Fantastic-Scientific Genre in Socialist Romania

When Romania became a socialist state in 1948, popular education in the spirit of scientific progress was reorganized based on practices already successful in the Soviet Union. Accordingly, two science magazines were published beginning in 1949: *Science and Technology for Young People* (*Ştiinţă şi tehnică pentru tineret*) and *Science and Culture* (*Știință și Cultură*). Dedicated to the education of youth and the general public, these magazines were designed to popularize science and technological advancements in the new people's republic. In 1954, these two publications were merged into the most important science magazine available in socialist Romania, *Science and Technology* (*Ştiinţă şi tehnică*). This magazine was more than simply a source of scientific information, as it offered a vast variety of content, including scientific curiosities, innovations, and other research achievements. More importantly, it provided its readers with access to science-fiction literature.

From October 1955 to April 1974, *Science and Technology* issued a special series called *The Scientific-Fantastic Stories Collection* (*Colecția "Povestiri științifico-fantastice"*). Selling more than 35,000 copies in the 1960s, this collection appeared in 466 individual issues, making it the longest existing collection of science-fiction short stories ever published in Romania. A result of Khrushchev's "thaw," it gave Romanian readers unrestricted access to translations of well-known Soviet authors for the first time. One of the first scientific-fantastic works to be published was the novel *An Expedition to the Bottom of the Sea*, by mainstream Stalinist writer Vladimir Nemtsov, which promoted the notion of fantastic-scientific literature. The Scientific-Fantastic Stories Collection, which targeted younger generations, initially published recent Soviet works, such as the serial publication of Ivan Yefremov's successful utopian novel, *Andromeda: A Space-Age Tale* (*Tumannost′ Andromedy*, 1957). It featured many other Soviet stories as well, including "The End of the Subterranean Town" by I. Kalnitsky and "The Indians" by V. Trenev. Although it continued to follow Soviet trends, the collection also gradually incorporated nineteenth-century American and Western authors, such as Nathaniel Hawthorne and Jules Verne.

As was the case with other sci-fi cultures, where the new literary forms appeared in stride with scientific and technological progress and the educational function of popular storytelling, Romanian authors adopted the so-called fantastic-scientific in their own particular way.[28] Romania's political "thaw" of the early 1970s made the publishing of Western and even American authors possible, gradually moving away from the influence of Soviet literature and culture. Many Romanian authors who were initially limited to the periodical publication in the Scientific-Fantastic Stories Collection later became acknowledged writers, published in the most important collections of the major Romanian publishing houses. Jules Verne's complete works were first translated and published as a special collection by the Ion Creangă publishing house, which specialized in children's books. Other classics of Western science fiction, such as the works of H. G. Wells and Edgar Rice Burroughs, were made available to the public along with recent Eastern European science-fiction masterpieces, including by the Strugatsky brothers and Stanisław Lem. Even Pierre Boulle's *The Planet of the Apes* (*La planète des singes*) was translated into Romanian in 1971, as well as several novels from Isaac Asimov's "robots" series, beginning with *Naked Sun* in 1975. Editura Tineretului (Youth Publishing House), renamed Editura Albatros (Albatross Publishing House) in 1971, created its own special science-fiction collection called the "Fantastic Club," where authors like Ray Bradbury, Stanisław Lem, and many young Romanian anticipation writers appeared alongside classics by H. G. Wells and E. R. Burroughs. One of the most esteemed publishing houses, Editura Univers, followed suit with its own collection of fantastic-scientific novels, publishing works by Michael Crichton, Isaac Asimov, and Herbert Franke.

When the Ceaușescu regime imposed economic and political autarky in Romania, however, it countered these positive developments. In 1974, the Science-Fiction Stories Collection was suppressed, and the publication of new materials was severely limited. As of 1982, *The Anticipation Almanac* (*Almanahul anticipația*) was published in yearly installments, with a much more politicized format. Appearing regularly until 1989, it provided one of the few Romanian outlets for anticipatory literature, which it combined with scientific information and political education. This new yearly almanac was similar to the Soviet biweekly *Knowledge Is Power*. While it never reached the popularity of the Soviet magazine, which had more than seven hundred thousand international readers by 1967, *The Anticipation Almanac* featured popular science, and each of its seven volumes included more than three hundred pages of scientific information and stories. The scientific information was written by various respectable academics; but special sections were dedicated to local, mostly younger fiction authors, like Alexandru Ungureanu, as well as well-known international science-fiction writers, like the 1983 issue that featured Arthur C. Clarke and Fredric Brown. The almanac thus preserved the spirit of scientific information and science-fiction literature working together, while also informing readers about the newest discoveries in all the fields of research.

The final edition of *Science and Technology*, put out by the Central Committee of the Communist Youth Union in 1989, opened with a glorification of the "Nicolae Ceaușescu era," which praised the contributions to scientific research made by Elena Ceaușescu, the president's wife. The issue also presented readers with some of the most advanced contemporary scientific issues, including water farming, the advent of the internet, and clean solar energy. More importantly here, it also featured an illustrated short story called "The Morning of a Faun" ("Dimineata de cerb") in the form of a comic strip; written and drawn by the young Romanian fantastic-scientific author Viorel Pîrligras, it recounts the story of a man who enjoys life but is revealed to be a robot who possesses emotions and a sense of morality.

Conclusion: Socialist Comics and the Fantastic-Scientific Imagination

In her work on anticipatory literature, Mircea Opriță, one of the most important historians of Romanian science fiction, argues that one of the major biases against anticipatory storytelling was that it was considered "childish," a literary field where "failed" authors and amateurs wrote fantasies for the uncultured.[29] And yet, magazines for children and youth in socialist Romania became a major instrument for popularizing science, and science fiction cultivated a public fascination for technological advancement and space travel. Organizations in charge of the education of children and youth, such as the National Council of the Pioneers youth organization, began publishing several youth magazines modeled upon those in the Soviet Union. *Firefly* (*Licurici*) was accordingly renamed *The Red Scarf* (*Cravata roșie*), and then, as of 1967, it appeared as *The Bold One* (*Cutezătorii*). This became the most important children's educational magazine of its time, exposing young readers to popular science and providing practical advice based on science. With several supplements published for various age groups—such as *Little Hedgehog* (*Arici*

Pogonici) for preschoolers and *Children's Evening Star* (*Luceafărul copiilor*) for high schoolers—the format of these periodicals also copied their Soviet counterparts, as their models, such as *Pioneer* (*Pioner*, also the name of the Romanian youth organization, the Pioneers) and *Spark of Light*, also made direct use of scientific facts and data in their rhetoric.

While they featured general information along with science and technology news, math problems, and brain teasers, many of these magazines published some of the first serialized science-fiction comics in Romania. These comics ranged from nationalistic depictions of national heroes to adventures and even westerns. They helped create an environment in which learning science was done through amusement. From 1969 to 1974, *The Bold One* published a monthly supplement called *Rocket of the Brave Ones* (*Racheta cutezătorilor*), in which scientific and technological discoveries were promoted in a comic series called *The Adventures of Minitehnicus*, showcasing the daring enterprises of a crew of children building their own space rocket. Once again, science and even space flight were presented as accessible to everyone.

This phenomenon deserves an in-depth analysis, but for the current discussion it suffices to observe that science-fiction themes and stories were continually used for educational purposes. In another series, called *Space-Logic* (*Spațiologica*), the narratives involved robots and took place in space, but presented children with factual information about science. *The Bold One* even offered a unique outlet for publishing special series of comics, some dealing with science-fiction subjects, that became extremely popular. The *Terra II* comics, for example, which appeared in 1968 with drawings by Albiu Stănescu and a story by Tudor Popescu, recounted the arrival of an extraterrestrial species called "Extra-Earthlings." In 1969, the magazine published another story about aliens visiting Earth. In *The Kidnapped Planet* (*Planeta răpită*), written by Horia Aramă, one of the most important representatives of the new generation of Romanian science-fiction authors, the aliens are presented as maleficent forces aiming to move our entire planet into another solar system using antigravity; they are, of course, defeated by a brave coalition of human scientists. Such comics also appeared for young people from other publishers. The almanac *Luceafărul copiiilo*, for example, featured a story based on the theory of relativity; in *Einstein's Dream* (*Visul lui Einstein*, 1982, scenario by Vasile Luca, drawings by Traian Marinescu), the great physicist encounters his own father, who has left Earth and traveled through space and time at the speed of light.

Horia Aramă—who created Radar Man, the first Romanian socialist superhero, in 1970—later wrote a remarkable story about the adventures of the first human mission to colonize Mars, with illustrations by Valentin and Liana Tănase. In 1980 and 1981, *The Bold One* issued a comic strip series on its final page, entitled *Meeting in Space* (*Intîlnire în spațiu*). The story of the Mars colony in the year 2010 describes a terraforming process, with bacteria accumulating in the valleys of Mars, very similar to the plot of *Red Planet* (2000, dir. Antony Hoffman, USA), in which Val Kilmer survives on the fourth planet in a similar way. There is also an enigma in the underground tunnels left behind by the Martians: it appears that some survivors are still alive, but it turns out that a robot left behind to guard the planet gave the humans a false impression. In 1982, this was followed by another comic titled *The Space Adventurers* (*Temerarii spațiului*), illustrated by the same cartoonists. The story, written by Sorin Stefănescu, is about an astronaut-to-be, a young teenager named Vlad, who illicitly boards a spaceship supervised by his brother and going on a rescue mission to save the cargo ship *Roza*. The latter had been lost in the vicinity of Jupiter's satellite in the year 2083. Vlad and the search team are confronted with strange phenomena and encounter an extraterrestrial species while on the rescue mission. In a nod to the first Soviet woman sent into space in 1963—former textile mill worker Valentina Tereshkova—Vlad and the rescue team finally manage to save Alma, a lost woman astronaut. These comics thus attend to the passion for space exploration and scientific discoveries that was earlier nurtured by the anticipation literature and science-fiction films made in socialist Romania. Hundreds of thousands of children were exposed to materials combining science, discoveries, innovative technologies, rockets, astrophysics, and robots, which offered them confidence in the future of a humanity driven by knowledge.

Some scholars believe that Romanian authors were more interested in the human condition and human psychology than in science, as Elaine Kleiner, for instance, claimed in an overview of Romanian science-fiction literature.[30] In fact, as stated by Ioan Albescu, the editor of the *Anticipation Almanac*, in his 1984 opening editorial, "anticipation literature" was always "engaged" and played a "militant" role in society. Whether denouncing the dangers of human alienation through excessive "technologizing" or promoting scientific achievements, these productions were not simply scientific fantasies. They were artistic creations intended as pathways to improve the lot of humankind and fulfill the dream of a future that seemed achievable.

Doru Aurel Pop is a professor of theater and television studies at the Babeș-Bolyai University in Cluj, Romania. He holds an MA in journalism and communication from Chapel Hill University, North Carolina, and a PhD from the Babeș-Bolyai University. He was a Fulbright and Ron Brown scholar in 1995–96 and 2000–2002. He has taught courses on European cinema and media studies at Bard College and Columbus State University. He has published several books on visual culture, media, and politics, as well as multiple essays on film studies. His publications include: *Romanian New Wave Cinema:*

An Introduction (McFarland, 2014); *The Age of Promiscuity: Narrative and Mythological Meme Mutations in Contemporary Cinema and Popular Culture* (Rowman & Littlefield, 2018); and *Romanian Cinema: Thinking Outside the Screen* (Bloomsbury, 2022).

Notes

1. Booker, *Historical Dictionary*.
2. Major, "Future Perfect?" 57–58; Fritzsche, *Science Fiction*.
3. *Cutezătorii* was a weekly popular magazine for children and teenagers, founded in September 1967 and issued until December 1989. Edited by the National Council of the Pioneer's Organization (Romania's youth organization), it included material on political education, encyclopedic and didactic articles, and scientific information, in addition to games and comics for younger audiences.
4. Hobana, "Introduction."
5. Manolescu, *Literatura S.F.*
6. Hobana, *Literatura*, 169.
7. Swirski, *Stanislaw Lem*, 21.
8. Krementsov, *Revolutionary Experiments*, 36–37.
9. Suvin, *Metamorphoses*, 12.
10. Khrushchev, *Documents*.
11. John F. Kennedy, "Moon Speech—Rice Stadium," 12 September 1962, retrieved 21 October 2023 from https://www.jfklibrary.org/learn/about-jfk/historic-speeches/address-at-rice-university-on-the-nations-space-effort.
12. Cadbury, *Space Race*.
13. Pascaru and Andreescu, *Stații interplanetare*.
14. Aronowitz, "The Politics of Science Wars," 209.
15. Andreescu, *Nave cosmice satelit*.
16. Andreescu and Dumitru, *Asaltul Cosmosului*; Andreescu, *Omul in spațiul cosmic*; Andreescu, *Spre Marte*.
17. Seed, *Under the Shadow*.
18. Evans, *Celluloid Mushroom Clouds*.
19. Râpeanu, *Filmat în România*.
20. Damian, "Pace."
21. Lukács, *The Destruction*, 472.
22. Lukács, *Goethe*, 176–77.
23. Lukács, *The Destruction*, 755.
24. Krementsov, *Stalinist Science*, 3–4.
25. Krementsov, *Revolutionary Experiments*, 20.
26. Shibarshina and Maslanov, "Science Communication."
27. Borissova and Malkov, "Russian Pendulum," 717–18.
28. Schwartz, "How *Nauchnaia Fantastika* Was Made."
29. Cited in Opriță, *Anticipația românească*.
30. Kleiner, "'Science Fantasy.'"

Bibliography

Andreescu, Dumitru. *Nave cosmice satelit*. Bucharest: Consiliul pentru Răspândirea Cunoștintelor Cultural-Științifice, 1962.

——. *Spre Marte: Ştiința invinge*. Bucharest: Editura Tineretului, 1966.

——. *Omul in spațiul cosmic: Ştiința Pentru Toți*. Bucharest: Editura Ştiințifică, 1974.

Andreescu, Dumitru, and Dumitru Ionescu. *Asaltul Cosmosului: Volum enciclopedic pentru copii*. Bucharest: Editura Tineretului, 1964.

Aronowitz, Stanley. "The Politics of Science Wars." In *Science Wars*, edited by Andrew Ross, 202–25. Durham, NC: Duke University Press, 2016.

Booker, Keith M. *Historical Dictionary of Science Fiction Cinema*. Lanham, MD: Scarecrow Press, 2010.

Borissova, Alexandra, and Dimitry Malkov. "Russian Pendulum: From Glorious Science Propaganda to Modest Public Engagement Initiatives." In *Communicating Science: A Global Perspective*, ed. Peter Brocks, Toss Gascoigne, Bernard Schiele, et al., 715–42. Canberra: The Australian National University Press, 2020.

Cadbury, Deborah. *Space Race: The Epic Battle between America and the Soviet Union for Dominion of Space*. New York: HarperCollins, 2006.

Damian, Laurențiu. "Pace: Un cuvânt pentru eternitate" [Peace: A word for eternity]. *Cinema*, 24(2) (1986).

Evans, Joyce A. *Celluloid Mushroom Clouds: Hollywood and the Atomic Bomb*. Boulder, CO: Westview Press, 1998.

Fritzsche, Sonja. *Science Fiction Literature in East Germany*. Oxford: Peter Lang, 2006.

Hobana, Ion. *Viitorul a început ieri*. Bucharest: Editura Tineretului, 1965.

——. *Literatura de anticipație: Autori, cărți, idei II*. Bucharest: Editura Eminescu, 1986.

Khrushchev, Nikita S. *Documents of the 22nd Congress of the PCSU*. Vol. 1. Report of the Central Committee of the CPSU delivered by N. S. Khrushchev, 17 October 1961. New York: Crosscurrents Press, 1961.

Kleiner, Elaine L. "'Science Fantasy' in the Cold War Era." In *Science Fiction Studies* 19(1) (1992): 59–68.

Krementsov, Nikolai L. *Stalinist Science*. Princeton, NJ: Princeton University Press, 1997.

——. *Revolutionary Experiments: The Quest for Immortality in Bolshevik Science and Fiction*. Oxford: Oxford University Press, 2014.

Lukács, György. *Goethe and His Age*. New York: H. Fertig. 1968.

——. *The Destruction of Reason*. Atlantic Highlands: Humanities Press, 1980.

Major, Patrick. "Future Perfect? Communist Science Fiction in the Cold War." *Across the Blocs: Cold War Cultural and Social History*, edited by Patrick Major and Rana Mitter, 71–96. London: Frank Cass, 2004.

Manolescu, Florin. *Literatura S.F.* Bucharest: Editura Univers, 1980.

Opriță, Mircea. *Anticipația românească: Un capitol de istorie literară* [Romanian anticipation: A chapter in literary history]. Cluj-Napoca: Editura Dacia, 1994.

Pascaru, Ion, and Dumitru Andreescu. *Stații interplanetare, gări cosmice*. Bucharest: Societatea pentru răspîndirea științei și culturii, 1962.

Râpeanu, Bujor T. *Filmat în România*. Bucharest: Editura Fundației Pro, 2004.

Schwartz, Matthias. "How *Nauchnaia Fantastika* Was Made: The Debates about the Genre of Science Fiction from NEP to High Stalinism." *Slavic Review* 72(2) (2013): 224–46.

Seed, David. *Under the Shadow: The Atomic Bomb and Cold War Narratives*. Kent, OH: Kent State University Press, 2013.

Shibarshina, Svetlana V., and Evgeny V. Maslanov. "Science Communication in the Soviet Union: Science as Vocation and Profession." *Social Epistemology* 34 (2019): 174–83.
Suvin, Darko. *Metamorphoses of Science Fiction: On the Poetics and History of a Literary Genre*. New Haven, CT: Yale University Press, 1977.
Swirski, Peter. *Stanislaw Lem: Philosopher of the Future*. Liverpool: Liverpool University Press, 2015.

CHAPTER 7

(Un)Healthy Tunes: Body, Mind, and Music on Socialist Television

Sandra Schnädelbach

Introduction

A reporter on the East German television program *Erleben Sie im Test: Wie beeinflußt uns Musik?* (Witness the test: How does music influence us? 1978, DDR F1) welcomes the audience in front of a bright, three-story building on a tree-lined street. It is the Research Institute of Suggestology in Sofia, Bulgaria. A few seconds pass, and the sound of classical music starts. The camera lingers on the building's façade against a backdrop of orchestral swells, then zooms through a window and into a room. Gathered here are the participants in a research institute experiment. They are listening to Joseph Haydn's *Symphony no. 67*, eyes closed, hands folded. They are studying Italian.[1]

In the 1970s, the research field of "suggestology" and its pedagogical application, "suggestopedia," as presented in this East German popular science television program were both novel. Building on a long history of Soviet medical research into hypnosis and suggestion, using stimuli beyond the threshold of perception to work on patients' unconscious, suggestology aimed to activate hidden resources within the brain via relaxation techniques and sensory, most prominently acoustic, stimulation. A Bulgarian research team, led by the physician and psychotherapist Georgi Lozanov (1926–2012), began its work in the mid-1960s but did not receive international attention until the following decade.[2] By the late 1970s, reports on suggestology that investigated the research and pedagogical approach, as well as the relationships of music, body, and mind, began to appear on television screens in the German Democratic Republic (GDR).

The preoccupation with music's influence on people was closely related to interest in the mind-body relationship in the health sciences, which had been flourishing in East and West since the 1970s.[3] Lozanov—remembered to this day as "the father of accelerated learning," especially in Bulgaria[4]—generated attention primarily with his spectacular promises of learning more easily and quickly, with the help of new scientific findings on the interplay between the body and the psyche.[5] Amid the "psycho-boom" and generally increased desire for self-realization of the 1970s, Lozanov's work aroused curiosity not only in the East but also in the West.[6] At the same time, however, mind-body science was perceived on both sides of the Iron Curtain as an important knowledge generator for psychological warfare.[7] Scientific progress had been a central factor in the Cold War from the start, and it referred not only to technological innovation but also to knowledge of the human psyche. Intelligence services in East and West alike were avidly interested in how to influence the mind as a means of political infiltration.[8]

After the end of the Cold War, historical research mainly focused on the military, technical, and political facets of the conflict.[9] Increasingly, however, historians have moved away from questions of state control and turned their attention to cultural products and different manifestations of the political. Cold War cultures—the plural is deliberate—have moved front and center. These cultures were by no means homogeneous, and should instead be understood as a "set of different, sometimes conflicting influences on culture," as Annette Vowinckel has pointed out.[10] This is all the more relevant given that the concept of the Cold War itself has recently been critically recast as a Western creation.[11] Acknowledging a plurality of Cold War cultures thus allows us to sharpen our perception of national specifics, without ignoring possible transfers and exchange processes between countries and blocs.[12]

Suggestopedia, for instance, was part of a significant European cultural tradition involving classical music; at the same time, it raised highly sensitive questions about influencing human will by studying suggestion via sound. Due to its concepts grounded in Soviet science and the flaws in its empirical design, suggestology was often evaluated as pseudoscience in Western academia. At the same time, it was carefully monitored by Western neurologists and psychologists as a possible source of brainwashing. Using audiovisual sources, this chapter sheds light not only on the aspirations but also and especially on the practices that were connected to suggestology in the Eastern Bloc—and that made it both attractive and explosive on the international stage of the Cold War.

The television programs on suggestology discussed in this chapter thus offer more than an excursion into the history of Soviet and Eastern European pedagogy.[13] In fact, they lead us to the core of the Cold War, as suggestopedia emerged as a special, applied field of much more fundamental scientific research in neurology and psychology. By focusing on GDR television programs on suggestology and the relationship between body, mind, and music of the late 1970s and 1980s, this chapter demonstrates that these seemingly marginal topics had significance far beyond national borders and were part of a scientific competition for psychological knowledge during the Cold War. It simultaneously exposes multiple connections between national and transnational influences and interdependencies. At a national level, the television programs shed light on the peculiarities of the socialist view of the body and health, which were built on materialistic approaches; and, at the same time, the programs were part of a widely ramified network of the era's transnational debates, reflecting the interconnections and similarities between the collectivist East and the individualist West.

Visualizing Science: Suggestology on GDR TV Screens

In East and West alike, audiovisual media like television played an important role in the production of scientific knowledge.[14] The programs discussed here were part of major East German science shows, which experienced a boom in the 1970s.[15] By this time, having a television set in the home was the norm in both Western Europe and the Eastern Bloc, and watching television became the most popular recreational activity.[16] From the very beginning, the rise of television was connected to the competitive structures of the Cold War, and this was particularly evident in a divided East and West Germany. From the race to be first to broadcast in the early 1950s to the fact that the people in the East were largely able to receive West German television and did so illegally, the new medium was marked by competition between systems.[17] In the GDR, television was further considered an important instrument of socialist education. In order to stay in the audience's good graces, however, East German television had to repeatedly implement program reforms. As variety and entertainment shows gained importance in the 1970s, the number of science and health advice shows, which enjoyed particular popularity, also increased.[18]

The thirty-minute show *Erleben Sie im Test: Wie beeinflußt uns Musik?*—introduced at the start of this chapter—was produced by (East) German Television Broadcasting (DFF) and shown during the 8 p.m. prime-time slot. DFF worked closely with the state-owned German Film Company (DEFA), which fulfilled numerous commissions for television shows. The program was part of the regular science series *Neue Fernseh-Urania* (New Urania television), named for Berlin's well-known Urania science education center, founded in 1888, which offered science-related readings, seminars, and discussions for lay audiences.[19] Initiated in the 1960s as a coproduction between Urania and the Academy of Sciences of the USSR, the series was relaunched in 1974 with an emphasis on popular science.[20] In television, such coproductions, cooperations, and program exchanges were well-established within the Eastern Bloc and functioned as proof of the good relationship between socialist neighbor states.[21] Along with sports and entertainment, educational shows were a central element of this international exchange; in the 1970s, for example, the GDR and Bulgaria jointly produced health education programs for cinema and television.[22] It is therefore unsurprising that *Erleben Sie im Test: Wie beeinflußt uns Musik?* picked up the new Bulgarian research field of suggestology and presented it as both a modern experimental science and a tool for socialist self-development through music.

The television program stages suggestopedic teaching as primarily an auditory experience. While the camera pans over adult students sitting in a semicircle, the soundtrack focuses on the voice of the teacher, who reads short sentences in Italian while Haydn's symphony plays (figure 7.1). The teacher's slow and declamatory reading style presents a distinctive speech rhythm and volume calibrated to a professional concept: what we see and hear are so-called "concert phases," and the teacher is not only a linguist but also a trained singer and violinist. This "set-up similar to that of attending a concert" was a core element of suggestopedia and was meant to secure a "quiet and intuitive perception," conducive to the "generation of moods, the emergence of associations, and the flashing of ideas."[23]

The camera makes suggestopedia visible as a specific physical practice: the television audience watches people reclining contentedly in comfortable armchairs, their heads leaning back or tilted to one side, their eyes closed and hands folded (figure 7.2). Guiding the way to relaxation and inner reflection, these "practices of contemplation," as I propose calling them, were based on contemporary research into the interplay of physical and mental processes, which were by no means applicable solely to pedagogy.

In an interview following the listening scene, Lozanov informs his audience that suggestopedia is about addressing "the unconscious factors" of learning and "tapping the mental-emotional reserves of the learner" via music.[24] According to the trained physician, sounds and melodies are able to activate unused areas of the brain and create a state of "concentrated mental relaxation."[25] He believes that music operates largely beyond conscious perception and is conveyed emotionally. It was this assumption that ultimately extended the scope of suggestology beyond language didactics, and it came to be seen as a route to a comprehensive investigation into the dependencies between the body and the psyche.

The mise-en-scène of the teaching scene unmistakably links listening practices to certain physical states and postures. These bodily practices were considered part of a scientific method that was based on sound technology and

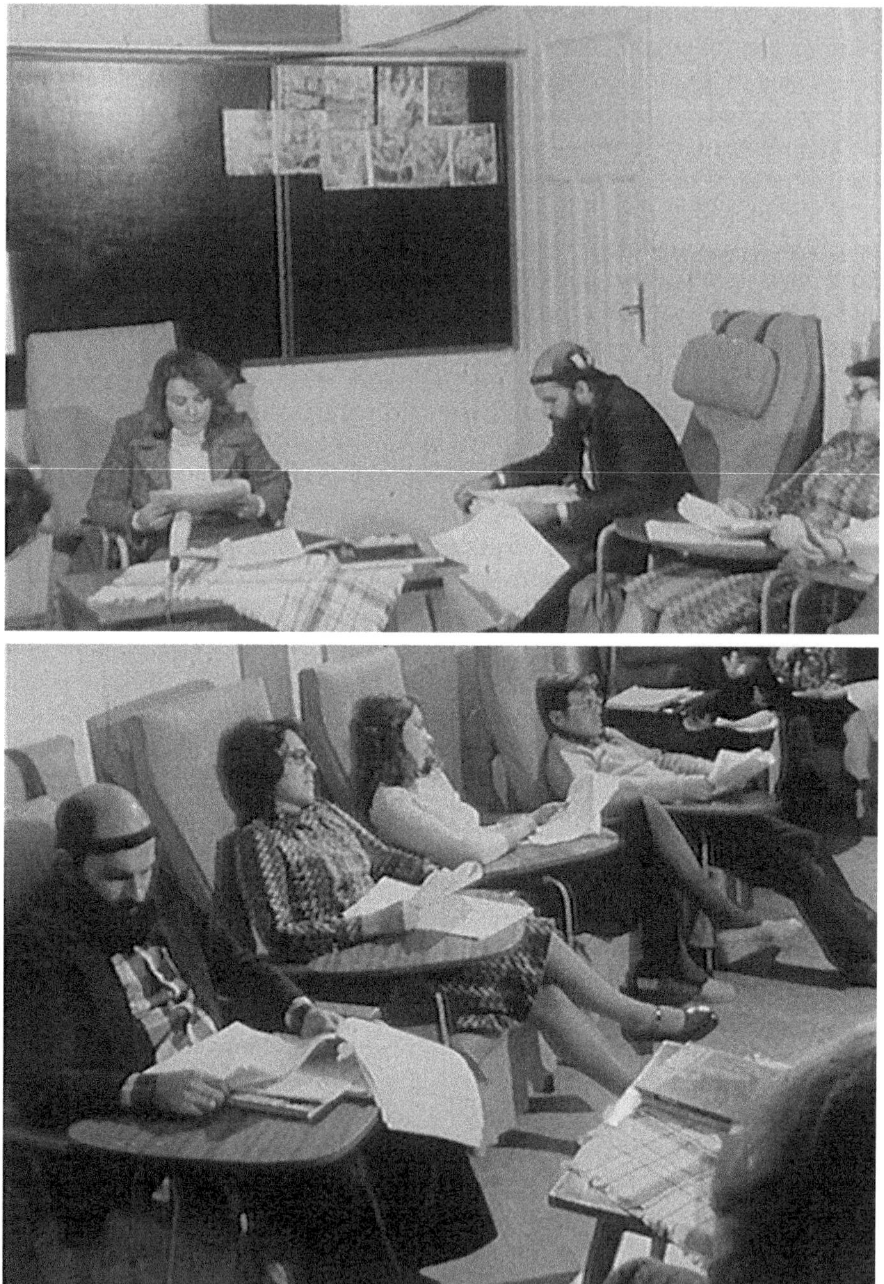

Figures 7.1. and 7.2. Suggestopedic teaching on GDR television. *Erleben Sie im Test: Wie beeinflußt uns Musik?* (Witness the test: How does music influence us? 1978, DDR F1). © Deutsches Rundfunkarchiv.

data collection. The camera shows not only the "technology center,"[26] a loudspeaker connected to the research laboratory, but also a student wearing a measuring device around his head, making it clear that the physical effects of listening were being evaluated via nerve and brain activity (figures 7.1 and 7.2).

A 1983 episode of the monthly science show *Umschau* (Review, DDR F1), which covered topics ranging from heart disease to robotics and was produced and broadcast by DFF from 1961 to 1991, presented suggestology even more explicitly as belonging to a complex, highly modern experimental project at the interface of health science and education. The popular show rendered a teaching scene with an even stronger focus on postures, with close-ups showing all students in the relaxation position. The interview with Lozanov—shot for *Erleben Sie im Test: Wie beeinflußt uns Musik?* in his office—was now taking place in the Suggestology Institute's research laboratory. The camera emphasizes the lab's extensive technical equipment: after a medium shot showing Lozanov in conversation, a long shot and extended pan over measuring equipment and individual devices, including a tape recorder and two television sets, follow (figure 7.3).

The staging of lab equipment in both science TV programs earmarked suggestology as ambitious psychological research that befitted other East German

Figure 7.3. The founder of suggestology, Georgi Lozanov, in his laboratory. *Umschau* (Review, 1983, DDR F1). © Deutsches Rundfunkarchiv.

research interests. *Erleben Sie im Test: Wie beeinflußt uns Musik?* presented suggestology as part of a broader research agenda investigating the effects of music on the human body and mind. The scenes that frame the report give us a vivid sense of this. The program begins with a performance by Metropol, at the time an up-and-coming East Berlin beat band, at a youth dance event. A young woman on the dance floor is asked by the reporter to take a blood pressure test; he then asks her to accompany him to the Department of Psychology at Berlin's Humboldt University for lab tests. Here, her respiratory, muscular, and skin conductance responses are measured under the influence of music. Like in the scenes at the Suggestology Institute in Sofia, the experimental, physiological dimension of the research is at the center, represented by numerous shots of the tests and data recording equipment (figures 7.4 and 7.5). With such references to "modern" tools of psychophysiology, the show characterizes suggestology as part of a vibrant field of research and situates it in a competition for scientific excellence that reached beyond the Eastern Bloc.[27]

Suggestology's Scientific Aspirations in the Cold War

The development of suggestology in the 1960s and 1970s was guided by a desire for scientific recognition on the international stage. The Suggestology Institute, formed around Lozanov, clearly exhibited a great ambition to generate international attention and raise the profile of Bulgarian research on psychophysiology beyond the Eastern Bloc. One foray in this direction was the "First International Symposium on Problems of Suggestology," held in Varna, Bulgaria, in June 1971. The six-day conference brought together researchers from various socialist states, including the GDR, as well as scientists from Canada, the United States, and several Western European countries, including the Federal Republic of Germany.[28] The event was held in cooperation with the Soviet Union's Ministry of Education. Its organizing committee included members of the Bulgarian Academy of Sciences—from medicine, physiology, psychology, and psychiatry—as well as someone from the Institute of Labor Protection and Industrial Diseases. Topics presented ranged from studies on the stress behaviors of train drivers, to the effect of suggestopedia on the cardiovascular system, to the learning behavior of children with mental disabilities.[29]

In accordance with Soviet psychology, suggestology was based on materialistic concepts of body-mind dynamics and built on the psychophysiological research on higher nervous activity of Ivan Pavlov (1849–1936).[30] Viewed as an alternative to Sigmund Freud's theory of the subconscious, Pavlov's findings were declared the only valid basis for psychology in Soviet countries in the 1950s.[31] Pavlov's physiological definition of consciousness did not con-

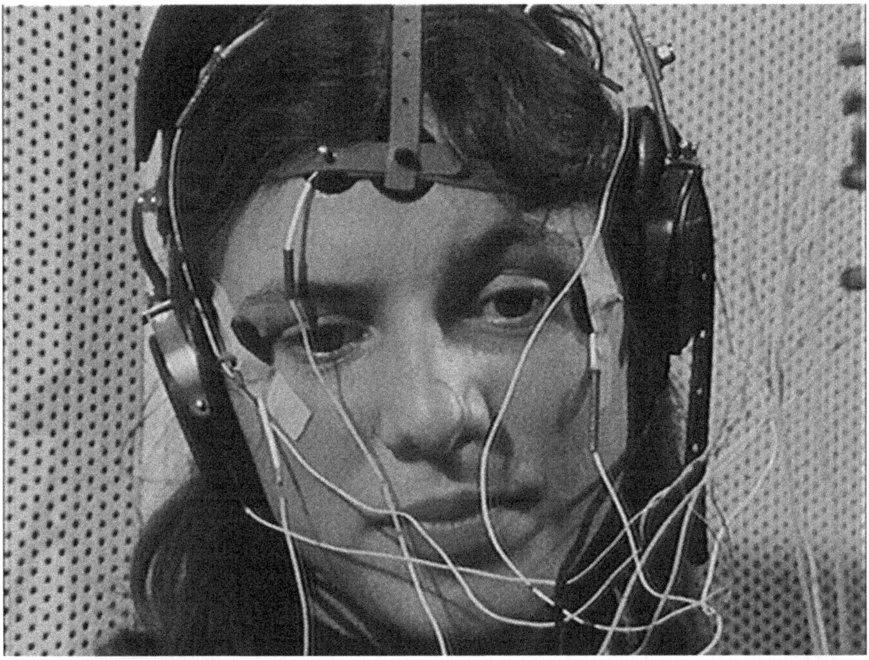

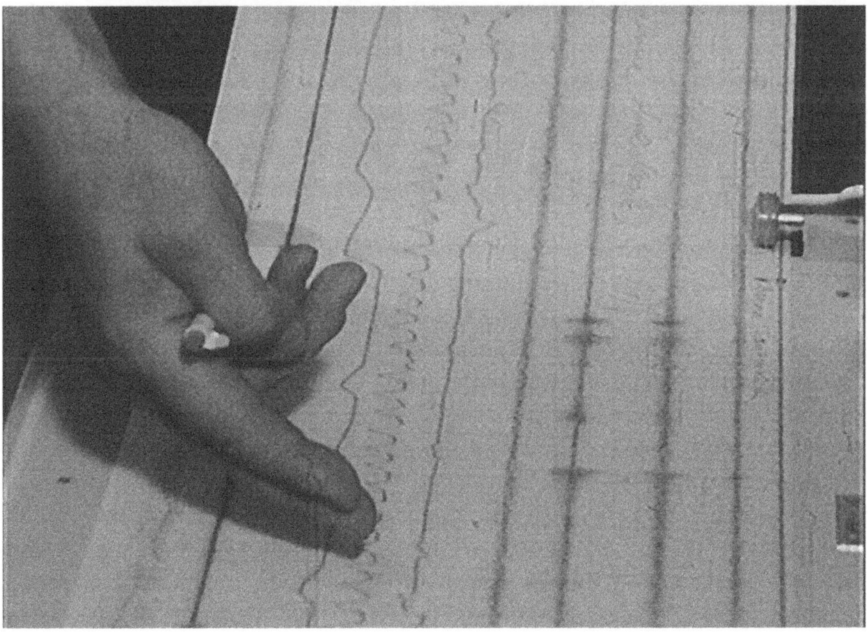

Figures 7.4. and 7.5. Measuring the effects of music on the body. *Erleben Sie im Test: Wie beeinflußt uns Musik?* (Witness the test: How does music influence us? DDR F1, 1978). © Deutsches Rundfunkarchiv.

sider the generation of subjective perceptions but instead investigated the excitability of brain areas. From the very beginning, Pavlov's team had paid special attention to sounds as sensory stimuli and the cause of psychophysical reactions.[32] As part of their research on conditioned—that is, learned but automated—responses, they experimented with trumpet sounds and the use of gramophones.[33]

Suggestology extended Pavlov's approach to pedagogy and brain performance. In line with Soviet psychology, Lozanov stressed that the "unconscious" was not to be confused with ideas of the "subconscious." Rather, "conscious" and "unconscious" described states of brain activity associated with different thresholds of excitability. Automatic responses, created by conditioning a person in a certain environment, were assigned to the unconscious.[34] The aim of suggestology was to psychophysically examine the unconscious and make it fruitful in educational contexts.

The concept of suggestion and the application of suggestive methods in medicine became more widely known in the second half of the twentieth century. Suggestology had grown in part from a body of psychological knowledge that had expanded since the mid-nineteenth century and was becoming hard to overlook in the 1970s. Suggestion was primarily used in the context of hypnosis techniques and had been in use in both East and West since the early twentieth century.[35] From the start, hypnosis had attracted criticism and suspicion among psychologists as well as the wider public since it was associated with the influence on a person's thoughts and feelings by a third party.[36] After 1945, research into hypnosis and related practices, like sleep therapy, primarily survived in Soviet and Eastern European medicine, where sleep therapy was in use until its shortcomings became visible and it fell into disrepute in the 1960s.[37]

With his research, Lozanov wanted to challenge such misgivings. By the early 1970s, he was already arguing that suggestology should not be confused with hypnosis. Suggestion, he claimed, did not occur in a passive state but was an active process of reflection that combined intuition and reason. The Bulgarian scientist was most interested in its potential for intellectual development and memory performance. The activation of the unconscious, he argued, created "conditions of control and feedback" that informed the individual's motivation and decision-making.[38] Cybernetics, with its emphasis on causal feedback loops involving biological and technological systems, offered an important source for these ideas. While this field of research was initially met with resistance in the GDR, from the mid-1950s onward it became an extremely popular point of reference in the scientific systems of the East as well as the West.[39]

A model of "telemetric registration" created at the Institute of Suggestology illustrates these references to cybernetics and the rising importance of new media. With the help of modern technology—specifically a television set connected to a laboratory—data-based feedback should enable the research team and the teachers to adapt suggestopedic methods to student responses and continuously improve the interaction with students.[40] Suggestology thus applied media theory of the time, which was often concerned with interactive ways of communication and which, as health advice programs on socialist television channels show, resonated well with the sociopsychological approaches of socialist medicine in general.[41]

All these elements reflect aspirations to establish suggestology within an international field of experimental psychophysiology beyond the Soviet zone. And indeed, while in his introductory lecture to the 1971 symposium Lozanov mentioned Pavlov, he refrained from placing his research in broader socialist contexts and emphasized his Western networks instead. Another interesting detail: the edited conference volume was partly published in English.

This endeavor to present suggestology as an internationally promising field of research was also evident in the television programs analyzed in this chapter. East German television not only boosted the perception of suggestology as a legitimate experimental science, however; it also gave suggestology a place within mental health care and psychosomatics, which were increasingly coming to the fore in both the East and the West.

International Trends in the Psycho-Boom

The emergence of suggestology on GDR television screens was closely connected to the growing importance of psychological knowledge in the second half of the twentieth century. Integral to this knowledge were questions about the psychophysical effects of music. As new dance styles became popular in both East and West in the 1950s, parents, politicians, pedagogues, and medical experts hotly debated the influence of music on the bodies and minds of young people. The discussions revolved primarily around (sexual) morality and were strongly shaped by Cold War rhetoric and efforts to distinguish socialist and nonsocialist societies.[42]

By the late 1970s, however, research increasingly shifted away from the dangers of popular music and toward music's potential for individual health and well-being. The East German television shows on suggestology thus appeared at a time when the health sciences were turning their attention to the therapeutic qualities of music.[43] This coincided with a broader international trend of paying more attention to mental health and, especially, psychosomatics.[44] With regard to Western nations, historians often refer to the "psycho-boom" of the 1970s and assume a delay in this development in the GDR and Eastern Bloc.[45] However, a look at GDR television programs reveals that the assumption that research developments in the East were simply delayed reactions to Western trends is misleading. Though the West differed significantly from the East in the breadth of its therapeutic and psychosomatic approaches,[46] the example of suggestology shows us that exploring the unconscious and the body-mind interconnection had been on the agenda in socialist countries since the 1960s.

In his interviews for the science television programs, Lozanov stressed that suggestopedia had its roots in therapeutic methods; in particular, suggestology was well-suited to East German approaches to psychotherapy. Psychotherapy in the GDR did not align with Western concepts and was instead built largely on materialist approaches like "rational psychotherapy," which—always focusing on higher nervous-system activity—relied on autogenic training and hypnosis.[47] As of the 1960s, however, a growing number of East German psychologists supplemented such materialist approaches by focusing on the psychological effects of social interactions and began including group therapy and music therapy in their treatment plans.[48] During this period, music therapy was also established as a subdiscipline in its own right and steadily gained prominence thereafter.[49]

East German psychology envisioned music not only as a means to regulate psychosomatic disorders but also as a path to self-reflection and personal growth.[50] The show *Erleben Sie im Test: Wie beeinflußt uns Musik?* reflects this as well; near the end, the narrator concludes, "Music: it influences not only our bodily functions but also our behavior toward our fellow human beings and ourselves."[51] The promise of suggestology was closely connected to the idea of self-development, which was popular in the West but—with its collectivist concern for the socialist personality—also firmly anchored in the Eastern Bloc.[52] In essence, suggestopedia employed practices that were also highly valued in the West in the 1970s. There, too, autogenic training, relaxation exercises, and biofeedback, which were used to create targeted effects on the autonomous nervous system, flourished.[53] On both sides of the Iron Curtain, psychophysical self-exploration was accompanied by a turn toward intuition, feelings, and relaxation, through which the individual was meant to fully develop their qualities.[54]

With its focus on music, suggestopedia tied in not only with current psychological trends but also with a traditional cultural practice of music reception. The method made use of classical music—i.e., a canon anchored primarily in Central Europe—whose reception was characterized by quiet aesthetic reflection, as advocated by European scholars of the Enlightenment.[55] Classical music was held in high esteem in the GDR, as it supported the country's claim to be the "better" Germany. Similarly, classical music anchored suggestology in western *Kultur* (high culture) and indirectly promoted the Soviet Union's Cold War search for its own connection to Germany's cultural heritage.[56] Socialist visions of classical culture were often presented in opposition to American popular culture on the Cold War's ideological battlefield. The myth of classical music reproduced by suggestology thus offered common ground: evoking an idea that had already resonated far beyond Germany or Europe in the nineteenth century, references to Haydn and others invoked the notion that music transcends emotional, interpersonal, and even societal conflicts.[57] These commonalities also made critics in the West, who held many of the same ideas, sit up and take notice; if suggestology was truly able to harness the power of music, it had to be taken seriously as a competitor in the scientific race.

Between Science and Pseudoscience: Evaluating suggestology in the West

The concept of suggestion, in particular, touched a nerve among Western psychologists and strategists of psychological warfare alike. In the 1970s, suggestology also received attention in Western countries, primarily from the US intelligence services. Since the 1950s, there had been fears in the United States of psychological manipulation of the population by opponents of the system, often referred to as brainwashing. This concept had first came to the fore in the 1930s, during the Stalinist show trials, but did not become widespread in the West until the conflict with China raised by the Korean War. In its original Chinese meaning, the idea of reeducating someone from a "false" to a Communist consciousness was associated with elements of Asian meditation traditions.[58]

As a method originating in the Eastern Bloc, suggestopedia was received by Western psychologists in the context of old and enduring concerns about the misuse of mental practices, and suggestology was easily conflated with hypnosis and states of passivity that touched upon questions of personal autonomy and freedom of choice.[59] Beyond the therapeutic setting, approaches like suggestology that promised insight into the malleability of human will quickly took on a political dimension.

The science programs on GDR television demonstrated the ambiguity inherent to suggestion. *Erleben Sie im Test: Wie beeinflußt uns Musik?* presents suggestopedia in a socialist context, as the voice-over confirms that "next to the painter [sits] the construction worker, next to the professor, the student," as is appropriate in an egalitarian socialist society.[60] The framing scenes also demonstrate the effects of music in a socialist community, as the voice-over concludes that music actually "gets into the legs"—i.e., stimulates bodily functions—and the camera presents astonishing scenes of kindergarten children who seem unstoppable as soon as a miners' march is played.[61] Whereas the political dimension remains unaddressed in this scene, the voice-over later becomes more explicit: music, it states, affects us not only as biological beings but as social ones. Expectations, knowledge, and experience are illustrated as various factors of individual reactions to music, and the program presents these as opportunities for socialist community building, which involve political and private, collective and individual experiences. Images from the Spanish Civil War of 1936 and of singing Soviet astronauts celebrate the unifying, immediate, positive effects of music, the causes of which elude our conscious reflection.[62]

The claim of unconscious influence through music, which was key to suggestopedia, is what rendered this pedagogical approach politically explosive. Although Lozanov tried to distinguish his approach to suggestion from old ideas about hypnosis, his references to the "unconscious" and the "hidden resources of the brain" left much room for interpretation.[63] According to *Erleben Sie im Test: Wie beeinflußt uns Musik?* in the suggestopedic learning of Italian, the vocabulary "infiltrated" the memory via music—a process that was staged as thoroughly positive. For Western critics of suggestology, however, Lozanov's terminology and the openness of his concepts were alarming and stoked the central anxieties of the Cold War.

It is therefore unsurprising that the Central Intelligence Agency (CIA) showed an interest in suggestology. A 1972 intelligence report on Soviet research into "controlled offensive behavior" that covered both hypnosis and suggestion contained a subchapter dedicated to suggestology. The report classified the new branch of research as quite powerful: "This method uses suggestion not as a means for some kind of mystical influence, not as some kind of abstract, 'vague' factor, but as a specific method for directly influencing the emotional world and intellectual activities, the entire personality, of the student."[64] The CIA addressed the exact aspects that Lozanov himself brought up and that were underlined in the television programs, which held that developing the socialist personality through music was central and allowed for better access to the self and other human beings.

Conversely, in the CIA's assessment, the educational and therapeutic components of suggestology were secondary to concerns about comprehensive manipulation of the personality, as was characteristic of the discourse on brainwashing in the United States.[65] The interpretative patterns of this discourse classified brainwashing as a form of "mental rape" with a remarkable impact. These ideas were propagated by popular Western feature films of the 1960s and 1970s that also prominently featured the use of music as a method of manipulation.[66] Since the nineteenth century, music had often been used in hypnotic procedures, as it was believed to have a direct psychophysical effect. Beginning in the 1950s, this sparked a widespread fear of musical brainwashing in the United States and, ironically, led some to suspect a communist conspiracy behind the rock 'n' roll hype of the time.[67]

The assessment of music as a danger to the mind was also made in the East, and Pavlov's psychological theories provided the ideal basis. Though Pavlov was generally not invoked in the West, due to his appropriation by Soviet science, East and West shared many commonalities, especially in the field of popular science.[68] Furthermore, the Medical Intelligence Division of the US Defense Intelligence Agency evidently considered the suggestological approach to be plausible, and noting its connection to meditative practices, use of music, and mental development goals, they sounded the alarm.[69] In fact, in the 1970s, the development of nonlethal weapons was closely connected to mind control research and became the preeminent concern in the Cold War arms race. In this decade, investigations into sound-based manipulation of mental states and brain capacities moved center stage, conveying the perceived importance of scientific excellence in psychology and neurology.[70]

Even beyond intelligence circles, suggestology initially received recognition by international experts in the West. The presentation of suggestology as experimental research at the interface of medicine, psychology, and pedagogy, as on GDR television, corresponded with the efforts of the Institute in Sofia to establish suggestology as a prestigious scientific project that would earn Eastern European research the attention and support of international organizations. A first step in this direction was made when a 1980 UNESCO commission reported that suggestology and suggestopedia were promising fields and encouraged further studies to validate and refine research findings.[71]

The reception of suggestology in pedagogical circles began in West Germany in the 1970s, but only picked up speed in the mid-1980s.[72] This delay can be attributed to the tentative nature of the rapprochement between Western sciences and research in the socialist states. Overall, the majority of Western researchers would simply ignore scientific events in the Eastern Bloc. Alternatively, accusations of pseudoscientific practices were quickly leveled, serving as a popular means of discrediting opponents on both sides of the Iron Curtain.[73]-

The late and suspicious reception of suggestology was not only due to Cold War dynamics, however, and reservations about the field ultimately proved to be justified. Over time, it became clear that the studies conducted at Lozanov's Suggestology Institute suffered from serious methodological flaws. The extraordinary results on foreign language learning presented by the institute could not be verified in experiments conducted by international researchers, suggesting that the results had either been deliberately enhanced or reflected inadequate scientific controls.[74]

The commercial success of suggestopedia, however, was not affected by the lack of scientific evidence. Lozanov worked to establish his name as a trademark and to have the suggestopedic method legally protected. Beginning in the mid-1970s, private institutes offering suggestopedic instruction were established in the United States and Europe; in West Germany, the "Lozanov Learning Institute" was housed in the Cologne School of Economics.[75] At the same time, American pedagogues and psychologists used Lozanov's theories to develop methods such as Superlearning or SALT (System of Accelerative Learning Techniques), which were successfully marketed internationally. Despite all this, the reputation of suggestology remained ambiguous: during their early years, Lozanov and his followers had often sympathized with parapsychology, thus raising suspicions in the West;[76] at the same time, in the 1980s Superlearning and the services of the Lozanov Learning Institute were presented in such established media outlets as the West German weekly newspaper *Die Zeit*.[77]

Thus, while suggestology remained a marginal phenomenon in the Western academic world, it was a success in the field of popular science and connected with interests and needs in both East and West. The two GDR TV programs on suggestology mentioned here aired in 1978 and 1983, helping fuel the popularity of the field in the late 1970s and early 1980s and bringing it to a mass audience. Here, the body practices of suggestology came to the fore visually, and the core of the method—namely the use of music—became visible to viewers. The programs fed the methods of suggestopedia into popular trends in knowledge of the body, health, and personality development, which also resonated in the West during that period.

Conclusion

The GDR television programs analyzed in this chapter reflect an effort to present suggestology as a serious experimental science. Image and, especially, sound emphasized the sensual qualities of suggestology and popularized certain bodily practices and techniques of self-development. These practices, though placed in a socialist context, showed considerable overlap with their Western counterparts in terms of practical design. The science programs enhanced the value of suggestology by presenting it as part of a vibrant medical investigation into the connections between the body, mind, and music. They also aided the aspirations of Lozanov and the Suggestology Institute in Sofia, Bulgaria, to achieve international scientific recognition. At the same time, references to the "power" of music and its untraceable effects on body and psyche brought to the fore precisely those aspects of the method upon which widespread mistrust and concern in the West was founded.

The television coverage of suggestology that began in the 1970s is linked to the various and complex evaluations that suggestology received in the East and West. While suggestopedia enjoyed little attention in Western research, it was taken seriously in intelligence circles and seen as a potential tool for Soviet infiltration. In their assumptions about human will and their desire to improve personal development, the rival powers of the Cold War were often closer to each other than was normally assumed. In the case of suggestopedia, this was in part because it built upon older cultural and scientific traditions and found support beyond systemic boundaries, especially in the German-speaking world. The high and quasi-spiritual status of classical music had a long history, and autogenic training and other psychosomatic relaxation practices were already in vogue in the early twentieth century. In this respect, the Cold War did not witness the advent of something unknown; rather, Cold War cultures drew extensively from historically rooted ideas and structures. Ambivalences and internal contradictions on both sides of the Iron Curtain are thus revealed here. On one hand, scientific self-depictions in the West did not admit to the myriad political and military motivations behind the research;[78] on the other hand, advances in the East were not limited to Moscow's socialist and anti-capitalist agenda but rather elicited targeted, increasingly commercial cooperation with other countries.

These television programs therefore reveal not only the characteristics of socialist psychology and psychosomatic medicine but also international trends that developed outside of and around the borders of the blocs. Suggestology was closely linked to a boom in psycho-knowledge as of the 1970s. In both East and West, therapeutic approaches were combined with practices aimed at enhancing individual well-being and personality development. The fact that these interests in "undiscovered mental resources" were shared across system boundaries was influential in the scientific revaluation and devaluation of suggestology over the decades.

Sandra Schnädelbach is a Gisela Eisenreich Fellow at Heinrich Heine University in Düsseldorf, Germany, where she explores the entanglement of visual history and the history of the body in East and West Germany. Her work focuses on cultural history, history of knowledge and science, media history, and the history of emotions. She received her PhD from the Free University Berlin and was a doctoral fellow at the Center for the History of Emotions at the Max Planck Institute for Human Development in Berlin. She was also a postdoctoral fellow at the University of Strasbourg, France. Her recent publications include *Entscheidende Gefühle: Rechtsgefühl und juristische Emotionalität vom Kaiserreich bis in die Weimarer Republik* [Decisive emotions: The sense of law and emotionality from the German Empire to the Weimar Republic] (Wallstein, 2020) and articles in the journals *Generus* and *View: Journal of European Television History and Culture*.

Notes

1. This project has received funding from the European Research Council (ERC) under the European Union's Horizon 2020 Research and Innovation Program (grant agreement No. 694817).
2. Edelmann, *Suggestopädie/Superlearning*, 34–46.
3. Harrington, *The Cure Within*, 27; Geisthövel and Hitzer, "Die Grenzen des Erfolgs."
4. Retrieved 16 March 2021 from https://en.wikipedia.org/wiki/Georgi_Lozanov.
5. Edelmann, *Suggestopädie/Superlearning*, 66–78.
6. Tändler, *Das therapeutische Jahrzehnt*, 90 and 174; Maasen et al., *Das beratene Selbst*, 35–36.
7. Krishnan, *Military Neuroscience*, 17–44.
8. Krishnan, *Military Neuroscience*; Needell, *Science, Cold War and the American State*, 1–8.
9. Siefert, "East European Cold War Culture(s)," 25.
10. Vowinckel et al., "European Cold War Culture(s)?" 5.
11. Vowinckel et al., "European Cold War Culture(s)?" 1–2; Siefert, "East European Cold War Culture(s)," 24.
12. Siefert, "East European Cold War Culture(s)," 24 and 31–35.
13. On communist educational programs in the Soviet Union and Italy, see Ettore Costa's chapter in this volume.
14. On the production and use of US science education films, see the chapter by Mario Schulze in this volume. Also see Bonah and Laukötter, "Visual Media and the Healthy Self," 19–23; Borge et al., "Introduction: The Science of Television."
15. Kreutz et al., *Von "AHA" bis "VISITE,"* 48–51.
16. Imre, *TV Socialism*, 8–9 and 19.
17. Gumbert, *Envisioning Socialism*, 14–29.
18. Vollberg, "'Because Every Recipient,'" 173–74 and 178–180; Kreutz et al., *Von "AHA" bis "VISITE,"* 48–51.
19. Schmidt-Lux, "Zwischen Partei und Publikum."
20. Kreutz et al., *Von "AHA" bis "VISITE,"* 51.
21. Beutelschmidt, *Ost—West—global*, 303–15.
22. Beutelschmidt, *Ost—West—global*, 336–48; Schwarz, "Vom Jahrmarktspektakel zum Aufklärungsinstrument," 31.
23. Lozanov, "Foundations of Suggestology," 84.
24. *Erleben Sie im Test: Wie beeinflußt uns Musik?* TC 10:22.
25. *Erleben Sie im Test: Wie beeinflußt uns Musik?* TC 10:24.
26. *Erleben Sie im Test: Wie beeinflußt uns Musik?* TC 10:22.
27. See also Mario Schulze's chapter on the role of audiovisual media in this scientific competition.
28. Lozanov, "Concluding Speech," 659.
29. See Ministry of Education and Research Institute of Suggestology, ed., *Problems of Suggestology*.
30. Early on, the theory and practice of Pavlov's reflexology had been represented on film by important Soviet filmmakers, such as Vsevolod Pudovkin (1893–1953). The imagery of his 1926 documentary *Mechanics of the Brain* became very influential in the following decades.
31. Busse, *Psychology in the GDR*, 59–66; Lubich, *Das Kreativsubjekt in der DDR*, 224.

32. Kennaway, "Musical Hypnosis," 272 and 280.
33. Pavlov, *Lectures on Conditioned Reflexes*, 170–81.
34. Lozanov, "Foundations of Suggestology," 71–72.
35. Geisthövel and Hitzer, "Leibseelische Komplexe," 35; Geisthövel, "Neurose," 52; Jensen, "Die Psychoanalyse," 63 and 67.
36. Jensen, "Die Psychoanalyse," 60 and 67.
37. Neuner, "Die Diktatur des Großhirns," 73; Balz, "Dietfried Müller-Hegemann," 217–18.
38. Lozanov, "Foundations of Suggestology," 72–75.
39. Tanner, "Die kontroverse Karriere der Kybernetik," 274.
40. Balevski, "Telemetric Registration," 213.
41. Schnädelbach, "Bad Vibes."
42. Poiger, *Jazz, Rock, and Rebels*, 31–70.
43. Jürgens, *Geschichte der ostdeutschen Musiktherapie*, 118–20; Lubich, *Das Kreativsubjekt in der DDR*, 227–30.
44. Geisthövel and Hitzer, "Die Grenzen des Erfolgs," 325–48.
45. Tändler, "Ratgeber oder Anleitungen," 371–73.
46. Tändler, "Ratgeber oder Anleitungen."
47. Maaz, "Zur Geschichte der Psychotherapie," 216.
48. Maaz, "Zur Geschichte der Psychotherapie."
49. Maaz, "Zur Geschichte der Psychotherapie," 227–30; Jürgens, *Geschichte der ostdeutschen Musiktherapie*, 118–20.
50. Maaz, "Zur Geschichte der Psychotherapie," 220.
51. *Erleben Sie im Test: Wie beeinflußt uns Musik?* TC 10:27.
52. See Tändler, "Ratgeber oder Anleitungen," 368–73.
53. Tändler, "Ratgeber oder Anleitungen," 367–75; Cocks, "Johannes Heinrich Schultz."
54. Tändler, "Ratgeber oder Anleitungen," 367–75.
55. On the education of concert audiences, see Weliver, *The Musical Crowd in English Fiction*, 5–7 and 56–58.
56. Gienow-Hecht, "Culture and the Cold War"; Siefert, "East European Cold War Culture(s)," 26–27.
57. Gienow-Hecht, "How Good Are We?" 276–79; Gienow-Hecht, "Culture and the Cold War in Europe," 402–9.
58. Kennaway, "Musical Hypnosis," 278–79.
59. Lozanov, "Foundations of Suggestology," 74–78.
60. *Erleben Sie im Test: Wie beeinflußt uns Musik?* TC 10:26.
61. *Erleben Sie im Test: Wie beeinflußt uns Musik?* TC 10:05 and 10:10.
62. For the rhetorical tools with which communists attributed political meaning to scientific information, see the chapter by Ettore Costa in this volume.
63. Lozanov, "Foundations of Suggestology," 74–78.
64. Defense Intelligence Agency, *Controlled Offensive Behavior*, 54.
65. Defense Intelligenge Agency, *Controlled Offensive Behavior*, 57.
66. Schwarzkopf, "Advertising, Emotions, and 'Hidden Persuaders,'" 175; Kennaway, "Musical Hypnosis," 279.
67. Schwarzkopf, "Advertising, Emotions, and 'Hidden Persuaders,'" 278–79.
68. Schwarzkopf, "Advertising, Emotions, and 'Hidden Persuaders,'" 279–81; Thiel and Walther, "'Pseudowissenschaft' im Kalten Krieg," 320–21.
69. Defense Intelligence Agency, *Controlled Offensive Behavior*, 57.

70. Krishnan, *Military Neuroscience*, 17–45.
71. UNESCO, *Final Report*.
72. Rupprecht S. Baur, "Lozanov"; Schiffler, *Suggestopädie und Superlearning*, 41.
73. Thiel and Walther, "'Pseudowissenschaft' im Kalten Krieg," 308–21.
74. Baur, "Lozanov," 43; Edelmann, *Suggestopädie/Superlearning*, 66–78; Schiffler, *Suggestopädie und Superlearning*, 13–21.
75. Miele, *Suggestopedia*, 87–96.
76. Ostrander and Schroeder, *Psychic Discoveries*, 268–81.
77. Anon., "Mit leichter Barockmusik."
78. On this topic, see Mario Schulze's chapter in this volume.

Bibliography

Television Programs

Erleben Sie im Test: Wie beeinflußt uns Musik? [Witness the test: How does music influence us?]. Dir. Ernst Hansch, DDR F1, 25 October 1978.

Umschau [Review] Dirs. Ulf Klaudius and Werner Schwarz, DDR F1, 23 August 1983.

Print Sources

Anon. "Mit leichter Barockmusik schnell ein Sprachgenie." *DIE ZEIT* (25) (1987).

Balevski, P. "Telemetric Registration of the Functional State of the Teacher-Student System in the Course of the Suggestopedic Process of Instruction." In *Problems of Suggestology: Proceedings of the I International Symposium on the Problems of Suggestology*, edited by the Ministry of Education and Research Institute of Suggestology, 211–18. Sofia: Nauka i Izkustvo, 1973.

Balz, Viola. "Dietfried Müller-Hegemann oder psychophysische Medizin à la Pawlow." In *Auf der Suche nach einer anderen Medizin: Psychosomatik im 20. Jahrhundert*, edited by Alexa Geisthövel and Bettina Hitzer, 211–21. Berlin: Suhrkamp, 2019.

Baur, Rupprecht S. "Lozanov—Wissenschaftler oder Idealist?" *Neues Lernen Journal* 1/4 (1986): 42–50.

Beutelschmidt, Thomas. *Ost—West—global: Das Sozialistische Fernsehen im Kalten Krieg*. Leipzig: Vistas, 2017.

Bonah, Christian, and Anja Laukötter. "Visual Media and the Healthy Self in the 20th Century: An Introduction." In *Body, Capital, and Screens: Visual Media and the Healthy Self in the 20th Century*, edited by Christian Bonah and Anja Laukötter, 13–40. Amsterdam: Amsterdam University Press, 2020.

Borge, Jessica, Tricia Close-Koenig, and Sandra Schnädelbach. "Introduction: The Science of Television. Television and It Importance for the History of Health and Medicine." *Gesnerus* 76 (2) (2019): 153–71.

Busse, Stefan. *Psychologie in der DDR: Die Verteidigung der Wissenschaft und die Formung der Subjekte*. Weinheim/Basel: Beltz, 2004.

Cocks, Geoffrey. "Johannes Heinrich Schultz oder vom steten Bemühen um entspannte Leistungsfähigkeit." In *Auf der Suche nach einer anderen Medizin: Psychosomatik im 20.*

Jahrhundert, edited by Alexa Geisthövel and Bettina Hitzer, 155–65. Berlin: Suhrkamp, 2019.

Defense Intelligence Agency. *Controlled Offensive Behavior—USSR (U)*. 1972. Retrieved 13 March 2020 from Central Intelligence Library, Electronic Reading Room: https://www.cia.gov/library/readingroom/document/cia-rdp96-00788r001300010001-7.

Edelmann, Walter. *Suggestopädie/Superlearning: Ganzheitliches Lernen—das Lernen der Zukunft?* Heidelberg: Asanger, 1988.

Geisthövel, Alexa. "Neurose oder eine vielgestaltige Diagnose zwischen Körper, Psyche und Gesellschaft." In *Auf der Suche nach einer anderen Medizin: Psychosomatik im 20. Jahrhundert*, edited by Alexa Geisthövel and Bettina Hitzer, 45–59. Berlin: Suhrkamp, 2019.

Geisthövel, Alexa, and Bettina Hitzer. "Die Grenzen des Erfolgs: Endgültige Etablierung und das Verschwinden einer großen Antwort (1970–2000)." In *Auf der Suche nach einer anderen Medizin: Psychosomatik im 20. Jahrhundert*, ed. Alexa Geisthövel and Bettina Hitzer, 325–48. Berlin: Suhrkamp, 2019.

———. "Leibseelische Komplexe: Zur Vor- und Frühgeschichte der psychosomatischen Medizin (1850/90–1945)." In *Auf der Suche nach einer anderen Medizin: Psychosomatik im 20. Jahrhundert*, edited by Alexa Geisthövel and Bettina Hitzer, 23–44. Berlin: Suhrkamp, 2019.

"Geschichte der Urania: Die Urania—das erste Science Center der Welt." 2018. Urania Berlin website. Retrieved 22 May 2020 from https://www.urania.de/die-urania/geschichte.

Gienow-Hecht, Jessica. "How Good Are We? Culture and the Cold War." In *The Cultural Cold War in Western Europe, 1945–1960*, edited by Giles Scott-Smith and Hans Krabbendam, 269–82. London: Routledge, 2003.

———. "Culture and the Cold War in Europe." In *The Cambridge History of the Cold War*, ed. Melvyn P. Leffler and Odd Arne Westad, 1:398–419. Cambridge: Cambridge University Press, 2010.

Gumbert, Heather L. *Envisioning Socialism: Television and the Cold War in the German Democratic Republic*. Ann Arbor: University of Michigan Press, 2014.

Harrington, Anne. *The Cure Within: A History of Mind-Body Medicine*. New York/London: W. W. Norton & Co., 2008.

Imre, Anikó. *TV Socialism*. Durham, NC/London: Duke University Press, 2016.

Jensen, Uffa. "Die Psychoanalyse oder Soma und Psyche auf der Couch." In *Auf der Suche nach einer anderen Medizin: Psychosomatik im 20. Jahrhundert*, edited by Alexa Geisthövel and Bettina Hitzer, 60–73. Berlin: Suhrkamp, 2019.

Jürgens, Petra. *Geschichte der ostdeutschen Musiktherapie: Entwicklung—Selbstverständnis—gesellschaftlicher und wissenschaftstheoretischer Kontext*. Frankfurt am Main: Peter Lang, 2007.

Kennaway, James. "Musical Hypnosis: Sound and Selfhood from Mesmerism to Brainwashing." *Social History of Medicine* 25(2) (2011): 271–89.

Kreutz, Anja, Uta Locher, and Doris Rosenstein, eds. *Von "AHA" bis "VISITE": Ein Lexikon der Magazinreihen im DDR-Fernsehen (1952–1990/91)*. Potsdam: Verlag für Berlin-Brandenburg, 1998.

Krishnan, Armin. *Military Neuroscience and the Coming Age of Neurowarfare*. London: Routledge, 2016.

Lozanov, Georgi. "Concluding Speech." In *Problems of Suggestology: Proceedings of the I International Symposium on the Problems of Suggestology*, edited by the Ministry of Education and Research Institute of Suggestology, 649–65. Sofia: Nauka i Izkustvo, 1973.

———. "Foundations of Suggestology." In *Problems of Suggestology: Proceedings of the I International Symposium on the Problems of Suggestology*, edited by the Ministry of Education and Research Institute of Suggestology, 71–87. Sofia: Nauka i Izkustvo, 1973.

Lubich, Barbara. *Das Kreativsubjekt in der DDR: Performative Art in Context*. Göttingen: V&R Unipress Verlag, 2014.

Maasen, Sabine, Jens Elberfeld, Pascal Eitler, and Maik Tändler, eds. *Das beratene Selbst: Zur Genealogie der Therapeutisierung in den 'langen' Siebzigern*. Bielefeld: Transcript, 2011.

Maaz, Hans-Joachim. "Zur Geschichte der Psychotherapie in der DDR." *European Journal of Mental Health* 6(2) (2011): 213–38.

Miele, Philip. H. *Suggestopädie: Der Weg zum leichten Lernen*. Cologne: Lozanov Learning Institute, 1983.

Ministry of Education and Research Institute of Suggestology, ed. *Problems of Suggestology: Proceedings of the I International Symposium on the Problems of Suggestology*. Sofia: Nauka i Izkustvo, 1973.

Needell, Allan A. *Science, Cold War and the American State: Lloyd V. Berkner and the Balance of Professional Ideals*. Amsterdam: Harwood Academic Publishers, 2000.

Neuner, Stephanie. "Die Diktatur des Großhirns: Die Lehren Pawlows im populärwissenschaftlichen Film der Stalin-Ära." In *Kamera! Licht! Aktion! Filme über Körper und Gesundheit 1915 bis 1990*, edited by Susanne Roeßiger, Uta Schwarz, and Stiftung Deutsches Hygiene-Museum, 71–77. Dresden: Sandstein, 2011.

Ostrander, Sheila, and Lynn Schroeder. *Psychic Discoveries behind the Iron Curtain*. New York: Bantam, 1971.

Pavlov, Ivan Petrovich. *Lectures on Conditioned Reflexes: Twenty-Five Years of Objective Study of the Higher Nervous Activity (Behaviour) of Animals*. Vol. 1. 3rd ed. New York: International Publishers, 1941.

Poiger, Uta G. *Jazz, Rock, and Rebels: Cold War Politics and American Culture in a Divided Germany*. Berkeley/Los Angeles: University of California Press, 2000.

Schiffler, Ludger. *Suggestopädie und Superlearning—empirisch geprüft*. Frankfurt am Main: Dieserweg, 1989.

Schmidt-Lux, Thomas. "Zwischen Partei und Publikum: Die Wissenschaftspopularisierung der Urania in der DDR." In *125 Jahre Urania Berlin*, edited by Ulrich Bleyer, Dieter B. Herrmann, and Otto Lührs, 198–213. Berlin: Westkreuz-Verlag, 2013.

Schnädelbach, Sandra. "Bad Vibes: Images of Communication, Emotional Balance and Health in East German Television, 1970s–1980s." *VIEW Journal of European Television History and Culture* 9(18) (2020): 97–111.

Schwarz, Uta. "Vom Jahrmarktspektakel zum Aufklärungsinstrument: Gesundheitsfilme in Deutschland und der historische Filmbestand des Deutschen Hygiene-Museums Dresden." In *Kamera! Licht! Aktion! Filme über Körper und Gesundheit 1915 bis 1990*, edited by Susanne Roeßiger, Uta Schwarz, and Stiftung Deutsches Hygiene-Museum, 23–39. Dresden: Sandstein, 2011.

Schwarzkopf, Stefan. "Advertising, Emotions, and 'Hidden Persuaders': The Making of Cold War Consumer Culture in Britain from the 1940s to the 1960s." In *Cold War Cultures: Perspectives on Eastern and Western European Societies*, edited by Annette Vowinckel, Marcus M. Payk, and Thomas Lindenberger, 172–90. New York/Oxford: Berghahn Books, 2012.

Siefert, Marsha. "East European Cold War Culture(s): Alterities, Commonalities, and Film Industries." In *Cold War Cultures: Perspectives on Eastern and Western European Soci-*

eties, edited by Annette Vowinckel, Marcus M. Payk, and Thomas Lindenberger, 23–54. New York/Oxford: Berghahn Books, 2012.

Tändler, Maik. *Das therapeutische Jahrzehnt: Der Psychoboom in den siebziger Jahren*. Göttingen: Wallstein, 2016.

———. "Ratgeber oder Anleitungen zur psychosomatischen Selbsthilfe." In *Auf der Suche nach einer anderen Medizin: Psychosomatik im 20. Jahrhundert*, edited by Alexa Geisthövel and Bettina Hitzer, 361–75. Berlin: Suhrkamp, 2019.

Tanner, Jakob. "Die kontroverse Karriere der Kybernetik." In *Auf der Suche nach einer anderen Medizin: Psychosomatik im 20. Jahrhundert*, edited by Alexa Geisthövel and Bettina Hitzer, 273–88. Berlin: Suhrkamp, 2019.

Thiel, Jens, and Peter Th. Walther. "'Pseudowissenschaft' im Kalten Krieg: Diskreditierungsstrategien in Ost und West." In *Pseudowissenschaft: Konzeptionen von Nichtwissenschaftlichkeit in der Wissenschaftsgeschichte*, edited by Dirk Rupnow, Veronika Lipphardt, Jens Thiel, and Christina Wessely, 308–42. Frankfurt am Main: Suhrkamp, 2008.

UNESCO. *Final Report: Expert Working Group on Suggestology and Suggestopedia*, ED/SCM/MMT, ED.80/WS/160, Sofia, 1980. Retrieved 17 March 2020 from UNESDOC Digital Library: https://unesdoc.unesco.org/ark:/48223/pf0000043398.

Vollberg, Susanne. "'Because Every Recipient Is Also a Potential Patient'—TV Health Programmes in the FRG and the GDR, from the 1960s to the 1980s." *Gesnerus* 76(2) (2019): 172–91.

Vowinckel, Annette, Marcus M. Payk, and Thomas Lindenberger. "European Cold War Culture(s)? An Introduction." In *Cold War Cultures: Perspectives on Eastern and Western European Societies*, edited by Annette Vowinckel, Marcus M. Payk, and Thomas Lindenberger, 1–20. New York/Oxford: Berghahn Books, 2012.

Weliver, Phyllis. *The Musical Crowd in English Fiction, 1840–1910: Class, Culture and Nation*. Basingstoke: Palgrave Macmillan, 2006.

PART III

Science in Public and Educational Media

CHAPTER 8

Science for Children and Adults

The Space Race in Italian Communist Party Publications, 1957–64

Ettore Costa

Introduction: The Land of Science and Socialism

The so-called space race—the competition between the Soviet Union and the United States to reach first the milestones of space exploration from 1957 to 1969—has left indelible traces not only in scientific labs, satellite networks, and museums but especially in the childhood memories of an entire generation of young boys and girls. Many children approached science for the first time through news of space accomplishments, pictures of satellites and astronauts, and the first rudimentary explanations of physics and cosmology. This was the case for US children, Soviet and Eastern European children, and even those caught in between: children of communist families living in the Western world. The Italian children's magazine *Pioniere* (Pioneer) reported on how Italian children experienced the space race and incorporated it into their daily lives. Italian children usually dress up for Carnevale, for example; children from communist families dressed as a Soviet rocket or as *Sputnik*, with headgear shaped like a satellite and a bodysuit painted with stars.[1] Across Italy, children formed youth groups to gather scientific information and engage in space-themed activities.[2] With the space race stimulating their imaginations, many children wrote short stories about interplanetary travel.[3] They also entered competitions to win prizes, such as a metallic model of a *Sputnik* satellite.[4]

Pioniere was not just any children's magazine, however, as it was published by the Italian Communist Party (Partito Comunista Italiano, PCI), staffed by party members, and distributed through party channels to the families of activists. *Pioniere* not only published propaganda; it also encouraged children to study and stay informed about scientific developments.[5] And yet, when children dressed up as *Sputnik*, they walked around town wearing the red star that symbolized the Soviet Union. The science-fiction comics they read featured space adventure narratives that doubled as morality plays about communist values. In focusing primarily on the scientific achievements of the Soviet Union, *Pioniere* also celebrated the socialist motherland. While *Pioniere* played a major role in distributing scientific information to children and encouraging their scientific education, it did not merely pass along scientific data; it also involved explicit and implicit biases. Understanding that science was not neutral, even when it was presented as such, is thus essential.

In this chapter, I will demonstrate how news of the Soviet Union's scientific achievements in the space race served to communicate the superiority of the socialist system to children and instill in their minds the conviction that socialism actually existed behind the Iron Curtain.[6] Reproducing the official propaganda of the Soviet government, *Pioniere* represented the Soviet Union as the land of science and socialism in an abstract and glossy way. This myth had huge emotional resonance, even for children, because it contrasted so vividly with their concrete daily existence. Their letters to *Pioniere* describe the poverty, social injustice, and lack of access to education still present in Italy, despite the incredibly high rates of growth associated with the Italian "economic miracle." Political mobilization is successful when political followers adopt the language of political innovators as their code for interpreting their grievances and aspirations in the real world.

Communist Science Discourse

This chapter analyzes the PCI's discourse on science from 1957 to 1964. During this period, Italian communists celebrated the Soviet Union as the vanguard of scientific progress, arguing that this demonstrated the superiority of the socialist system. I examine the kind of scientific information Italian communists communicated in their periodicals, their framing of scientific issues, and the role science played in their worldview. This topic has only recently received some attention.[7] Along with *Pioniere*, I will examine *Rinascita* (Rebirth), another PCI periodical, aimed at adults. In addition to the scientific content passed on by the communist publications, I analyze the rhetorical tools used to imbue this scientific information with political significance, focusing on the illocutionary and perlocutionary roles of these scientific articles—that is, both the writers' objectives and the effect of the texts on their audience.[8] Finally, I examine whether the texts produced by Italian communists confirmed or subverted the conventions of Italian culture and the discourse of the international communist movement.

Analyzing the PCI's media output on science is highly relevant to understanding scientific literacy and popularization in Italy during the Cold War. After World War II, the PCI played a major role in fostering the integration of subaltern industrial and rural workers into Italian democracy and society, while also being the dominant cultural influence among intellectuals and artists.[9] Examining the PCI's framing of science and scientific issues is thus essential to appreciating how science was understood in both high and popular culture in Italy. Focusing on the publications that Italian communists produced specifically for children further reveals the deep penetration of communist ideas of science into Italian society. Communist leaders did not engage with scientific topics solely for propaganda. Since its inception, the worldwide labor movement—like other movements of the oppressed—had stressed the importance of self-improvement and education because the world of tomorrow was to be a world of New Men, guided by a more humane morality.[10] The materials that communists produced for children were thus doubly pedagogical in character, serving to instill the right values in a new generation of party activists and citizens that would fight for, and maybe even build, socialism.[11]

Another interesting dimension to the scientific information shared by media like *Pioniere* is the part played by the PCI in the international communist movement. A transnational examination shows that much of its culture and rhetoric was dependent on sources and ideas from the Soviet Union. Most Western European countries depended on US and, to a lesser degree, British sources for scientific information and ideas about science. In Western countries with large communist movements, such as Italy and France, we can see a penetration of Soviet scientific ideas, information, and topics, however. We must therefore first examine the role science played in Soviet communist ideology from 1957 to 1964.

After the death of Joseph Stalin in 1953, the Soviet Union desperately needed to distance itself from the system of endemic terror and intense exploitation of industrial and agricultural workers he had engendered. The process of distancing is known as de-Stalinization. At the 20th Congress of the Communist Party of the Soviet Union (CPSU) in 1956, Nikita Khrushchev, the general secretary of the CPSU and the most prominent figure among Stalin's successors, denounced many Stalinist crimes against the Soviet population. That same year, Soviet tanks quashed a popular revolution against communism in Hungary. Without democratic consent, the new Soviet leadership enacted internal reforms to enshrine an "implicit social contract" with the population, which would involve redistribution of the economic surplus generated by technological progress, political quietism, and low intensity in labor.[12]

Khrushchev made scientific progress and technological innovation the center of his strategic vision as well as the defining feature of the new ideological turn of Soviet communism.[13] While the celebration of science had been foundational to socialism since the nineteenth century, as highlighted by the leader of the October Revolution Vladimir Lenin, it was now emphasized more urgently in an effort to distance Soviet communism from Stalinism and stress the promise of future prosperity.[14] The Soviet system was believed to fit the new model of "Scientific-Technological Revolution," in which the power of science would be used to drastically increase production. As historian Susan Reid explains, "The promised abundance that would win the Cold War for socialism was to be attained by harnessing the achievements of modern science and technology."[15] Communists argued that the October Revolution of 1917 had turned the backward Russian Empire into a leading industrial power. The socialist world could counterbalance imperialist capitalism and would soon outstrip its production. The Soviet Union presented a model of modernization, scientific progress, and peace to Western workers and decolonized nations. As historian Federico Romero writes, "The competition was thus focused on the terrain of economic and scientific progress, assimilating communism to the ability to produce abundance and solve the material problems of the population."[16]

Scientific progress and its radical social consequences were thus central to the Cold War, especially in the late 1950s and 1960s. According to leading Cold War historian Odd Arne Westad, the essence of the Cold War was a "competition . . . for the society of the future, and there were only two fully modern versions of it": the Soviet Union and the United States.[17] The space race was the most visible signifier of the scientific competition between the two blocs. The USSR celebrated the first victories. Its first artificial satellite, *Sputnik I*, shocked Western societies in October 1957. The launch of the first living being, the dog Laika, into space in November 1957 coincided with the fortieth anniversary of the Soviet revolution. The flight of the first man in space, Yuri Gagarin, in April 1961 coincided with the CPSU Congress that promised the transition to communism in twenty years. And the flight of Valentina Tereshkova, the first woman in space, in June 1963 gave legitimacy to the claim that the socialist world was doing more for gender equality than the capitalist West.

The period of the space race marks what historian Santi Fedele calls the "autumn" of the Soviet myth, as the attraction of Soviet communism around the world drew strength less from its "hard" elements—revolutionary violence and radical social transformation—and more from an emphasis on "soft" elements, such as the success of the Soviet Union in producing scientific progress and prosperity.[18]

The ideological turn undertaken by Khrushchev greatly helped communist parties outside the Soviet Union. The connection to the USSR was a source of both identity and prestige for the worldwide communist movement, but the revelation of the crimes of Stalinism had been damaging. The new focus on science therefore provided an opportunity for renewal. Since the days of the antifascist struggle and postwar democratic reconstruction in Italy, Palmiro Togliatti, leader of the PCI, had worked to make Italian communists an integral component of Italian democracy; he would continue this work until his death in 1964, a few months before Khrushchev's fall. In 1956, Togliatti defined a new strategy for the PCI, placing a stronger emphasis on the "Italian road to socialism." To what extent Italian communists were independent from or dependent on the Soviet Union and whether the break with the Stalinist recent past was stronger than continuity are highly contentious historical topics.[19] But, ultimately, denouncing Stalinism required recasting, not recanting. Togliatti shifted the terms of Soviet and communist legitimacy as soon as he returned from the 20th Congress of the PCSU in 1956.[20] He praised the "new civilization that experiences continuous progress regarding labor productivity, technology, science, culture."[21] He predicted that the socialist system would outperform capitalism in technological innovation and solutions to social problems, and also produce New Men through education and leisure time. These themes were reiterated for the fortieth anniversary of the Soviet revolution, which, according to Togliatti, had made a backward nation into the vanguard of progress.[22] After the flights of the first Soviet cosmonauts, an editorial in *Rinascita*—expressing directly or indirectly the opinion of Togliatti—commented that the "scientific leadership of the world" had passed from the West to the socialist system.[23] In the ensuing years, the party press would continue to follow the lead of the highest echelons of international and Italian communism.

Reaching Children and Adults

As a mass party with deep social penetration, the PCI published different publications for different audiences. In this chapter, I focus on two periodicals. *Rinascita* was a monthly—as of 1962, weekly—magazine that dealt with the most important political, economic, and cultural topics of the day. Since PCI leader Togliatti was also the editor, the periodical gave the ideological lead to adult activists and voters. I give more attention to a second periodical, *Pioniere*, published from 1950 to 1962 as a weekly magazine for children and from 1963 to 1966 as a weekly supplement to the PCI newspaper, *L'Unità*. Children's education was a key concern for communists, and large and well-funded parties like the Italian and French Communist Parties could devote entire organizations and periodicals to children. *Pioniere* started as the publication of the Associazione Pionieri Italiani (Association of Italian Pioneers), an organization tied to the PCI that provided scouting experiences and youth services for the children of communist activists. Similar organizations existed across the socialist world.[24]

The existence of separate pedagogical, recreational, and cultural institutions for the children of communists was not atypical in postwar Europe. European societies were usually divided into separate subsections, defined by strong cultural, social, and religious identities—a phenomenon known as "pillarization," from the Dutch case. For example, there could be a Catholic pillar, a Protestant pillar, a bourgeois pillar, or a working-class pillar, depending on the country. The term "pillar" refers to vertical integration, since each subsection of society developed a whole array of institutions that regulated most aspects of daily life. Members of different pillars prayed at different churches (or no church), bought their food from different shops, read different newspapers and books, spent their free time at different clubs, and voted for different parties. In the Italian case, the bourgeois and Catholic pillars had children's periodicals of their own. This pillarization would start to dissolve in the 1960s with the spread of mass culture, consumerism, and Americanization.[25] A common trend of Cold War propaganda and often the biggest challenge to specific ideological and cultural discourses—whether it was among Italian communists, or French Gaullists, or German Christian Democrats, as the chapters by Stein and Uelzmann show—was the hegemonic assault of US culture. Indeed, the communists needed to form their youth services and publications in order to protect their separate identity, since parish priests often influenced children's education and rites of passage, even in communist families, and Christian Democrats controlled public education. In turn, conservative authorities sabotaged communist attempts to run children's activities; reading *Pioniere* could thus mean excommunication, and there was harassment and occasional legal persecution.

This is also why, unlike other chapters in this book, cinema is not the focus here. The PCI had to rely on more traditional means, including its press and in-person meetings, as producing and distributing movies involved much higher costs and obstacles. More importantly, Christian Democrats and the Church had a privileged relationship with the movie industry, with a grasp that was tighter than what Uelzmann shows for Christian Democrats in West Germany. As Ermanno Taviani explains, the PCI did not aim to build a counterpart to the movie industry—a countercinema, which it would have had to fund—but rather to support freedom of speech and quality movies against more vulgar productions and Catholic censorship. This was not least because communist leaders, particularly Togliatti, had a very traditional conception of culture that did not include cinema.[26] Making independent movies would have been impossible anyway, and early attempts were stamped out.

In the early 1950s, the PCI had tried to make its own propaganda movies and distribute movies from communist countries but found it impossible because of a continuous struggle with the censorship board. It was so bad that the party had to shut down its production and distribution company because of the impossibility of projecting films, documentaries, or newsreels. Even the projection of still images, technically legal, was met with regular police harassment. Only starting in the early 1960s did the PCI slowly return to audiovisual propaganda.

Gianni Rodari, an activist and journalist for the PCI, developed the idea for *Pioniere* and served as its first editor. This experience would allow him to become the most important Italian children's author of the twentieth century, and many of his books were translated across the socialist world. The main driving force of the magazine would be Dina Rinaldi, its editor from 1953 to 1962, however.[27] Although party leaders were still committed to serious art and traditional literature and were suspicious of new forms of mass communication, especially comics, Rinaldi wanted to avoid a top-down model of education and propaganda. Having started her political activism during the war, she worked to make communist publications more approachable and interesting for women and children. As one comic writer put it, *Pioniere* needed to avoid a "mechanical transposition of ideological values" and political propaganda.[28] In her weekly column, Rinaldi presented the goal of the magazine as assisting in the intellectual, moral, and sentimental education of children to make them good citizens and good communists. The magazine rejected the idea of childhood as a time of innocent disengagement from reality; instead, children were supposed to be part of the grown-up world, just in a way accessible to them.[29] The coverage of the space race demonstrates this commitment in action.

"The First Country to Storm the Heavens"

I will now turn my analysis to the two-part rhetorical strategy employed by *Pioniere* and *Rinascita* in transmitting scientific information. The first rhetorical step was to identify scientific success with socialist success. The successes of the *Sputnik* satellites, Laika the dog, the *Lunik* launches, and cosmonauts Gagarin and Tereshkova were used as means to argue the superiority of the Soviet system. The second step involved connecting the superiority of the socialist system with the potential to improve the living conditions of Italian workers. This step was more difficult to achieve than the first, because it involved connecting abstract, remote events with the ordinary lives of Italian adults and children.

The main rhetorical move by Soviet and Italian communists was to create a metonymic association of *Sputnik* with the Soviet system. As a PCI journalist wrote, "Sputniks are the products of the most developed social system and thus propagandize its principles."[30] *Pioniere* denied that space travel was about national prestige while simultaneously comparing *Sputnik* with Soviet gold medals at the Olympics.[31] The strength of the metonym was enhanced by the fact that the first *Sputniks* were launched around the celebration of the fortieth anniversary of the Soviet revolution, which legitimized the use of the topos of temporality, or the procession from past to future. Socialism had turned archaic Russia into the most prominent scientific state in the world, and the future thus belonged to socialism. Alongside regular allusions to Christopher Columbus—also useful to legitimize the national credentials of the PCI—*Pioniere* told its readers in enthusiastic terms that a new era of humanity had begun.[32] As a contested concept, the future was a focal point of political passions and ideas during the Cold War, with different ideologies proposing different visions for the future and associating these with people's hopes and fears.[33] A common expression was "the future has already begun," echoing the German title of a popular book by Robert Jungk. The implication that socialism and the Soviet Union were the "future" (or "tomorrow") was made explicit by a headline celebrating the anniversary of the October revolution: "The Future Has Already Begun with October."[34]

Italian communist leaders employed the trope that the *Sputniks* and later Soviet space achievements proved the superiority of the socialist system in combatting reformists and other supporters of a mixed economy, who believed capitalism was still vital and could be steered to achieve social goals.[35] An article by the deputy leader of the PCI, Luigi Longo, demonstrates the use of space achievements as a rhetorical tool. In order to break the association of socialism with the violence of Stalinism and the repression of the Hungarian Revolution of 1956, Longo used a series of metaphors to emphasize the fragility of capitalism and the solidity of socialism. The *Sputniks* had swept away the arguments of social democrats and critics of the Hungarian invasion "like fog in the sun," he wrote, and their defenses were nothing more than "a castle of words, illusions, and deceits."[36] The success of *Sputnik* also enabled a reversal of the traditional hierarchy of civilization, by associating Eastern Europe with progress and the Western world with backwardness:

> Where are now the prophets of doom, after just one year during which the first country of socialism, the USSR, opened the roads of heaven to humanity, and the countries of the capitalist world, which believed themselves to be the most advanced, the most educated, the most civilized, were left to watch, astonished, surprised, envious?[37]

Finally, the failed predictions of social democrats and capitalist apologists served to reinforce the claim that Marxism was the only science of society that provided objective knowledge. Longo presented the science of Marxism and the science of *Sputnik* as an indissoluble dyad, with each reinforcing the other.

Italian communists often used the dialectic between master and slave as a metaphor for the liberation struggle as well as for the relationship between human beings and nature.[38] Both Togliatti and Rinaldi castigated conservative thinkers and writers who celebrated the contemplation of nature or expressed pessimism about the future. They argued that the appropriate approach put humankind in the position of master able to subjugate nature.[39] Italian communists argued that conservative ideology depicted nature as static, by projecting the idea of a static social order over nature, with the status quo depicted by conservative ideologies as perfect. In contrast, dialectical materialism depicted nature as a dynamic product of past factors that is always changing and malleable. *Pioniere* hailed *Sputnik I* as a demonstration of the power of humankind: "A socialist state, the Soviet people, were the first to achieve this conquest that demonstrates how the things that surround us, like the entire universe, are not immutable, they are open for man to know and discover."[40]

The Italian communists celebrated scientific progress with rhetoric arguing that the increase in human power was good in itself, and that Soviet achievements in space were a victory of rationality over backwardness and superstition.[41] The notion that scientific progress would automatically yield social progress was rejected, however. A journalist in *Rinascita* warned that, while humankind's victory over nature was good, the system under which this victory took place would determine its consequences.[42] Good scientific tools could still be misused by capitalists and warmongers, as in the case of nuclear weapons. This was a key point for the communists, since they could not concede that technological progress would be enough to solve the problems of workers and bring about universal prosperity without dismantling capitalism. Drawing a distinction between scientific power and its use allowed the communists to condemn capitalism. The use of science was thus ultimately a political responsibility, and while capitalist nations could make important discoveries and deliver some benefits, only under socialism could science serve all humankind. Rinaldi, writing in *Pioniere*, and another journalist for *Rinascita* both alternated between the epic register and irony to contrast the immense potential of science for all humankind and the fact that science under capitalism was associated with fear of destruction and privilege for the few.[43] The relationship between human power and scientific and social progress was also central to one of Togliatti's most important speeches, on the "destiny of man," in which he asserted that man had to be not just the master of nature but also the master of social developments.[44]

In 1959, the launches of the Soviet Luna program—nicknamed *Luniks*—were the first to reach the moon, generating particular enthusiasm and supplying another opportunity to celebrate the Soviet Union. Italian communists regularly used the trope of the contradiction between Earth and space, but

Rinaldi tied the success of the Soviet Union in space to its achievements on Earth, since only under a fair socialist system were the great achievements of space exploration possible. She wrote:

> The Soviet Union is the first country to storm the heavens, but it is also the first country where slavery, ignorance, and unemployment were fought and overcome. Forty years ago, it became the first country to give all its children schools and books to study, which armed them with knowledge, which helped them become what Soviet men and women are today.
>
> And isn't this an equally wonderful fact? Heaven was stormed, but men were also given everything they needed to live and progress.[45]

Rinaldi repeated a typical communist trope that the failures of the US space program were due to the nation's regressive social structure. Togliatti concurred that capitalism hindered scientific progress.[46] Alessandro Natta—the PCI's general secretary in the 1980s—criticized the Italian ruling class for not understanding how *Lunik II* revealed the failure of Italian education.[47] Rinaldi in particular, however, appealed to emotion as well as reason. That Soviet children received free books and schooling, enabling them to do great things in life, was particularly resonant for the many Italian children who had to leave school early—like a child named Otello, who had his letter published in the same issue as Rinaldi's text above.[48] This emotion was translated directly into a call to action: young people should not wish to live in the Soviet Union, but were instead called to organize and fight to improve things in Italy.[49] Communist propaganda thus used a simple salvation narrative, contrasting current and past miserable conditions with the paradisiacal situation to be brought by technical innovation and political action. The fact that the same narrative was available to both communists and neo-colonialists, as Stein's chapter shows, demonstrates its flexibility but also the strength of the modernization paradigm across the political spectrum.

The contrast between celestial and terrestrial matters could also be used to change perspective and reframe issues. *Pioniere* argued, for example, that space exploration highlighted the contradictions and injustices on Earth.[50] Speaking to the adult readers of *Rinascita*, mathematician Lucio Lombardo Radice, a key communist expert on science and education, made the same argument: in demonstrating the enormous potential and power available to man, the *Luniks* made the persistence of poverty, ignorance, inequality, and war on Earth absurd and anachronistic.[51] Speaking about Tereshkova's flight, another PCI politician commented that it produced a humiliating contrast to the patriarchal laws still in force in Italy.[52] Rodari employed a trope that would become widespread in the political discussion about the space race: interest in space exploration should not mean ignoring problems on Earth. Finally, he used another trope, which dealt with identification and representation, writing that he was happy that the first man-made object to touch the Moon had been Soviet, because every one of its components had been made in a factory owned by the workers.[53] In contrast, US rockets were made by for-profit firms without benefiting the scientists and workers who actually built them. The success of the Soviet Union, Rodari argued, was therefore more resonant for ordinary workers. It is thus clear that Italian communists employed every opportunity to show not just that the Soviet Union was superior but that socialism was particularly well-suited for science.

New Men (and Women)

As a revolutionary ideology, Communism proclaimed the need to forge not only a new society, but also a New Man, who was able to think and feel more clearly and live in harmony with others. Antonio Gramsci, the most important intellectual of the PCI, had already invoked the need for moral reform in Italy in his prison notebooks.[54] This aspiration tied into the trope that, as science opened up a new era, human nature needed to open up and adapt to the space age. Scientific understanding was considered essential for developing the morality and sensitivity of the New Man and moving him away from superstition and fear. Connecting opportunity and risk through the trope of urgency, Lombardo Radice stressed that the space age offered the opportunity to forge a New Man, but also argued that the New Man was necessary for the survival of humankind in the new era. In *Rinascita*, he explained to adult readers that the development of science over the centuries had raised philosophical and cultural questions, and that the man that created satellites and synthetic materials could not think like the man who believed in the Ptolemaic system.[55] The education of the New Man of tomorrow was also central for *Pioniere*, which presented Soviet achievements in space by highlighting their moral value as well as their scientific significance. In the new era, men would become freer, gaining mastery of the world and themselves. Only by educating the children of today with the right values would it be possible to prepare the men of tomorrow.[56]

The trope of the New Man influenced the celebration of Yuri Gagarin, whose space flight marked the apex of enthusiasm for science among Italian communists. In the USSR, Gagarin was the subject of hagiographical descriptions, displacing Aleksey Stakhanov as the epitome of the communist New Man.[57] *Pioniere* emphasized the celebration not of the individual but of the society that had supported and educated him, because Gagarin would have remained a mere farmer without the Soviet revolution. Only socialism had allowed him to cultivate his capacity and personality to the fullest: "His journey into the cosmos has been therefore a great victory for that socialist society that has broken the path of life and the future for all its Gagarins."[58] Using Gagarin as a plural noun to refer to all people coming from humble beginnings allowed Italian communist writers to create a point of identification for Italian children. In this case, the rhetoric was completely successful, as evidenced by letters from readers of *Pioniere*. A girl from Genoa wrote that she felt as if she had been on Gagarin's spaceship, while a boy from Rome wrote that he envied Gagarin not only because he was the first man in space, but because he had had the possibility to continue his studies.[59] Rinaldi emphasized this point, using the trope that socialism and science were two sides of the same rationality.[60] Only by eliminating slavery and superstition had the Soviet Union been able to develop the technology to go into space.

The PCI's rhetoric associated the emancipation of women with the Soviet Union as well, since the nation had actually instituted gender equality instead of only paying lip service to it, as in the capitalist world. Togliatti and other PCI leaders used the role of women in Soviet science, especially space exploration, as evidence of socialist gender equality.[61] Maria Antonietta Macciocchi, a communist journalist and activist, wrote in *Rinascita* that the Italian establishment failed to value female scientists.[62] Rinaldi was particularly passionate about the issue of women in science as well, and she always encouraged girls to become involved in scientific and technological activities and games.[63] For Rinaldi, acquiring a scientific mentality was a necessary step for the full emancipation of women, especially from the backward customs and ideologies that relegated them to a subordinate position. When young girls wrote to her about how their grandparents imposed their superstitious beliefs and traditional gender roles on them, Rinaldi responded that science could teach women to believe in their abilities and reject fatalism.[64] The contrast between the fatalistic and superstitious past and a modern and enlightened age was expressed through the analogy of the old and young generations. Many young girls responded well to the appeal of science. A girl named Anna wrote a short essay about how Gagarin's flight stimulated her interest in science. She wanted to be the first woman astronaut, both out of vanity and "to make it clear (as many do not believe) that nowadays women have the same abilities as men and could have arrived in space first."[65]

By the time Valentina Tereshkova went into space, Dina Rinaldi was no longer editor of *Pioniere*. However, the new editor still encouraged girls to pay attention to science. One child wrote that she could not go to space because she was a girl; but the editor wrote that Valentina Tereshkova had proved the opposite.[66] *Rinascita* published a cartoon in which Tereshkova flew over a crowd of admiring women from every profession and race, while some men and older women glared with disapproval from the sidelines.[67] The generation gap was once again used as a metaphor for the conflict between tradition and progress. *Rinascita* published the comments of two women about the flight of Tereshkova, one of whom wrote that she wanted women to excel in the same fields as males but admired that Tereshkova had not sacrificed her femininity, because the equality of the sexes did not erase their identity.[68] It is also possible to discern some resistance to the messaging around Tereshkova in the other published comment, which was from Anna Banti, a key author of twentieth-century Italian novels. Banti argued that true gender equality should minimize the distinction between male and female cosmonauts; as a result, the idea that Tereshkova's flight was a triumph for all women was "childish jubilation."[69] Banti was suspicious of space exploration, arguing that the intelligence and money spent on the space race should have been used to solve social problems on Earth. Not everyone embraced the pro-science approach put forward in *Pioniere*. Two young girls, for example, complained that *Pioniere* was a boys' magazine because it had too much science fiction. The editor answered that girls were not limited to traditionally feminine subjects; Tereshkova had demonstrated that women could be equal to men in science, and many girls appreciated the articles of *Pioniere*.[70]

Peace was a central concern in Soviet propaganda, especially when used to argue that the US and capitalist worlds were ruled by warmongers. Italian communists stressed that the world of tomorrow would be a world of peaceful men, spearheaded by the peaceful science of the Soviet Union. Conversely, science in the hands of capitalists could be used for war. *Pioniere* oscillated between celebrating even US space experiments and mocking the failure of US rockets or the smaller size of their satellites.[71] One comic went even further, with a storyline centered around the US space program led by the Nazi "Von Bran [*sic*]."[72] The artists were not subtle in exploiting text and iconography to create connections between Wernher von Braun, a former Nazi aerospace engineer and the leading scientist of the US space program, and Nazism. His comic counterpart shouted "sig-ail [*sic*]" and spoke of "inferior beings." The comic also made reference to the Italian Resistance; in one strip, Von Bran used torture to make an alien talk, but the alien was smuggled outside the base using the same clandestine techniques used by Italian women working for the Resistance during World War II.[73] As the chapter by Uelzmann shows, West Germany had enormous trouble in dispelling the Nazi image, in part because the communists brought it up again and again.

The point of the comics was to demonstrate that, although the US space program employed the rhetoric of impartial science, it was still imbued with militarism and imperialism, as the comics showed its plans to conquer other planets with H-bombs and its disregard for the lives of its astronauts. The topic of peace was received well by *Pioniere*'s readers. A young reader wrote a science-fiction story in which a federation of planets boycotted Earth because of its warmongers, but the aliens approved the Soviet space program as opening a new era of peace.[74] Another reader wrote that space conquest required that humankind be united and nuclear war avoided.[75] Of course, even children understood that *Sputnik I* could mean peace because the Soviet Union was now strong enough to force others to accept peace.[76] Most readers, adults and children alike, thus seemed to accept communist discourse on the promise of new society, new men, and new global peace.

Limitations and Side Effects of the Communist Discourse on Science

Despite the strong celebration of Soviet science, the communist discourse on science had its weaknesses. One flaw of the Soviet and communist conception of science was that its celebration of human mastery over nature blurred the line between science and magic. One Bruno, a reader of *Pioniere*, wrote a short text describing a future with no hard work or inconveniences, just like a fairy tale, with fairies using magic wands to make anything appear or disappear.[77] These puerile dreams might be forgiven in a child, but not in the general secretary of the CPSU: Khrushchev's fascination with science had less to do with rigorous study and more with Promethean dreams of reshaping nature completely, as demonstrated by his fascination with Trofim Lysenko's pseudoscientific promises to multiply agricultural production beyond the limits of scientific genetics.[78]

Dino Platone, *Pioniere*'s scientific expert, reported on the scientific achievements of the Soviet Union, including space travel, cultivation of newly reclaimed land, modification of rivers, and automatic factories.[79] Platone also enthusiastically described a series of scientific projects that Soviet scientists proposed to undertake once all nations agreed on world peace. These included digging a hole to the center of the earth, transferring icebergs to dry land in North America, and creating a new sea in the Sahara. Other plans included creating a giant mirror in space to reflect sunlight and abolish nighttime, making deserts cultivable, and melting polar ice.[80] Rinaldi praised the proposal for a dam on the Bering Strait, which would modify oceanic streams and make Northern Siberia cultivable, writing: "Once again it is shown that man's work can transform nature and, by transforming it, improve himself and his own existence."[81] *Pioniere* also celebrated Soviet plans to use H-bombs to create artificial lakes, irrigate deserts, and produce electricity.[82] The United States had similar plans, for example the infamous Project Plowshare, but they had nothing to envy from the Soviet projects.[83]

The biggest flaw of miraculous Soviet science was that its miracles failed to materialize. The continuous promises and exaltation of science eventually generated cynicism among the Soviet population.[84] "And while the citizens might have rejoiced about Sputnik and nuclear reactors," said one official's wife, "who cares about Sputniks if you cannot produce shoes, underwear, or washing machines of high quality and the right mix?"[85] At some point, there were surely more genuine believers in Soviet science in Italy than in the Soviet Union.

Another problem for Italian communists was the actual prosperity generated by the postwar Italian economic miracle. Parallel to what Uelzmann's chapter demonstrates for West Germany, Christian Democratic politicians in Italy stressed the miraculous economic growth that the country was experiencing after fascism, war, and destruction. Although it heightened social contradictions due to the lack of a general plan and social provisions, the rapid pace of growth ended misery for millions. Thus, while the PCI continued to encourage social integration and the nationalization of the masses through political participation, the "acquisitive model" of Italian economic growth promoted consumerism and individualism, contradicting socialist values and fracturing the working class.[86] The PCI was torn between fighting for higher salaries and mass consumerism, on one hand, and promoting austerity in customs and moralism against material consumption, on the other. As capitalism started to deliver tangible results in material abundance, communists grew ambivalent toward the miraculous power of science, since it seemed to deliver better results under capitalism than communism. In 1957, Togliatti mocked those who believed that technology alone could create a "universal Cockaigne."[87] While he was referring to capitalists, it could also apply to Khrushchev's excessive promises. Thus, the identification of science with socialism was important, but not without resistance and inconveniences.

Conclusion

This chapter has analyzed the discourse on science produced by the Italian Communist Party in the two magazines it published for children and for adult activists. While the registers and tropes employed were tailored to the respective audiences, the discourse used in each was inspired by the same ideological conception of science and served the same strategic goals. In weaving this discourse, Italian communists employed a series of rhetorical tools to produce an identification between scientific success and Soviet socialism and encourage readers to accept Soviet science as a future-oriented solution for their daily problems. Letters that children wrote to *Pioniere* indicate their warm reception of the communist science discourse. This conclusion seems to be confirmed by later developments, as many of the children reading *Pioniere* in the late 1950s would have become cadres and activists by the 1980s, when—despite the developments of Eurocommunism—the majority of PCI members still considered the Soviet Union the model of an ideal society.[88]

This case is particularly interesting because it offers a clear-cut example of the transnational circulation of ideas connecting and adapting to local conditions. Many of the tropes and arguments used by Italian communists were of Soviet origin, in contrast to those used by the rest of the Western world, which depended on US ideas and sources. The impetus to recenter communist ideology around scientific progress came from the need of Soviet leadership to move away from the association of communism with Stalinist repression and exploitation. Usually, foreign state propaganda would not be so readily accepted. The Soviet model was successful among Italian communists and their children, however, because it connected to their daily experiences and struggles, be it exploitation in factories, the lack of access to secondary education, or bad living conditions. As Stein points out in her chapter, this propaganda connected with deprived people by both stressing sympathy for their suffering and offering concrete solutions to it. Celebrating the Soviet Union was thus a way for a particular subsection of Italian society—the communist pillar—to express its identity and advance its social demands.

The scientific achievements of the Soviet Union peaked between 1957 and 1964, with the space race providing a prime opportunity to celebrate the Soviet model. The Sputniks showed that human beings had immense power at their disposal, which they could use to reshape nature and society. The *Luniks* forced many to reconsider their perspective on human problems. The flights of Gagarin and Tereshkova seemed to prove that the Soviet Union was creating a new society and the New Man. Italian communists celebrated the socialist system that made it possible for Gagarin to study and travel into space—promising the same for children in the future communist society. Although met with some resistance, Tereshkova offered a new role model for women, with science countering old superstitions and female subjugation.

The celebration of Soviet science among Italian communists was not without its limitations and drawbacks, of course. Promoting a foreign nation as the guide of scientific progress could mar a genuine desire to promote scientific education among young people with a paternalistic attitude; this is also evident in the neocolonial case analyzed by Stein. As we saw above, communist discourses on science also often gave the impression of discussing a wish-granting genie. This issue was not confined to communists, however; there is always the risk, when praising the power of science for laymen, to turn into a magic show instead of a rigorous discipline, as Stein's chapter also shows. Nevertheless, it created the impression of an even greater contrast with reality given the unfulfilled promise of technological prosperity in the Soviet Union.

Meanwhile, Italian communists had their own problems reconciling with the prosperity produced by capitalism, leading some to become suspicious of consumerism and technological products. In any case, by the late 1960s, the

US space programs surpassed those of the Soviet Union. Concurrently, Western societies were developing their own skepticism of scientific rationality and technology, due to an increasing concern for the environment and human health, and about scientific warfare and social alienation. The space race, which had started with a celebration of science's benefits, thus ended with widespread skepticism and some open hostility.[89] For all his faults, Khrushchev had found a way to revive the myth of communism after Stalin. The decline of the identification between scientific progress and socialism left the ideology of the Soviet Union and the international communist movement with an empty center that would never be filled.

Ettore Costa is assistant professor at the Scuola Superiore Meridionale in Naples. He received his PhD at the University of Rome, La Sapienza. His doctoral research focused on socialist internationalism in Europe at the beginning of the Cold War. The focus of his research is the relationship between socialism and science, on which he worked at the Centre of European Research at the University of Gothenburg, Sweden, and the New Europe College, in Bucharest. His publications also address topics such as human rights, European integration, and Euroscepticism, as well as the European Union's science policies.

Notes

1. "Costumi," *Pioniere*, 25 January 1959. Most articles in *Pioniere* were published anonymously; some were attributed to pseudonymous authors, so I will often have to write "*Pioniere* says ..." instead of mentioning a specific journalist.
2. Rinaldi, "Ufficio postale," 22 December 1957.
3. Marcello S., "Sul pianeta verde," *Pioniere*, 24 November 1957; Mariangela T., "Destinazione Giove," *Pioniere*, 12 January 1958. (These children's surnames have been removed to protect their privacy.)
4. "I 100 vincitori," *Pioniere dell'Unità*, 21 November 1963.
5. Rinaldi, "Ufficio postale," 3 November 1957.
6. Marchioro, "Il Pioniere," 83.
7. See Bassi, "'Solo con il socialismo'"; Costa, "Whoever Launches the Biggest Sputnik."
8. Skinner, "Meaning and Understanding."
9. De Luna, "Partiti e società"; Gualtieri, *L'Italia*, 21–22.
10. The concept of the New Man refers to the idea that radical political transformation would require a change in human nature. The new human beings would have a higher consciousness, be less selfish, better educated, and in better physical and mental health. While it was part of official Soviet ideology, it is also common among major religions and in utopian revolutionary ideology to the left and the right. NB: At the time, Italian used the term *uomo* (man) to mean both adult males and human beings

as a whole; this ambiguity is preserved in the text, since "man" used to have the same connotations in English.

11. Franchini, *Diventare grandi*, 7.
12. Graziosi, *L'URSS dal trionfo al degrado*, 169.
13. Graziosi, *L'URSS dal trionfo al degrado*, 212–19.
14. Priestland, *The Red Flag*.
15. Reid, 2005. "The Khrushchev Kitchen," 290.
16. Romero, *Storia della guerra fredda*, 143. NB: All translations from Italian are my own.
17. Westad, *The Cold War*. See also Pivato and Pivato, *I comunisti sulla luna*, 45.
18. Fedele, *L'autunno del mito*, 87. See also: Degl'Innocenti, *Il mito di Stalin*, 175.
19. Cafagna, *C'era una volta*; Gozzini and Martinelli, *Storia del Partito Comunista Italiano*; Sorgonà, *La svolta incompiuta*.
20. Fedele, *L'autunno del mito*, 18–19.
21. Togliatti, "La relazione."
22. Fedele, *L'autunno del mito*, 65–66; Togliatti, "Per la unità"; Togliatti, "1917–1957."
23. "Il dibattito di Mosca," *Rinascita*, November 1961. This *translatio imperii* was an old trope for Togliatti; see, e.g., his "L'umanità al bivio" and "Per una sinistra europea."
24. Franchini, *Diventare grandi*, 10–13.
25. Pombeni, "Il sistema dei partiti."
26. Taviani, "Il Cinema Di Propaganda."
27. Franchini, *Diventare grandi*, 1–2.
28. Marchioro, "'Il Pioniere,'" 72.
29. Franchini, *Diventare grandi*, 159.
30. Boffa, "Il lancio del III Sputnik."
31. Rinaldi, "Ufficio postale," 18 March 1962; Rosser, "Gli 'Sputnik' dell'atletica."
32. Ruschi, "Destinazione Luna."
33. Andersson, *The Future of the World*, 17.
34. "Una nuova epoca aperta ai millenni," *L'Unità*, 7 November 1957.
35. Togliatti, "Alternativa democratica e comunisti"; Togliatti, "Chi fa di più per il popolo?"
36. Longo, "La realtà."
37. Longo, "La realtà."
38. "L'uomo crea le stelle," *Pioniere*, 27 October 1957; Franchini, *Diventare grandi*, 40.
39. Togliatti, "Guardando il futuro"; Rinaldi, "Ufficio postale," 2 August 1959.
40. "L'uomo ha conquistato lo spazio." *Pioniere*, 20 October 1957.
41. Franchini, *Diventare grandi*, 39; Pivato and Pivato, *I comunisti sulla luna*, 17; Lombardo Radice, "Il posto della scienza."
42. Giuseppe Boffa, "Il XXI Congresso."
43. Luporini, "La scienza"; Rinaldi, "Ufficio postale."
44. Palmiro Togliatti, "Il destino dell'uomo."
45. Rinaldi, "Ufficio postale," 25 October 1959.
46. Palmiro Togliatti, "A ciascuno il suo."
47. Natta, "Attualità e urgenza."
48. "Che cosa ne pensano," *Pioniere*, 25 October 1959.
49. Rinaldi, "Ufficio postale," 27 September 1959.
50. "L'anno memorabile della conquista dello spazio," *Pioniere*, 28 May 1961.
51. Lombardo Radice, "Un nuovo metro."
52. Spinella, "Per la libertà di tutti."
53. Rodari, "La terra e la Luna."

54. Gramsci, *Quaderni*, 2108.
55. Lombardo Radice, "Un nuovo metro"; Lombardo Radice, "Il posto della scienza."
56. "Lunik: Scienza e civiltà," *Pioniere*, 4 October 1959.
57. Pivato and Pivato, *Comunisti sulla luna*, 55.
58. Lannutti, "La vita di Yuri."
59. Rinaldi, "Ufficio postale," 30 April 1961; Rinaldi, "Ufficio postale," 7 May 1961.
60. Rinaldi, "Ufficio postale," 7 May 1961.
61. Pivato and Pivato, *Comunisti sulla luna*, 74.
62. Macciocchi, "Il peso della tradizione conservatrice."
63. Rinaldi, "Ufficio postale," 29 December 1957; Franchini, *Diventare grandi*, 46.
64. Rinaldi, "Ufficio postale," 8 December 1957 and 13 July 1958.
65. "Radiotraguardo." *Pioniere*, 4 June 1961.
66. Amico del giovedì, "Posta," *Pioniere dell'Unità*, 27 June 1963.
67. "Uomini e donne dinanzi al volo di valentina," *Rinascita*, 29 June 1963.
68. Della Pergola, "La vittoria di Valentina."
69. Banti, "L'emancipazione non c'entra."
70. Amico del giovedì, "Posta," *Pioniere dell'Unità*, 5 November 1964.
71. Colazingari, "Film illustrati"; "Il futuro è già cominciato," *Pioniere*, 21 September 1958.
72. Argilli and Berti, "Chiodino e Giacinto."
73. Argilli and Berti, "Chiodino e Giacinto," *Pioniere*, 1 March 1959 and 29 March 1959.
74. Maria Grazia B., "Gli 'Stati Solari,'" *Pioniere*, 19 January 1958.
75. Rinaldi, "Ufficio postale," 22 February 1959.
76. "Che cosa ne pensano," *Pioniere*, 18 October 1959.
77. Bruno M., "La vita, il lavoro dell'uomo nell'anno duemila," *Pioniere*, 29 March 1959.
78. Graziosi, *L'URSS dal trionfo al degrado*, 245.
79. Platone, "L'uomo e l'universo."
80. Moralevic, "Scienza o fantascienza."
81. Rinaldi, "Ufficio postale," 17 April 1960.
82. Atomo, "Il misterioso mondo degli atomi," *Pioniere*, 16 June 1957.
83. Kirsch, *Proving Grounds*.
84. Graziosi, *L'URSS dal trionfo al degrado*, 263.
85. Josephson, *Would Trotsky Wear a Bluetooth?* 294.
86. Spagnolo, *Sul memoriale di Yalta*.
87. Costa, "Whoever Launches the Biggest Sputnik," 102.
88. Ignazi, *Dal PCI al PDS*, 153.
89. Costa, "The Western European Left and the First Moon Landing."

Bibliography

Andersson, Jenny. *The Future of the World: Futurology, Futurists, and the Struggle for the Post Cold War Imagination*. New York: Oxford University Press, 2018.

Argilli, Marcello, and Vinicio Berti. "Chiodino e Giacinto." *Pioniere*, 22 February 1959.

Banti, Anna. "L'emancipazione non c'entra." *Rinascita*, 29 June 1963.

Bassi, Giulia. "'Solo con il socialismo comincia un'epoca nuova nella scienza': Mito e ambivalenze del PCI sul discorso scientifico sovietico (1949–1969)." In *Scienziati e guerra fredda: Tra collaborazione e diritti umani*, edited by Elisabetta Bini and Elisabetta Vezzosi, 2020, 37–64. Rome: Viella.

Argilli, Marcello, and Vinicio Berti. "Chiodino e Giacinto." *Pioniere*, 1 March 1959.

——. "Chiodino e Giacinto." *Pioniere*, 29 March 1959.

Boffa, Giuseppe. "Il lancio del III Sputnik la più audace impresa scientifica dei nostri tempi." *Rinascita*, May 1958.

——. "Il XXI Congresso e gli sviluppi teorici e politici dell'azione dei comunisti sovietici." *Rinascita*, March 1959.

Cafagna, Luciano. *C'era una volta . . . : Riflessioni sul comunismo italiano*. Venice: Marsilio, 1991.

Carioli, Stefania. "Racconti e giornali illustrati per l'infanzia d'ispirazione marxista nell'Italia del secondo dopoguerra." *Società di Politica Educazione e Storia* 10 (2019): 155–74.

Colazingari. "Film illustrati." *Pioniere*, 2 March 1958.

Costa, Ettore. "Whoever Launches the Biggest Sputnik Has Solved the Problems of Society? Technology and Futurism for Western European Social Democrats and Communists in the 1950s." *History of European Ideas* 46(1) (2020): 95–112.

——. "The Western European Left and the First Moon Landing: The Fall of Scientific Enthusiasm and the Ebb of Socialism." *International History Review* 44(6) (2022): 1347–68.

Degl'Innocenti, Maurizio. *Il mito di Stalin: comunisti e socialisti nell'Italia del dopoguerra*. Manduria: P. Lacaita, 2005

Della Pergola, Paola. "La vittoria di Valentina." *Rinascita*, 29 June 1963.

De Luna, Giovanni. "Partiti e società negli anni della ricostruzione." In *Storia dell'Italia Repubblicana*, vol. 2, edited by Francesco Barbagallo. Turin: Einaudi 1995.

Fedele, Santi. *L'autunno del mito: La sinistra italiana e l'Unione Sovietica dal 1956 al 1968*. Milan: Franco Angeli, 2016.

Franchini, Silvia. *Diventare grandi con il Pioniere, 1950–1962: Politica, progetti di vita e identità di genere nella piccola posta di un giornalino di sinistra*. Florence: Firenze University Press, 2006.

Gozzini, Giovanni, and Renzo Martinelli. *Storia del Partito Comunista Italiano*. Vol. 7. *Dall'attentato a Togliatti all'VIII Congresso*. Turin: Einaudi, 1998.

Gramsci, Antonio. *Quaderni del carcere*. Edited by V. Gerratana. Turin: Einaudi, 1975.

Graziosi, Andrea. *L'URSS dal trionfo al degrado: Storia dell'Unione Sovietica, 1945–1991*. Bologna: Il Mulino, 2011.

Gualtieri, Roberto. *L'Italia dal 1943 al 1992: DC e PCI nella storia della Repubblica*. Rome: Carocci, 2006.

Ignazi, Piero. *Dal PCI al PDS*. Bologna: Il Mulino, 1992.

Josephson, Paul R. *Would Trotsky Wear a Bluetooth? Technological Utopianism under Socialism, 1917–1989*. Baltimore: Johns Hopkins University Press, 2010.

Kirsch, Scott. *Proving Grounds: Project Plowshare and the Unrealized Dream of Nuclear Earthmoving*. New Brunswick, NJ: Rutgers University Press, 2005.

Lannutti, Giancarlo. "La vita di Yuri." *Pioniere*, 7 May 1961.

Lombardo Radice, Lucio. "Un nuovo metro." *Rinascita*, October 1959.

——. "Il posto della scienza nella cultura italiana." *Rinascita*, December 1959.

Longo, Luigi. "La realtà e l'attualità della prospettiva socialista." *Rinascita*, March 1958.

Luporini, Cesare. "La scienza e il destino umano." *Rinascita*, December 1959.

Macciocchi, Maria Antonietta. "Il peso della tradizione conservatrice." *Rinascita*, March 1961.

Marchioro, Michela. "'Il Pioniere,' settimanale di tutti i ragazzi d'Italia." *L'Almanacco* 29–30 (1998): 71–95.
Moralevic, Juri. "Scienza o fantascienza." *Pioniere*, 12 March 1961.
Natta, Alessandro. "Attualità e urgenza della riforma scolastica." *Rinascita*, November 1959.
Pivato, Franco, and Marco Pivato. *I comunisti sulla luna: l'ultimo mito della rivoluzione russa*. Bologna: Il Mulino, 2017.
Platone, Dino. "L'uomo e l'universo." *Pioniere* supplement, 20 March 1960.
Pombeni, Paolo. "Il sistema dei partiti dalla Prima alla Second Repubblica." In *L'Italia contemporanea dagli anni Ottanta a oggi*. Vol. 3: *Istituzioni e politica*, edited by Simona Colarizzi, Agostino Giovagnoli, and Paolo Pombeni, 307–31. Rome: Carocci, 2014.
Priestland, David. *The Red Flag: A History of Communism*. New York: Grove Press, 2009.
Reid, Susan E. 2005. "The Khrushchev Kitchen: Domesticating the Scientific-Technological Revolution." *Journal of Contemporary History* 40(2) (2005): 289–316.
Rinaldi, Dina. "Ufficio postale," weekly column in *Pioniere*. Cited issues: 8 December 1957; 3 November 1957; 22 December 1957; 29 December 1957; 13 July 1958; 22 February 1959; 27 September 1959; 25 October 1959; 2 August 1959; 17 April 1960; 30 April 1961; 7 May 1961; 18 March 1962.
Rodari, Gianni. "La terra e la Luna." *Pioniere*, 27 September 1959.
Romero, Federico. *Storia della guerra fredda: l'ultimo conflitto per l'Europa*. Turin: Einaudi, 2009.
Rosser. "Gli 'Sputnik' dell'atletica." *Pioniere*, 24 November 1957.
Ruschi, Berto. "Destinazione Luna." *Pioniere*, 3 November 1957.
Sorgonà, Gregorio. *La svolta incompiuta: Il gruppo dirigente del Pci dall'8. all'11. congresso (1956–1965)*. Rome: Aracne, 2011.
Skinner, Quentin. "Meaning and Understanding in the History of Ideas." *History and Theory* 8(1) (1969): 3–53.
Spagnolo, Carlo. *Sul memoriale di Yalta: Togliatti e la crisi del movimento comunista internazionale (1956–1964)*. Rome: Carocci, 2007.
Spinella, Mario. "Per la libertà di tutti." *Rinascita*, 29 June 1963.
Taviani, Ermanno. "Il cinema di propaganda: il caso del Pci." In *Propaganda, cinema e politica 1945–1975*, edited by Ermanno Taviani. Archivio Audiovisivo Del Movimento Operaio e Democratico, Annale 11, 2008.
Togliatti, Palmiro. "L'umanità al bivio." *Rinascita*, November 1951.
———. "La relazione di Togliatti al Comitato centrale sui lavori del XX Congresso del PCUS." *L'Unità*, 15 March 1956.
———. "Per la unità e la compattezza del movimento comunista." *Rinascita*, October–November 1957.
———. "1917–1957." *Rinascita*, November 1957.
———. "Alternativa democratica e comunisti." *Rinascita*, January 1958.
———. "Chi fa di più per il popolo: Il collettivismo o il capitalismo sociale?" *Rinascita*, April 1958.
———. "Guardando il futuro." *Rinascita*, November–December 1958.
———. "Per una sinistra europea." *Rinascita*, March 1959.
———. "A ciascuno il suo." *Rinascita*, February 1960.
———. "Il destino dell'uomo." *Rinascita*, 30 March 1963.
Westad, Odd Arne. *The Cold War: A World History*. New York: Basic Books, 2017.

CHAPTER 9

"Effective Films": Science Film and Education in East Germany

Sophia Gräfe and Kerrin von Engelhardt

Introduction

Given the surge of interest in the interdisciplinary history of scientific media currently emerging in a number of disciplines—including media studies, cultural studies, and the histories of science, medicine, psychology, and education—the comparatively small number of scholarly studies on science films in the former German Democratic Republic (GDR) is remarkable.[1] Compared to other national contexts, the history of East German science film has received hardly any scholarly attention.[2] The research that has been done so far is often too specific to characterize a distinctive audiovisual culture.[3] And yet, East German institutions and individuals devoted a great deal of attention to the aesthetics and didactics of educational films and promoted research into an effective use of films in schools, universities, and popular science. Similarly, GDR scientists continued making research films for experimental purposes and tried to keep up with international technical standards. The production of this "useful films" is difficult to track today.

Problematic access to archives and sources of GDR science films is a major reason for this difficulty. When the East German state institutions were dissolved after the German reunification in 1990, the transfer, archiving, and preservation of films and records posed a significant challenge. For many years post-1990, it remained unclear where the responsibility for establishing and maintaining science film archives should lie. For instance, the film stock of the central body of science film in East Germany, the Institut für Film, Bild und Ton (IFBT, Institute für for Film, Image, and Sound), was not transferred to the German Federal State Archive. A large part of the ISFA film collection was stored at the Hochschule für Technik und Wirtschaft Berlin (HTW, University of Applied Sciences), pending systematic cataloging. The Federal Film Archive holds individual film copies, which have been deposited by smaller institutions. This collection is also awaiting cataloging and restoration. The complex institutional network of GDR science film outlined in this chapter contributes to an ongoing discussion, but there is no established canon of films or literature that we can rely on.

This chapter has two goals. It first reviews the institutions, actors, discourses, and collections relevant to East German science film. This overview is intended to provide potential points of entry for future studies of science film in the GDR. As a preliminary contribution to the interdisciplinary history of science film in the GDR, the chapter also highlights the knowledge politics of these visual media by tracing the various approaches taken to optimize the use of film, from Pavlovism to information theory to cybernetics. As the chapter reveals, educational films occupied a distinguished position within the audiovisual landscape of the GDR. Government institutions focused on aesthetics and didactics in science films for the most effective use of filmic images to optimize learning processes in a wide range of academic disciplines. Their commitment to this goal can be traced through a rich body of literature produced by a powerful organizational apparatus. It also led to the emergence of a distinct field of educational media research that explored the educational impact of film. The second part of the chapter will then examine GDR research films and analyze less-known structures and archives dedicated to this genre, which existed in the shadow of educational film.

A Way into GDR Science Film History

The East German journal *Bild und Ton* (Image and Sound) is a good starting point from which to explore the history of the science film in the GDR. Originally licensed in 1948 by the occupying Soviet Military Administration, the richly illustrated magazine served as a resource for both amateurs and professionals, providing information on major technical developments in the fields of science film and photography until 1992.[4] A selection of its articles on the theory, practice, and history of filmmaking—as well as paratext, such as advertisements and event announcements—sheds light on the topics that were especially relevant to the field.

A particular emphasis on education is apparent in the early years of the journal's existence. In a late-1949 issue, its founder and editor in chief Albert

Wilkening (1909–90) advocated for the implementation of educational films in schools—"Unterrichtsfilme in die Schule!" (Educational films in schools!)—and called for the professionalization of the struggling film industry.[5] After all, it would support the country's ongoing reconstruction efforts after World War II; as he wrote, "[. . .] the war and its aftermath have robbed many of our young people of the education they need to ensure the ongoing development of our economy."[6] Only two journal issues later, a mildly indignant response to Wilkening's criticism painted a completely different picture. Werner Hortzschansky (1898–1982), whose name would reappear in the institutional history of science films in the GDR until the 1960s,[7] advertised the achievements of the Zentrale Bildstelle (ZBS, Central Film Office), founded in July 1946, in compliance with Soviet Command No. 170. Under his direction, the ZBS had restored to working order the five Landesbildstellen (state film offices) and 182 Kreisbildstellen (district film offices) in the Soviet Occupation Zone, which had been battered during World War II.[8] Hortzschansky boasted that, in collaboration with the newly formed DEFA-Studio für populärwissenschaftliche Filme (DEFA Studio for Popular Science Films) in Berlin-Babelsberg, these film offices were busily producing new educational films in response to the pressing challenges of the day.[9] These new films—sometimes still referred to as *Kulturfilme* (cultural films), a term discredited by the Nazis—were intended for use in schools and universities.[10] They dealt with key problems of the time, such as infectious diseases like tuberculosis and diphtheria, and pests such as the housefly, the potato beetle, and the spruce bark beetle, which were looked at in terms that went beyond the merely biological.[11]

In August 1950, the Zentralinstitut für Film und Bild in Unterricht, Erziehung und Wissenschaft (ZFB, Central Institute for Film and Image in Teaching, Education, and Science) was founded. The ZFB collaborated with the Pädagogisches Zentralinstitut (Central Pedagogical Institute) and the Deutsches Zentralinstitut für Berufsbildung (German Central Institute for Vocational Training) to coordinate the technical and organizational supply of audiovisual teaching aids for all schools and universities in the GDR. Schulbildstellen, or film offices assigned to each school, taught teachers and students how to operate photographic and film equipment, while also exercising political control. Their media repositories, in fact, served as an interface between state educational policy and the classroom. As one teacher, Gerd Maier, reported in an article for *Bild und Ton* in 1955, these film offices also impacted the cultural life of districts and neighborhoods, as parents were invited to the district film office auditorium, where they were shown the very same educational films in the evenings.[12] Furthermore, Hochschulbildstellen (university film offices) circulated copies of films available for instructional use at all GDR universities as well as supporting scientists in producing their own research and educational films. Pedagogical institutes also had their own film departments, where they conducted research on the psychological and pedagogical dimensions of film.[13]

Without approval from the central authority—which was renamed in 1954, 1962, and ultimately in 1970 as the Institut für Film, Bild und Ton (IFBT, Institute for Film, Image, and Sound)—no film could be circulated, whether among universities, schools, or the general public.[14] Hortzschansky and his colleagues initially justified this by referring to the efficiency—made necessary by the scarcity of resources—of having a central authority manage all teaching aids, "from kindergarten to university."[15] Upon closer examination, however, the ideological slant of this centralization becomes clear: every citizen, including the rural population, was to become aware of certain timely or urgent matters—even, and perhaps especially, those not visible to the naked eye. Progress, peace, and the efficient use of resources were thus ideals embodied by state-sponsored filmmaking and distribution.[16] It was this industry's mission to direct the professional and social path of each individual toward participating in the socialist collective.[17]

To accomplish this ambitious task, approaches to film design (*Gestaltung*) were published regularly in *Bild und Ton*. Wilkening intended to explore how to make socialist cinematography as rational and effective as possible. In 1950, he proclaimed that physiological and psychological research could be productive in studying film reception and reported that "research abroad" had already begun to explore this new perspective.[18] In the 1950s, Wilkening and other GDR educators applied the theories of Anton Makarenko (1888–1939) and Ivan Pavlov (1849–1936) to develop their approach to learning, explicitly distancing themselves from the behaviorist approach of B. F. Skinner (1904–90). After the 1953 Pavlov Conference, held in East Germany, a so-called Pavlov Commission was established at the Deutsche Pädagogische Zentralinstitut (DPZI, German Central Pedagogy Institute) in 1954, to ensure that "the fundamental findings of Pavlovian physiology would become an integral component of the general education of every future citizen of the German Democratic Republic."[19] Accordingly, Wilkening referred to Pavlov's experiments to declare that "conditioned reflexes" are an intrinsic principle of film.[20] In 1952, Wilkening published an editorial "Pawlows bedingte Reflexe und der Film" (Pavlov's conditioned reflexes and film).[21] Distancing himself from the "mystic" notion of film as an illusion propagated by so-called "dream factories," he instead understood film as a natural, physiological process that forged temporary associations between different places, times, images, and sounds by means of cinematic montage.[22]

The 1951 educational film *Bedingte Reflexe (Pawlow)* (HF 36, Conditioned reflexes [Pavlov]) may be regarded as an emblematic example of Wilkening's understanding of a "reflexological" film aesthetic.[23] German physiologist Emil von Skramlik (1886–1970) supervised the production of this short film, which

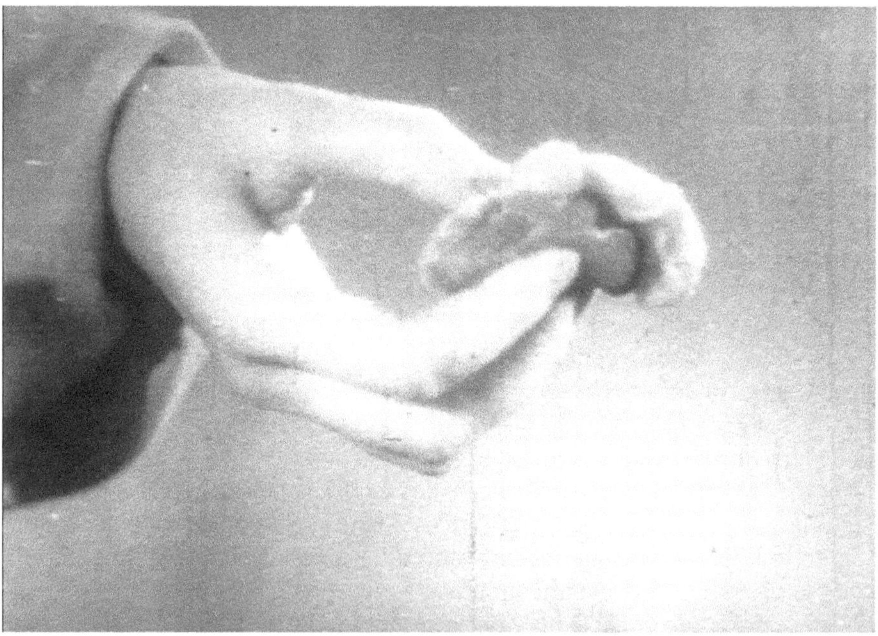

Figures 9.1 and 9.2. *Bedingte Reflexe* (Pawlow) (HF 36, Conditioned reflexes [Pavlov], 1951). © Library & Collection Medical Humanities of the Institute for the History of Medicine and Ethics in Medicine at Charité.

is a re-edit of footage from the Moscow Studio for Popular Science Film. The silent film intercuts shots of a dog that has been prepared with a saliva probe, with images of food, a test tube or a ticking metronome. Sound stimulus and salivation are thus semantically linked in the film, without requiring commentary.

It soon became clear, however, that Pavlov's principles were only partially applicable to education. As pedagogical theory became more subject-oriented,[24] Pavlov was criticized for his reduction of complex processes to biology and for the biologistic determinism inherent in his reflex theory.[25] A resolution was therefore passed to disband the DPZI's Pavlov Commission in 1957, just a few years after its formation.[26] Nevertheless, as we shall see below, the goal of controlling learning processes through conditioning was nevertheless still evident in parts of the GDR's cybernetics and computer science programs as late as the 1980s.

In 1963, Wilkening shifted gears. He now sought to apply the new field of information theory to film aesthetics—by, for instance, establishing a guideline of one hundred bits per second as the upper limit of consumable information. According to Wilkening, the information presented had to be efficiently allocated to the various film techniques, such as sequence shots or close-ups, so that the most essential information could be conveyed to the audience.[27]

"Effective Films"

What significance do these considerations have for our understanding of science film practice in the GDR? Even beyond the publications of an influential author like Wilkening, a preoccupation with creative ways of using films "effectively" for all spheres of education is evident from a large number of sources. Journal articles, handouts, compendia, and dissertations focusing on the "effective" use of film reveal a densely woven network of discourse in the GDR, in which the implementation of political guidelines through media—rooted in an economical theory of education and psychology—was negotiated. Manuals for teachers were produced as introductory textbooks and supplements to individual films.[28] Various institutions in the field of GDR science film also published articles on educational film in their journals and newsletters. The archive of the Research Library for the History of Education (BBF) in Berlin preserves so-called *Pädagogische Lesungen* (pedagogical lectures), which report on best practice methods for the "effective" use of educational films. Furthermore, unpublished dissertations on related topics can be found in university libraries. Drawing on these sources, we can reconstruct the main debates that took place about binding guidelines for the production of "effective" educational films.

Although the GDR educational system largely followed the Soviet model, pedagogical research from non-socialist countries, including the United States, was also taken into consideration. Numerous references to international studies in dissertations from the 1970s and 1980s indicate that East German academics and educators alike were interested in the empirical research and tangible data on the "effectiveness" of educational films.[29] For example, in his dissertation, East German Karl Steuding quotes Russian scholar L. V. Chashko, including his 1961 paper *The Educational Film as a Vehicle for Increasing Teaching Effectiveness*, and Walter Arno Wittich's and Charles Francis Schuller's 1953 overview *Audiovisual Materials: Their Nature and Use*, which originated from the United States.[30]

East German research also focused on locating suitable settings for pedagogical experiments and developing methods for analyzing film-based instruction. A range of methods were tested and applied: in some cases, researchers simply used questionnaires to assess the impact and benefits of educational films.[31] Others made use of complex pedagogical experiments involving control groups and statistical analysis. Researchers compared and statistically evaluated the increase in students' knowledge after the test screening of films.[32]

Programmed Lessons

In the mid-1970s, empirical research design became the distinguishing characteristic of GDR educational film didactics. Central was the work of pedagogue Ewald Topp, who led the DPZI from 1964 to 1968 and was appointed deputy head of the Akademie der Pädagogischen Wissenschaften der DDR (APW, Institute for Teaching Aids at the Pedagogical Sciences Academy of the GDR) in 1971.[33] He was concerned with employing film as an "effective medium for programming lessons."[34] In 1973, Topp analyzed the use of audiovisual teaching aids, or "modern information and educational resources," from the perspective of economic educational principles.[35] In his view, films should be part of every school's basic technical equipment in order to support the "programming of teaching and learning processes as a component of socialist educational theory and the socialist school."[36]

In addition to technical support, the concept of programming included the control and automation of learning processes.[37] Predefined lesson plans for programmed instruction were intended to support teachers and enable independent study on the part of the students. According to Topp, films, photographs, audiovisual sequences, projection slides, and television programs contained a "high degree of materialized work" and could therefore replace the actual didactical labor of teachers and students.[38] This approach was informed by the principles of cybernetics, which had been introduced into educational materials in the GDR beginning in the 1960s.[39] In order to increase rationalization, Topp encouraged that research be done into the sequencing of learning processes and the efficacy of teaching aids, which was implemented through state-run research plans.

Komplexfilm *vs.* Fragmentfilm

The GDR educational films strongly reflect the growing role of science in its educational system.[40] The visual style of most of these films forefronts an explicit and specific scientific character, which in GDR educational policy was linked to the principle of *Parteilichkeit* (adherence to the party line).[41] Audiovisual teaching aids and other teaching tools were thus meant to increase the efficiency of learning and educational performance while ensuring that the knowledge transmitted was solidly framed in socialist ideology.[42]

A variety of film formats were examined for this purpose. In their influential 1976 handbook *Audiovisuelle Lehrmittel* (audiovisual teaching aids), Rolf Fuchs and Klaus Kroll distinguished between *Komplexfilm* (complex film) and *Fragmentfilm* (fragment film).[43] According to the authors, the *Komplexfilm* presents "a sequence of learning steps, leading from single phenomena to more general statements or from theoretical insights to practical examples for application."[44] This type of film was considered an effective introduction to or summary of a topic.

In contrast, the fragment film required the comment of a teacher during the projection: "The Fragmentfilm is limited to one learning step or a short sequence of learning steps. The very term 'fragment' indicates that it only ever concerns one basic element within a didactic-methodological unit."[45] A fragment film often featured formal, factual infographics and scientific notations and was thus considered as part of the teacher's instruction or as an accompaniment to classroom discussions.[46]

Produced in the 1970s for tenth-grade biology classes, the film *Replikation der DNS* (DNA replication), provides a good example of the fragment film genre. Using images of a three-dimensional DNA model in combination with schematic animation and photographs of microscopic images, it illustrates the process of DNA replication dynamically, yet without sound. The film also employs multiple repetitions of the image sequence, making it easier to didactically integrate into classroom instruction. With the additional montage of microphotographs, it evokes an aesthetic of scientific factuality.

The design principles of socialist realism were especially relevant for complex films, as Hortzschansky proclaimed as early as in a 1951.[47] To achieve "effective" indoctrination with their viewers, complex films were interspersed with visual references to everyday life and filmed in clearly identifiable locations. Complex films also occasionally used music to evoke emotions, as explained by Bernd-Rainer Denecke in his 1986 dissertation on image-sound

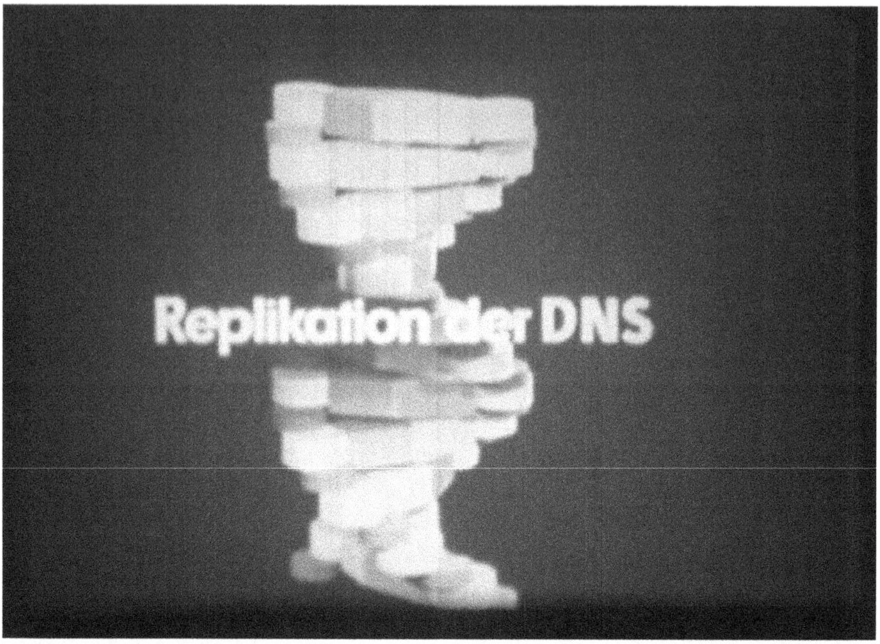

Figures 9.3 and 9.4. *Replikation der DNS* (KF 96, DNA replication, 1974). © Bundesarchiv-Filmarchiv.

design in teaching aids.[48] Denecke stated that on an auditory level, sound, and especially music, can have associative, suggestive effects by linking unrelated visual material. Manipulation of the soundtrack's volume can also induce physical reactions, such as "accelerated breathing, increased heart rate, elevated body temperature, and perceptible flushing,"[49] which are reminiscent here of Pavlov's reflex theory.[50] By contrast, the fragment film dispensed with such supplementary elements, sometimes omitting sound altogether.[51]

Although no music was used in *Die Atmung des Menschen* (Human respiration), it is an excellent example of a complex film. Produced at the DEFA Studio for Documentary Films, it was released in 1982 for use in eighth-grade biology classes. The silent film primarily centers on shots of young athletes on a playing field, all wearing identical jerseys, intercut with filmed X-ray images of lungs and abstracted animated images of rib cage movements. These are followed by shots of a classroom, where a window is opened, followed by a staged scene in which lung volume is measured. This four-minute film frames biological knowledge in a sociopolitical context. It promotes physical fitness (sports and abstinence from tobacco consumption) and regular fresh-air activities; the health of the respiratory organs is to be maintained to prevent fatigue during mental work.

Besides the didactic "efficiency" of educational films, researchers also examined the frequency of their use, yet with sobering results. For instance, Günter Heichel's 1975 cross-analysis of biology classes at East German secondary schools found that only 6 to 23 percent used audiovisual teaching aids.[52] He attributed this low percentage primarily to a scarcity of resources: if the same subject were taught at the same time at every school, there may not be enough copies of the film available for everyone.[53] Heichel also noted that, even in the 1970s, many schools still lacked the necessary equipment.

This reveals a fundamental problem: despite extensive studies on the benefits of educational films, GDR teachers rarely made use of them. Efforts were undertaken to counteract this, such as with the annual awards for the aforementioned *Pädagogische Lesungen* that reported on best teaching practices.[54] In 1986, Hans-Jürgen Richter, head of the district film office in Beeskow, calculated the frequency of borrowing such films for chemistry classes between 1979 and 1985. His analysis revealed the reason for the low number of films on loan and their insufficient use: the teachers' lack of knowledge and commitment.[55] Even a contest for best practice reports had once again brought a grievance to light.

This episode underscores the need to critically examine the archival records of the GDR institutions that once administered educational films—institutions that were primarily run by political authorities and may have embellished the reality in their reports. Future research must therefore incorporate continuous

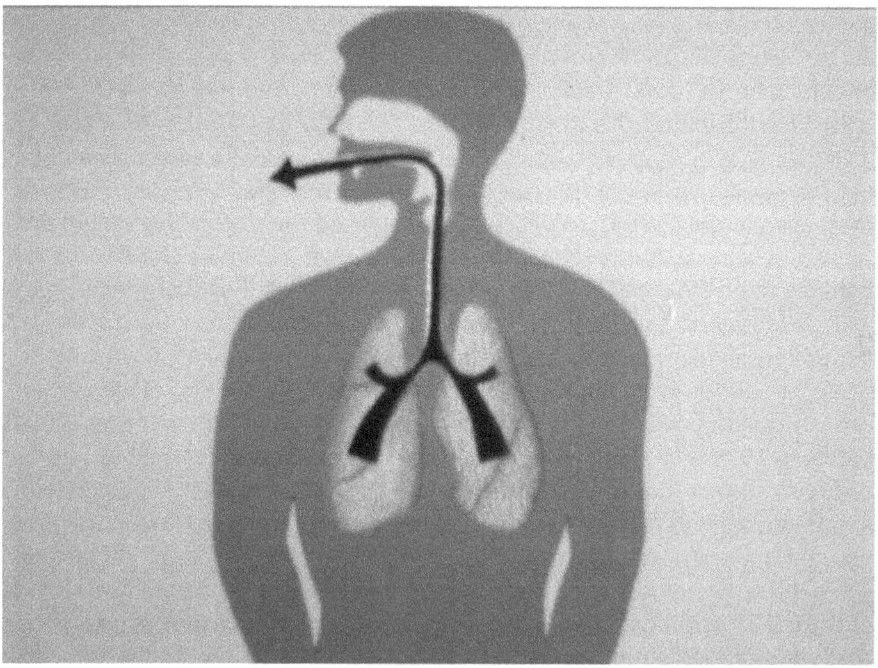

Figures 9.5 and 9.6. Figures 5–6: *Die Atmung des Menschen* (TF-1040, Human respiration, 1982), © Bundesarchiv-Filmarchiv.

comparison of theory and practice to minimize distortions in our view of the history of educational films in the GDR.[56]

Research Films—Off the Radar?

In contrast to films used in education and popular science, the GDR research film had slipped somewhat off the radar. The GDR had no centralized institution for these ephemeral films. Today, research films reside in specialist collections or the estates of individual researchers; one example is the archive of Berlin's Museum für Naturkunde, which houses the film collection of behavioral biologist Günter Tembrock (1918–2011).[57]

In the GDR, as elsewhere, research films were produced as a tool and source of knowledge of everyday laboratory research.[58] Due to their ontology these films were most often not intended for a larger audience; in some cases, they even contained confidential research data and remained unpublished.[59] In the 1950s, Albert Wilkening had tried to create more visibility for the research film genre by requesting reports on applied cinematography (*angewandte Kinematographie*) for *Bild und Ton* in 1958.[60] While the responses he received cannot be considered representative of the profile of all research films in the GDR, they do illustrate how they were perceived in a wider scientific context.

One of the initial responses was by a pedagogue who described how his working group had transitioned from using written documentation to using film to record classroom teaching. Where previously it had required up to twelve observers sitting in a classroom and taking shorthand notes to record the interactions between teachers and students, these interactions could now be documented on film "without the slightest mistake."[61] Another author, Johannes Haase, based at the Deutsche Hochschule für Körperkultur (DHfK, German College for Physical Culture) in Leipzig, presented his use of film in attempts to identify the ideal movements for certain sports. Experts in biomechanics used measurement markers and precision timers to determine the timing and spatial vectors of the movements. The recordings of "ideal" movement sequences would then be projected—for instance, on the wall of a gym—and used for training and teaching purposes.[62] Additional reports from experts in technical and material science fields followed, providing the readers of *Bild und Ton* with knowledge about research films that went well beyond the biology and medical films to which they were accustomed.

In the GDR and internationally, it was common practice to reuse research film footage and integrate it into educational films. University film offices and the DEFA Studio for Popular Science Films teamed up with researchers to reedit film footage.[63] The DEFA studio was more than just the central resource for popular science films. It produced documentary films of all kinds, which were shown before the feature film program at movie theaters—including at the DEFA *Zeitkinos*, located in popular places such as the Leipzig main station and Berlin-Friedrichstraße.[64]

The studio was also commissioned by the DZL, and later the Institut für Film, Bild und Ton, to produce a series of more advanced educational films in collaboration with individual researchers. Such films depicted scientific research practices, announced research results, and presented topics that were complex or difficult to convey in a university classroom setting. Some of these were also intended to replace dangerous or overly elaborate experiments; in a 1953 brochure instructing teachers in the use of teaching aids, for instance, the film *Bedingte Reflexe (Pawlow)* (Conditioned reflexes [Pavlov]) is proposed as a substitute for actual animal experimentation.[65]

In many respects, medical films constituted a unique category. Time-lapse and slow-motion images, surgical images, invasive endoscopic imaging techniques, X-ray and infrared images, as well as long-term film records of various conditions had been used as medical documentation, as well as for diagnostic purposes.[66] As early as 1950, the DEFA Studios had established a special subunit for medical and hygiene films under the leadership of Fritz Dick.[67] Medical films were taken out of university filmmaking facilities for organizational purposes. For instance, at the Humboldt University Berlin, the medical films section was separated from the central University Film Office in the late 1950s and revamped as the Charité Film Studio in the tradition of the Medizinisch-kinematographische Institut für Unterricht und Forschung (Medical-Cinematographic Institute for Teaching and Research), which had existed at the Berlin-based Charité university hospital from 1923 to 1931.[68] From 1959 on, the Charité hospital had its own film production group whose archive is one of the rare extant collections of medical teaching and research films.[69]

The Humboldt University Pädagogisches Labor (Pedagogical Laboratory) also had a specialized focus.[70] Under the direction of Hans-Georg Heun, it merged with the University Film Office in 1979 to form the Zentralinstitut für audiovisuelle Lehr- und Lernmittel (ZAL, Central Institute for Teaching and Learning Resources).[71] A selection of films and the correspondence of the Central Institute are now accessible via the Humboldt University archive in Berlin. The rare holdings of such local film institutes are among the few archival testimonies of research films in the GDR.

Transnational Archives

After the dissolution of the East German state in 1990, all rights to the GDR's educational and research films were transferred to the Federal Republic of

Germany (FRG). However, this was not accompanied by a physical transfer of existing science film collections. The film stock of the IFBT, until then the most important body of science film in East Germany, was not given a home in a state archive in the post-reunification period. A large portion of the GDR educational film collection was stored at the University of Applied Sciences (HTW) in Berlin, pending systematic cataloging.[72] Additionally, the German Federal Archive maintains individual film copies, which have been deposited by smaller and regional institutions; this collection also awaits cataloging and restoration and has barely been used for research. Other partial holdings are scattered across such diverse locations as the libraries of the Universities of Oldenburg, Rostock, and Freiberg, and the Wende Museum in the United States, making it virtually impossible to gain an overview. To learn more about science films in the GDR, one must paradoxically turn one's gaze abroad. Beginning in the early 1950s, individual proponents of East German science films joined the International Scientific Film Association (ISFA).[73] Indeed, in 1958 Werner Hortzschansky was elected to be the first vice president of the Ständiges Komitee für den höheren Unterrichtsfilm (Standing Committee on Film for Higher Education) at the ISFA's 12th Congress in Moscow.[74]

In 1947, the international association was conceived by the renowned scientific film pioneer Jean Painlevé (1902–89) and founded in partnership with the Scientific Film Association in London. It was partially funded by UNESCO, which was working to promote scientific collaboration, as well as the dissemination and appreciation of scientific research and technology, in postwar societies. Announcements and proceedings of individual ISFA meetings were published in the journal *Nature* well into the 1960s. The annual conferences featured sessions on research films, educational films, and popular science films and largely focused on establishing standards for film production and design. Accordingly, the GDR's Nationale Vereinigung für den Wissenschaftlichen Film (NVWF, National Association for Scientific Film), founded in 1963, sought to align its departments and standards with these international conventions. In contrast to the West German Institut für den wissenschaftlichen Film (IWF, Institute for Scientific Film), the NVWF did not produce its own films, serving instead as a platform for institutions and other members to exchange experiences.[75] It joined the ISFA in 1963 and convened ISFA meetings that were held, for instance, in Dresden (1969) and Jena (1982).[76]

The ISFA conferences not only provided the GDR with an opportunity for specialized discussion and to showcase their achievements but also offered East German representatives an official avenue for making contact with colleagues in the West, like at the IWF. Although the science film institutions in East and West Germany had both emerged from the earlier Reichsanstalt für Film und Bild in Wissenschaft und Unterricht (RWU, Reich Institute for Film and Image in Science and Education), the absence of a cultural agreement hindered official working relations between the two postwar Germanys.[77]

The international ISFA meetings are of particular significance to the historiography of GDR science films: the discussion papers, session reports, and film lists that emerge from these ISFA conferences documented historical debates on all subgenres of science film—from popular science film to educational film and research film.[78] At one point, the ISFA additionally maintained a central film collection, called the Cinémathèque des films scientifiques, and a specialized library in Brussels,[79] whose history and holdings have until now remained unexplored. Research into the materials preserved there could open up the potential for future comparative studies that would, in turn, offer a more balanced analysis of a broad, international set of sources and help overcome what have hitherto been limited, one-sided accounts—including, for example, of the role of ideology in films used in science and education.

Conclusion: Critical Historiography of GDR Science Film

The widely ramified institutional network of GDR science film and the opaque archival situation create certain historiographical conundrums. Because relevant archives and sources are scattered so widely and haphazardly, the study of educational films in the GDR urgently needs sources that facilitate locating and reviewing film holdings.

It is true that organizational hubs such as the IFBT, publications like *Bild und Ton*, and individuals such as Wilkening, Hortzschansky, and Topp played a significant role in the development and direction of the field. However, we concur with sociologist Rainer Lepsius that the challenge is to "extrapolate from this institutional description the consequences for the lives of individuals, the peculiarities of social relations, the degree of conformity, and the diverse forms of adaptation, evasion, and avoidance of norms."[80] Thus, research on the everyday experience of science and education in the GDR remains a pressing task.

Several methodological considerations for future work in this field of research emerge from our analysis. To write impactful histories of science films in the GDR, it will first be necessary to confront the archival landscape. In the future, these collections should be consolidated into a common digital registry. Secondly, efforts should be made to establish a joint science film archive that would include films of both East German and West German provenance. Lastly, the written sources of former GDR institutions that are accessible today predominantly reveal a rhetoric of success, clearly motivated by fulfilling the state's grand plans and loyalty to the party line.[81] In studying historical science films, it will thus be essential to retain a critical distance from their ostensible optimism. Discrepancies between scientific policy and educational reality surface in several ways in GDR sources, and their undertones, even when barely perceptible, deserve greater attention. This challenge can be met through a more intensive oral history of East German realities.[82]

Sophia Gräfe holds a BA in media culture from the Bauhaus University in Weimar and an MA in cultural history and theory from the Humboldt University Berlin. She is a research assistant at the Cluster of Excellence "Matters of Activity" at Humboldt University. From 2018 to 2023, she was a research associate in the DFG-Heisenberg Project "Transdisciplinary Networks of Media Knowledge" at the Philipps-University Marburg. Her research interests include media history and history of science, useful film, surveillance film, natural history image archives, and film and animal studies.

Kerrin von Engelhardt (née Klinger) studied fine arts at the Bauhaus University Weimar and cultural history/folklore studies, art history, and philosophy at Friedrich Schiller University Jena. She received her doctorate in the history of sciences with a dissertation on mathematics education around 1800. She is currently leading the sub-project "The Myth of Scientific Neutrality: The School Educational Film in the Cold War" as part of the BMBF-funded collaborative project "Educational Myths—A Dictatorship and Its Afterlife" at the Humboldt University Berlin.

Notes

1. For an introduction into interdisciplinary studies of science film and the *Gebrauchsfilm* (useful film) history, see Hediger, "'Dann sind Bilder also nichts!'"; Streible, Roepke, and Mebole, "Nontheatrical Film"; Acland and Wasson, *Useful Cinema*; Wellmann, "Cinematography, Seriality, and the Sciences"; Zimmermann, "Analyse nicht-fiktionaler Filmformen"; and Waller, *Beyond the Movie Theater*. See also the online bibliography on nontheatrical film compiled by Chen and Gräfe, "Mapping the Nontheatrical Field."
2. For previous studies on national film histories, with a focus on educational film, see, e.g., Masson, *Watch and Learn*; Orgeron, Orgeron, and Streible, *Learning with the Lights Off*; Dahlquist and Frykholm, *The Institutionalization of Educational Cinema*; and Geoff, *Academic Films for the Classroom*.
3. Film historians studying the GDR have hitherto primarily dealt with feature films and children's films, whereas those studying German educational and science films have mostly looked at the early years of the twentieth century or on medical and hygiene films in the Nazi era. Examples include: Bonah and Laukötter, "Screening Diseases"; Laukötter, "How Films Entered the Classroom"; Friedland, "'. . . doch erscheint in seiner Denkschrift"; Annegarn-Gläß, *Neue Bildmedien revisited*; Weiss, "Pedagogy,

Professionalism and Politics"; Kühn, *Unterrichtsfilm im Nationalsozialismus*; and Schmidt, *Medical Films, Ethics and Euthanasia*. Recently, however, researchers have begun to examine the science and educational film history of the GDR. See Karin Kneile-Klenk's foundational study of the depiction of National Socialism in East German school films and television (*Der Nationalsozialismus in Unterrichtsfilmen*), as well as Schwarz, "Vom Jahrmarktspektakel zum Aufklärungsinstrument"; Schluss and Jehle, *Videodokumentation von Unterricht*; Klinger, "Functional Bodies on Scientific Educational Film"; and Anja Laukötter's noteworthy "'One Feels So Much in These Times.'" For a broader history of science in the GDR, see Schleiermacher and Pohl, *Medizin, Wissenschaft und Technik*.

4. On the historical importance of *Bild und Ton*, see also Josephine Diecke's chapter in this volume.
5. In addition to being the editor in chief of *Bild und Ton*, Albert Wilkening was also a production manager at the DEFA Studios and, from 1969 to 1976, the general director of DEFA. He later also held a professorship at the German Film Academy (Deutsche Hochschule für Filmkunst, now Film University) in Potsdam-Babelsberg. In 1966, he coedited a compendium on film: Wilkening, Baumert, and Lippert, *Film: kleine Enzyklopädie*.
6. Wilkening, "Unterrichtsfilme in die Schule!" 353.
7. On Hortzschansky, see Hagemann and Obst, "Werner Hortzschansky 1889 bis 1982."
8. On institutions of science film production and distribution during the NS era, see Annegarn-Gläß, 71–78.
9. See Hortzschansky, "'Unterrichtsfilme in die Schule!'"
10. On the question of distinguishing documentary and educational films from the concept of the *Kulturfilm*, see Gebhardt, "Was unterscheidet den Unterrichtsfilm," and Juliane Scholz's chapter in this volume. For an introduction into the history of the *Kulturfilm*, see Reichert, *Kulturfilm im Dritten Reich*. See also Knopfe, "Kalendarium," 299.
11. See Brandes, "Unterrichtsfilme in die Schule."
12. See Maier, "Welche Bedeutung hat die Bildstelle."
13. See Hortzschansky, *Das deutsche Zentralinstitut für Lehrmittel*.
14. See a list of all major institutions for research and educational film in the GDR at the end of this chapter. For an extensive overview of film institutions in the GDR, see Jordan, *Film in der DDR*.
15. Hortzschansky, *Das deutsche Zentralinstitut für Lehrmittel*, 3.
16. See Hortzschansky, "Neue Perspektiven."
17. On the formation of socialist citizens, see Andresen. *Sozialistische Kindheitskonzepte*.
18. Wilkening, "Ist die Filmkunst eine Wissenschaft?"
19. Zabel, "Zur Geschichte des Deutschen Pädagogischen Zentralinstituts," 148–51.
20. In Pavlov's famous experiment, a dog's salivation was elicited to measure the temporary association made between hearing a bell ring and the expectation of being fed. See Pavlov, *Conditioned Reflexes*. On Pavlov's experimental design, see also Todes, *Pavlov's Physiology Factory*.
21. Wilkening, "Pawlows bedingte Reflexe."
22. Wilkening, "Pawlows bedingte Reflexe," 160.
23. On reflex theory and film aesthetics, see Vöhringer and Hagner, "Vsevolod Pudovkins Mechanik des Gehirns"; Nowak, "Mit Pavlov ins Kino." See also Killen, "Cinema, Conditioning, and the Cold War Subject."
24. See Schmidt, "Erziehungsbedingungen in der DDR."

25. On the history of academic psychology in the GDR, see Busse, "Von der Sowjetunion lernen," 213–19.
26. Busse, "Von der Sowjetunion lernen," 220.
27. See Wilkening, "Filmgestaltung und Informationsverarbeitung."
28. The most popular examples were: Schwarze, *Technik der Arbeit*; Fuchs and Kroll, *Audiovisuelle Lehrmittel*.
29. See Geißler and Wiegmann, "Sowjetisierung."
30. Steuding, *Untersuchung zur Bestimmung*, 5–6.
31. See, e.g., Denecke, "Pädagogisch-psychologische Grundlagen."
32. See Steuding, *Untersuchung zur Bestimmung*; Küster, "Untersuchungen zur Wirksamkeit."
33. See Zabel, "Zur Geschichte des Deutschen Pädagogischen Zentralinstituts," 292.
34. Topp, *Zur Funktion, Nutzung und Weiterentwicklung*, 27.
35. Topp, *Zur Funktion, Nutzung und Weiterentwicklung*, 8.
36. Topp, *Zur Funktion, Nutzung und Weiterentwicklung*, 26.
37. See Hoffmann-Ocon and Horlacher, "Technologie als Bedrohung oder Gewinn."
38. Topp, *Zur Funktion, Nutzung und Weiterentwicklung*, 26.
39. On cybernetics in the GDR, see Segal, *Kybernetik in der DDR*; Dittmann, *Kybernetik steckt den Osten an*. See also Zabel, "Die Lehrmaschine und der Programmierte Unterricht," 128–29.
40. On "scientism" in the GDR, see Schmidt-Lux, *Wissenschaft als Religion*.
41. See Geißler, "Zur pädagogischen Diskussion"; von Engelhardt, "'Wissenschaftlichkeit' ohne 'Parteilichkeit'?"
42. See Zabel, "Die Lehrmaschine und der Programmierte Unterricht"; von Engelhardt, "Educational Film in East Germany."
43. See Fuchs and Kroll, *Audiovisuelle Lehrmittel*, 80. On the typology of educational film, see also Gebhardt, "Populärwissenschaftlicher Film," 124.
44. Gebhardt, "Populärwissenschaftlicher Film," 124.
45. Gebhardt, "Populärwissenschaftlicher Film," 81. In his 1972 dissertation on the use of film in chemistry classes, Karl Steuding similarly differentiated between the educational film as a "demonstration tool" and as a "working tool" in the classroom. See Steuding, "Untersuchung zur Bestimmung."
46. See Gebhardt, "Inhalt und Gestaltung des Unterrichtsfilms," 24.
47. On the visual culture of socialism, see Köhring and Rüthers, Ästhetiken des Sozialismus. On *Komplexfilms* see Hortzschansky, *Unterrichtsfilme und Lichtbilder*, 15.
48. Denecke, "Pädagogisch-psychologische Grundlagen." See also von Engelhardt, "Complex Associations."
49. Denecke, "Pädagogisch-psychologische Grundlagen," 84.
50. See Denecke, "Pädagogisch-psychologische Grundlagen," 79–88.
51. See Gebhardt, "Inhalt und Gestaltung des Unterrichtsfilms," 35.
52. Heichel, "Analyse des Ausstattungsstandes."
53. See Heichel, "Analyse des Ausstattungsstandes," 93.
54. See Wähler and Hanke, "'Pacemakers Report.'"
55. See Richter, *Zum wiederholten Einsatz*.
56. This historiographical method was recently highlighted by a collection of online essays edited by Pilz and Schätz, in the special issue of *Research in Film and History*, "Educational Film—Histories and Historiographies of Practice."
57. The Tembrock film collection contains research films, e.g., from laboratory research

on animal behavior at the Zoological Institute of the Humboldt University of Berlin. On Tembrock's use of films, see Gräfe, "Red Foxes in the Filing Cabinet." It also contains educational films on biology topics from the GDR and the previous Nazi era.
58. See Wilkening, "Der Film im Dienste der Wissenschaft."
59. See Barényi, "Angewandte Kinematographie."
60. Wilkening, "Angewandte Kinematographie."
61. See Bohnsack, "Moderne Technik hilft," 31. On procedures for audiovisual recording of school lessons in the GDR, see Reh and Jehle, "Visual history of education"; Jehle, *Politische Bildung und Erziehung*.
62. See Haase, "Die Bedeutung der Kinematographie."
63. See Puppe, "Film und Bild in Wissenschaft und Forschung"; Gussmann, "Das populärwissenschaftliche DEFA-Filmstudio."
64. See N., "Das DEFA-Zeitkino, Berlin-Friedrichstraße"; Knopfe, *Der populärwissenschaftliche Film*, 304.
65. See ZFB, *Lehrmaterial für die Durchführung*, 15–16.
66. For an introduction into medical film and photography by East German authors, see Boy, Luhmann, and Schweinitz, *Foto- und Filmtechnik in der Medizin*. See also Wilkening, Baumert, and Lippert, *Kleine Enzyklopädie Film*, 337–50, and Lustermann, "*Foto- und Filmtechnik im Dienste der Wissenschaft*," as well as Schweinitz, "*Aufgabenstellung, historische Entwicklung*."
67. See Dick, *Über die Anfänge des Medizinfilms nach 1945*; Knopfe, *Der populärwissenschaftliche Film*, 299.
68. See Friedland, "Doch erscheint in seiner Denkschrift."
69. See "Schriftwechsel mit dem Prorektorat für Forschungsangelegenheiten" of the Humboldt-Universität zu Berlin, Universitätsarchiv der Humboldt-Universität zu Berlin, HUUA Rek.02 — Rektorat 1945–1968, 0293: 1–3. In the 1970s, the newly formed Akademie für Ärztliche Fortbildung der DDR (AfÄF, Academy for Continuing Medical Training) began acting as DEFA's main commissioner for educational medical films, for instance producing the the black-and-white film series *Wegweiser Gesundheit* (Guide to health, GDR 1974–84, 43 episodes).. From 1978 to the mid-1980s, a large number of medical films were commissioned by the German Hygiene Museum in Dresden entitled *Du und Deine Gesundheit* (You and your health); see Osten, "'Wer lässt sich schon gerne belehren?'"
70. See the files of the *Sektion Pädagogik*, Universitätsarchiv der Humboldt-Universität zu Berlin, HUUA Sek. Päd., 3662 and 3659.
71. See Heun, "Die Organisation und Leitung der Arbeit"; Heun, "Die Organisation und Leitung . . . und an anderen Hochschulen in der DDR."
72. See Ruedel and Koerber, "The Materiality of Heritage"; von Engelhardt and Ruedel, "Collecting and Using Audiovisual Educational Aids."
73. Werner Hortzschansky was elected as the first vice president of the Standing Committee on Film for Higher Education in 1958, at the XII Congress of the International Scientific Film Association (ISFA) in Moscow. See Kneile-Klenk, *Der Nationalsozialismus in Unterrichtsfilmen*, 56–57.
74. Se Kneile-Klenk, *Der Nationalismus in Unterrichtsfilmen*, 57.
75. See K.-H. Kretschmar, *Information über Entwicklung und Zielstellung der Nationalen Vereinigung des Wissenschaftlichen Films und Fernsehens der DDR (NVWF)*, AMH-Conference, Göttingen, June 1990, Göttingen: Institut für den Wissenschaftlichen Film Göttingen, 1990, 1. Private archive of Werner Große.

76. See Schweinitz, "Betrachtungen vor dem AICS/ISFA-Kongress."
77. See Große, *Filme für die Wissenschaft*, 51. The writings of Werner Große, a member of the IWF from 1975 to 2011, and IWF records—archived at the Leibniz Information Centre for Science and Technology and University Library (TIB) in Hanover, Germany—both depict individual examples of these transnational exchanges. Also see Mario Schulze's chapter in this volume. For an overview of recent studies on individual science films produced by the IWF, see Sattelmacher, Waltenspül, and Schulze, *Research Film*.
78. See, e.g., the ISFA papers in the film and file collection of the former IWF, housed in the TIB in Hanover.
79. See Goldsmith, "The International Scientific Film Association."
80. Lepsius, "Die Institutionenordnung als Rahmenbedingung," 18.
81. See also Jessen, "Diktatorische Herrschaft als kommunikative Praxis."
82. See Obertreis and Stephan, "Erinnerung, Identität und 'Fakten.'"

Bibliography

Former East German Institutions for Research and Educational Films

Zentralbildstelle (ZBS) (1946–1950)
Zentralinstitut für Film und Bild in Unterricht, Erziehung und Wissenschaft (ZFB) (1950–1954)
Deutsches Zentralinstitut für Lehrmittel (DZL) (1954–1962)
Sektion Unterrichtsmittel am Deutschen Pädagogischen Zentralinstitut (DPZI) (1962–1970)
Institut für Unterrichtsmittel an der Akademie der Pädagogischen Wissenschaften der DDR (APW) (1970–1990)
Institut für Film, Bild und Ton (IFBT) (1970–1990)
Nationale Vereinigung für den Wissenschaftlichen Film (NVWF) (1962–1991)

Archives, Collections, and Libraries with sources related to East German Science Film

BBF—DIPF Leibniz Institute for Educational Research and Information (Berlin), Research Library for the History of Education
Charité University Hospital, Library & Collection Medical Humanities (BSMH)
Humboldt University Berlin, university archive and library
Leibniz Information Centre for Science and Technology and University Library (TIB) (Hanover), media collection of the West German Institute for Scientific Film (IWF)
Museum für Naturkunde Berlin, Tembrock Film Collection, archive
The Wende Museum (Culver City, California), media library
University of Applied Sciences (HTW) (Berlin), media archive
University of Freiberg, library
University of Oldenburg, library
University of Rostock, library
Werner Große (Göttingen), private collection

Print Sources

Acland, Charles R., and Haidee Wasson, eds. *Useful Cinema*. Durham, NC: Duke University Press, 2011.
Andresen, Sabine. *Sozialistische Kindheitskonzepte: Politische Einflüsse auf die Erziehung*. Munich/Basel: Ernst Reinhardt Verlag, 2006.
Annegarn-Gläß, Michael. *Neue Bildmedien revisited: Zur Einführung des Lehrfilms in der Zwischenkriegszeit*. Bad Heilbrunn: Klinkhardt, 2020.
Barényi, Béla. "Angewandte Kinematographie in Wissenschaft und Technik: Eine Erwiderung der Filmindustrie." *Bild und Ton* 8(17) (1959): 224.
Bohnsack, Siegfried. "Moderne Technik hilft der pädagogischen Lehre und Forschung." *Bild und Ton* 12(2) (1959): 30–32.
Bonah, Christian, and Anja Laukötter. "Screening Diseases: Films on Sex Hygiene in Germany and France in the First Half of the 20th Century—An Introduction." *Gesnerus: Swiss Journal of the History of Medicine and Sciences* 1 (2015): 5–14.
———, eds. *Body, Capital, and Screens: Visual Media and the Healthy Self in the 20th Century*. Amsterdam: Amsterdam University Press, 2020.
Boy, Manfred, Heinz Lühmann, and Jürgen Schweinitz. *Foto- und Filmtechnik in der Medizin*. Halle (Saale): VEB Fotokinoverlag 1960.
Brandes, H. "Unterrichtsfilme in die Schule." *Bild und Ton* 3(3) (1950): 95–97.
Busse, Stefan. "'Von der Sowjetunion lernen': Pawlowisierung und Psychologie." *Psychologie und Geschichte* 1(2) (1998): 150–73.
Chen, Hongwei Thorn, and Sophia Gräfe. "Mapping the Nontheatrical Field: A Useful Teaching Bibliography." *JCMS Teaching Media Online Dossier: Teaching Non-theatrical Media*, edited by Tanya Goldman, Joseph Clarke, and Hongwei Thorn Chen, and Sophia Gräfe. Retrieved on 11 August 2023 from https://quod.lib.umich.edu/cgi/t/text/idx/j/jcms/18261332.0060.607/—mapping-the-nontheatrical-field-a-useful-teaching?rgn=main;view=fulltext.
Dahlquist, Marina, and Joel Frykholm, eds. *The Institutionalization of Educational Cinema: North America and Europe in the 1910s and 1920s*. Bloomington: Indiana University Press, 2019.
Denecke, Bernd-Rainer. "Pädagogisch-psychologische Grundlagen der Bild/Ton-Gestaltung audio-visueller Lehr- und Studienmittel." PhD diss., Pädagogische Hochschule "K.F.W. Wander," Dresden, 1986.
Dick, Fritz. "Über die Anfänge des Medizinfilms nach 1945." *Betriebsgeschichte* 3, Babelsberg 1984, 45–47.
Dittmann, Frank, ed. *Kybernetik steckt den Osten an: Aufstieg und Schwierigkeiten einer interdisziplinären Wissenschaft in der DDR*. Berlin: Trafo-Verlag, 2007.
Felsmann, Klaus-Dieter, and Bernd Sahling. *Deutsche Kinderfilme aus Babelsberg: Werkstattgespräche, Rezeptionsräum*. Berlin: DEFA-Stiftung, 2010.
Friedland, Alexander. "'. . . doch erscheint in seiner Denkschrift die Bedeutung des klinischen Films für den Unterricht allzustark betont': Zur Geschichte des Medizinisch-kinematographischen Instituts der Charité 1923–1931." *Medizinhistorisches Journal* 52(2/3) (2017): 148–79.
Fuchs, Rolf, and Klaus Kroll. *Audiovisuelle Lehrmitte*. 2nd ed. Leipzig: Fotokinoverlag, 1982.
Gebhardt, Walter. "Was unterscheidet den Unterrichtsfilm von den anderen Filmarten?" *Bild und Ton* 5(1) (1952): 2–3.

——. "Inhalt und Gestaltung des Unterrichtsfilms als Mittel hohe Lernergebnisse zu erreichen." *Unterrichtsmittel Information* 3(6) (1968): 16–39.

——. "Populärwissenschaftlicher Film contra Unterrichtsfilm?" *Film, Fernsehen, Filmerziehung* 6(2) (1969): 108–34.

Geißler, Gert. "Zur pädagogischen Diskussion in der DDR 1955 bis 1958." *Zeitschrift für Pädagogik* 38(6) (1992): 913–40.

Geißler, Gert, and Ulrich Wiegmann. "Sowjetisierung: Zur Geschichte west- und ostdeutscher Narrative über die Schule und Pädagogik in der SBZ/DDR." *Zeitschrift für Pädagogik* 1(supplement) (2023): 106–26.

Geoff, Alexander. *Academic Films for the Classroom: A History*. Jefferson, NC: McFarland, 2010.

Goldsmith, Maurice. "The International Scientific Film Association." *Nature* 188 (1960): 372.

Gräfe, Sophia. "Red Foxes in the Filing Cabinet: Günter Tembrock's Image Collection and Media Use in Mid-century Ethology." *Berichte zur Wissenschaftsgeschichte* 45(1–2) (2022): 55–86.

Große, Werner. *Filme für die Wissenschaft: Die Epoche des wissenschaftlichen Films in Göttingen*. Göttingen: Göttinger Tageblatt, 2012.

Gussmann, Manfred. "Das populärwissenschaftliche DEFA-Filmstudio Babelsberg." *Bild und Ton* 9(5) (1956): 122–23.

Haase, Johannes. "Die Bedeutung der Kinematographie für die Sportwissenschaft." *Bild und Ton* 12(4) (1959): 125–26.

Hagemann, Werner, and Joachim Obst. "Werner Hortzschansky 1889 bis 1982: Sein Wirken in Erkenntnis der Werte von Lichtbild und Film im Unterricht." In *Wegbereiter der neuen Schule*, edited by Gerd Hohendorf, 143–49. Berlin: Volk und Wissen, 1989.

Hediger, Vinzenz. "'Dann sind Bilder also nichts!' Vorüberlegungen zur Konstitution des Forschungsfelds 'Gebrauchsfilm.'" *Montage AV* 14(2) (2005): 11–22.

Heichel, Günter. "Analyse des Ausstattungsstandes, des Ausnutzungsgrades, des methodischen Einsatzes von Unterrichtsmitteln und der erzieherischen Einflußnahme bei der Arbeit mit Unterrichtsmitteln im Biologieunterricht der allgemeinbildenden polytechnischen Oberschulen der DDR." PhD diss., Martin-Luther-Universität Halle-Wittenberg, 1975.

Heun, Hans-Georg. "Die Organisation und Leitung der Arbeit mit audiovisuellen Lehr- und Lehrmitteln an der Humboldt-Universität zu Berlin und an anderen Hochschulen in der DDR." In *Neodidagmata XX, Poznati 1991*, 131–36. Poznan: Adam Mickiewicz University Press, 1991.

——. "Die Organisation und Leitung der Arbeit mit audiovisuellen Lehr- und Lehrmitteln an der Humboldt-Universität zu Berlin." *Wissenschaftliche Zeitschrift der Humboldt-Universität, Gesellschaftswissenschaftliche Reihe* 35(7) (1986): 620–25.

Hoffmann-Ocon, Andreas, and Rebekka Horlacher. "Technologie als Bedrohung oder Gewinn: Das Beispiel des programmierten Unterrichts." *Jahrbuch für Historische Bildungsforschung* 20 (2014): 153–56.

Hortzschansky, Werner. "Unterrichtsfilme in die Schule! Eine Erwiderung der Zentralbildstelle." *Bild und Ton* 3(2) (1950): 33–34.

——. *Das deutsche Zentralinstitut für Lehrmittel, seine Aufgaben, Organisation und Arbeitsweise*. Schriftenreihe des Deutschen Zentralinstituts für Lehrmittel 11. Berlin: Volk und Wissen, 1957.

———. *Unterrichtsfilme und Lichtbilder in der Schularbeit.* Schriftenreihe Film und Bild 1, Zentralinstitut für Film und Bild in Unterricht, Erziehung und Wissenschaft. Berlin: Volk und Wissen, 1951

———. "Neue Perspektiven der Unterrichtsfilm- und Lichtbildarbeit," *Bild und Ton* 3(9) (1950): 275.

Jehle, May. *Politische Bildung und Erziehung im Unterricht: Kontrastive Fallstudien anhand von Videoaufzeichnungen politischen Fachunterrichts in Ost-, West- und Gesamtberlin von 1978 bis 1993.* Bad Heilbrunn: Verlag Julius Klinkhardt, 2022.

Jessen, Ralph. "Diktatorische Herrschaft als kommunikative Praxis: Überlegungen zum Zusammenhang von 'Bürokratie' und Sprachnormierung in der DDR-Geschichte." In *Akten. Eingaben. Schaufenster: Die DDR und ihre Texte. Erkundungen zu Herrschaft und Alltag*, edited by Alf Lüdtke and Peter Becker, 57–75. Berlin: Akademie Verlag, 1997.

Jordan, Günter. *Film in der DDR: Daten, Fakten, Strukturen.* Potsdam: Filmmuseum Potsdam, 2009.

Killen, Andreas. "Homo pavlovius: Cinema, Conditioning, and the Cold War Subject." Grey Room 45 (fall 2011): 42–59.

Klinger, Kerrin. "Functional Bodies on Scientific Educational Film During the Cold War." *Medienimpulse* 54(4) (2016): 1–18.

Kneile-Klenk, Karin. *Der Nationalsozialismus in Unterrichtsfilmen und Schulfernsehsendungen der DDR.* Weinheim: Beltz, 2001.

König, Ingelore. *Arbeiten mit DEFA-Kinderfilmen.* München: KoPäd, 1998.

Köhring, Alexandra, and Monica Rüthers, eds. Ästhetiken des Sozialismus: *Populäre Bildmedien im späten Sozialismus.* Vienna/Cologne/Weimar: Böhlau, 2018.

Kühn, Michael. *Unterrichtsfilm im Nationalsozialismus: Die Arbeit der Reichsstelle für den Unterricht/Reichsanstalt für Film und Bild in Wissenschaft und Unterricht.* Mammendorf: Septem Artes, 1998.

Küster, Jürgen. "Untersuchungen zur Wirksamkeit von audio-visuellen Hochschulunterrichtsmitteln im Prozeß der Festlegung des Wissens, durchgeführt im Forschungsbereich der Methodik des Physikunterrichts." PhD diss., Pädagogische Hochschule "Theodor Neubauer," Erfurt/Mühlhausen, 1976.

Laukötter, Anja. "How Films Entered the Classroom: Emotional Education of Youth through Health Education Films in the US and Germany, 1910–1930." *Osiris* 31 (2016): 181–200.

Lepsius, Mario Rainer. "Die Institutionenordnung als Rahmenbedingung der Sozialgeschichte der DDR." In *Sozialgeschichte der DDR*, edited by Hartmut Kaelble, Jürgen Kocka, and Hartmut Zwahr, 17–30. Stuttgart: Klett Cotta, 1994.

Lustermann, Erich. "Foto- und Filmtechnik im Dienste der Wissenschaft." *Bild und Ton* 3(1) (1950): 363–67.

Maier, Gerd. "Welche Bedeutung hat die Bildstelle für die Schulen?" *Bild und Ton* 5(8) (1955): 140.

Masson, Eef. *Watch and Learn: Rhetorical Devices in Classroom Films after 1940.* Amsterdam: Amsterdam University Press, 2012.

N., N. "Das DEFA-Zeitkino, Berlin-Friedrichstraße." *Bild und Ton* 4(3) (1951): 86.

Nowak, Lars. "Mit Pavlov ins Kino: Die Orte der Tiere im sowjetischen Montagefilm der 1920er Jahre." *Tierstudien* 6 (2014): 71–83.

Obertreis, Julia, and Anke Stephan. "Erinnerung, Identität und 'Fakten': Die Methodik der Oral History und die Erforschung (post)sozialistischer Gesellschaften (Einleitung)." In *Erinnerung nach der Wende: Oral History und (post)sozialistische Gesellschaften*, edited by Julia Obertreis and Anke Stephan, 9–26. Essen: Klartext Verlag, 2009.

Osten, Philipp. "'Wer lässt sich schon gerne belehren?' Die Entstehung der Filmreihe 'Du und Deine Gesundheit.'" In *KAMERA! LICHT! AKTION! Filme über Körper und Gesundheit 1915 bis 1990*, edited by Susanne Roeßiger and Uta Schwarz, 50–70. Stiftung Deutsches Hygiene-Museum. Dresden: Sandstein, 2011.

Orgeron, Devin, Marsha Orgeron, and Dan Streible, eds. *Learning with the Lights Off: Educational Film in the United States*. Oxford/New York: Oxford University Press, 2012.

Pavlov, Ivan. *Conditioned Reflexes: An Investigation of the Physiological Activity of the Cerebral Cortex*. Translated and edited by G. V. Anrep. London: Oxford University Press, 1927.

Pilz, Katrin, and Joachim Schätz, eds. "Educational Film—Histories and Historiographies of Practice." Special issue, *Research in Film and History* (July 2023). Retrieved October 2023 from https://film-history.org/issues/issue-5-educational-film-practices.

Puppe, Kurt. "Film und Bild in Wissenschaft und Forschung." *Bild und Ton* 10, no. 5 (1957): 123–25.

Reh, Sabine, and May Jehle. "Visual history of education—audiovisuelle Unterrichtsaufzeichnungen aus der DDR." In *Qualitative Videoanalyse in Schule und Unterricht*, ed. Michael Corsten et al., 348–70. Weinheim, FRG: Beltz Juventa, 2020.

Reichert, Ramón, ed. *Kulturfilm im Dritten Reich*. Wien: Synema, 2006.

Richter, Hans-Jürgen. *Zum wiederholten Einsatz von Unterrichtsfilmen im Fach Chemie Erfahrungen und Hinweise*. Beeskow: APW, 1980 (BBF/DIPF: PL 5756).

Ruedel, Ulrich, and Martin Koerber. "The Materiality of Heritage: Moving Image Preservation Training at HTW Berlin." *Synoptique* 6, no. 1 (2018): 91–96.

Sattelmacher, Anja, Sarine Waltenspül, and Mario Schulze, eds. Focus Section: Research Film, *Isis* 112, no. 2 (2021): 291–378.

Schleiermacher, Sabine, and Norman Pohl. *Medizin, Wissenschaft und Technik in der SBZ und DDR: Organisationsformen, Inhalte, Realitäten*. Husum, FRG: Matthiesen, 2009.

Schluss, Henning, and May Jehle, eds. *Videodokumentation von Unterricht—Historische und vergleichende Zugänge zu einer Quellengattung der Unterrichtsforschung*. Wiesbaden: Springer VS, 2013.

Schmidt, Hanns-Dieter. "Erziehungsbedingungen in der DDR." In *Sozialisation und Entwicklung von Kindern vor und nach der Vereinigung*, ed. Gisela Trommsdorff, 34–35. Opladen, FRG: Leske und Budrich, 1996.

Schmidt, Ulf. *Medical Films, Ethics and Euthanasia in Nazi Germany: The History of Medical Research and Teaching Films in the Reich Office for Educational Films/Reich Institute for Films in Science and Education, 1933–1945*. Husum, FRG: Matthiesen Verlag, 2002.

Schmidt-Lux Thomas. *Wissenschaft als Religion. Szientismus im ostdeutschen Säkularisierungsprozess*. Würzburg, FRG: Ergon-Verlag, 2008.

Schwarz, Uta. "Vom Jahrmarktspektakel zum Aufklärungsinstrument: Gesundheitsfilme in Deutschland und der historische Filmbestand des Deutschen Hygiene-Museums Dresden." In *Kamera! Licht! Aktion! Filme über Körper und Gesundheit 1915 bis 1990*, ed. Susanne Rößiger, 12–49. Dresden: Sandstein, 2011.

Schwarze, Günter. *Technik der Arbeit mit audiovisuellen Unterrichtsmitteln*, 3rd ed. Berlin: Volk und Wissen, 1963.

Schweinitz, Jürgen. "Betrachtungen vor dem AICS/ISFA-Kongress." *Bild und Ton* 35, no. 7 (1982): 205–6.

———. *Aufgabenstellung, historische Entwicklung und Arbeitsmethoden der medizinischen Kinematographie*. Babelsberg, GDR: Deutsche Hochschule für Filmkunst, 1967. Diploma thesis.

Segal, Jérôme. *Kybernetik in der DDR: Begegnung mit der marxistischen Ideologie*. Dresden: Technische Universität, 2001. Ph.D. dissertation.

Steuding, Karl. "Untersuchung zur Bestimmung der pädagogisch-methodischen Wirksamkeit von Unterrichtsfilmen im Unterrichtsprozeß des Faches Chemie der Zehnklassigen Allgemeinbildenden Polytechnischen Oberschule." Erfurt-Mühlhausen, GDR: Pädagogische Hochschule "Dr. Theodor Neubauer," 1972. Ph.D. dissertation.

Streible, Dan, Martina Roepke, and Anke Mebole. "Introduction: Nontheatrical Film," *Film History* 19, no. 4 (2007): 339–43.

Todes, Daniel Philip. *Pavlov's Physiology Factory*. Baltimore MD: Johns Hopkins University Press, 2002.

Topp, Ewald. *Zur Funktion, Nutzung und Weiterentwicklung der technischen Grundausstattung der Oberschulen der DDR*. Berlin: Volk und Wissen, 1973.

Vöhringer, Margarete, and Michael Hagner. "Vsevolod Pudovkins Mechanik des Gehirns – Film als psychophysiologisches Experiment." *Bildwelten des Wissens* 2, no. 1: Bildtechniken des Ausnahmezustands (2004): 82–92.

von Engelhardt, Kerrin. "'Wissenschaftlichkeit' ohne 'Parteilichkeit'? Der naturwissenschaftliche Unterrichtsfilm der DDR." *Zeitschrift für Pädagogik* 69, Supplement (2023): 40–55.

——. "Complex Associations: On the Emotional Impact of Educational Film in the German Democratic Republic (1950–1990)." In *Teaching with Light Projection—Media Performance Histories*, ed. Nelleke Teughels and Kaat Wils, 245–267. Turnhout, Belgium: Brepols, 2023.

——. "Educational Film in East Germany (GDR) in 1950–1990, in Perspective of the 'Grammar of Schooling'." In *Research in Film and History*, Special issue: Educational Film—Histories and Historiographies of Practice, ed. Katrin Pilz and Joachim Schätz, July 2023: Retrieved 11 August 2004 from https://film-history.org/issues/text/educational-film-east-germany.

von Engelhardt, Kerrin, and Ulrich Ruedel. "Collecting and Using Audiovisual Educational Aids from East Germany." In *Collecting Educational Media: Making, Storing and Accessing Knowledge*, edited by Peter Carrier and Anke Hertling, 111–25. New York/ Oxford: Berghahn Books, 2022.

Wähler, Josefine, and Maria-Annabel Hanke. "'Pacemakers Report': GDR Pedagogical Innovators and the Collection of Pädagogische Lesungen, 1952–1989." *Paedagogica Historica* 58(1) (2020): 66–83.

Waller, Gregory A. *Beyond the Movie Theater: Sites, Sponsors, Uses, Audiences*. Berkeley: University of California Press, 2023.

Weiss, Sheila F. "Pedagogy, Professionalism and Politics: Biology Instruction during the Third Reich." In *Science, Technology and National Socialism*, edited by Makus Renneberg and Mark Walker, 184–96. Cambridge: Cambridge University Press, 1994.

Wellmann, Janina, ed. "Cinematography, Seriality, and the Sciences," Special Issue. *Science in Context* 24(3) (2011): 311–464.

Wilkening, Albert. "Filmgestaltung und Informationsverarbeitung." *Bild und Ton* 16(12) (1963): 353–54.

——. "Angewandte Kinematographie in Wissenschaft und Technik." *Bild und Ton* 11(11) (1958): 281.

——. "Der Film im Dienste der Wissenschaft." *Bild und Ton* 10(1) (1957): 1.

——. "Pawlows bedingte Reflexe und der Film," *Bild und Ton* 5(3) (1952): 160.

——. "Ist die Filmkunst eine Wissenschaft?" *Bild und Ton* 3(12) (1950): 361.

——. "Unterrichtsfilme in die Schule!" *Bild und Ton* 2(12) (1949): 353.

Wilkening, Albert, Heinz Baumert, and Klaus Lippert, eds. *Kleine Enzyklopädie Film*. Leipzig: VEB Bibliographisches Institut, 1966.

Zabel, Nicole. "Zur Geschichte des Deutschen Pädagogischen Zentralinstituts der DDR." PhD diss., Technische Universität Chemnitz, 2009.

——. "Die Lehrmaschine und der Programmierte Unterricht: Chancen und Grenzen im Bildungswesen der DDR in den 1960er und 1970er Jahren." *Jahrbuch für Historische Bildungsforschung* 20 (2014): 123–52.

Zentralinstitut für Film und Bild in Unterricht, Erziehung und Wissenschaft, ed. *Lehrmaterial für die Durchführung der "Fachseminare für die Einweisung aller Lehrkräfte, in die Methoden der Verwendung von Unterrichtsfilmen, Lichtbildern und Magnettonbändern."* Schriftenreihe Film und Bild 3. Berlin: Volk und Wissen, 1953.

Zimmermann, Yvonne. "Analyse nicht-fiktionaler Filmformen." In *Handbuch Filmanalyse*, edited by Malte Hagener and Volker Pantenburg, 443–56. Wiesbaden: Springer VS, 2018.

CHAPTER 10

German Technology and Education for "Young Nations"

Cold War Politics and Aesthetics of Development Aid in Two West German Government PR Films of 1961

Jan Uelzmann

Introduction

This chapter analyzes the image of development aid in two West German public relations (PR) films produced by the Deutsche Wochenschau newsreel company under the direction of the Federal Press Office: *Zu Gast in Bonn: Ein Bericht aus dem politischen Leben der Bundeshauptstadt* (Visiting Bonn: A report on political life in the federal capital, 1961, dir. A. W. Uhlig); and *Koblenzer Straße 99–103: Der Alltag des Auswärtigen Amts* (Koblenzer street 99–103: Day-to-day at the Foreign Office, dir. Heinrich v. Tiedemann), released in 1961 and revised in 1963.[1] It explores how the films' progressive rhetoric enlists recent economic, technological, and educational advances made by the Federal Republic of Germany (FRG). In a filmic dialogue across the Iron Curtain, the films favorably compare the FRG to the socialist states with which it competed for influence in decolonized countries. While the films adopt an enlightened and technocratic tone in touting the FRG's advances in these areas, their message is ultimately undermined by the persistence of racist and colonial modes of thinking.

During the early 1960s, in an age preceding the rise of television and determined by escalation in Cold War tensions, films like these played a crucial role in the PR arsenal of Konrad Adenauer, West Germany's first federal chancellor (1949–63). Adopting the style and production methods of the newsreel industry but focusing on a single topic, the PR films consisted of documentary-style narratives of twenty to forty-five minutes that centered on political events, such as state visits or celebrations in Bonn, the FRG's provisional capital. These films promoted the Adenauer administration's political successes during election season and, in more general terms, documented the country's status as a thriving democracy and integral partner in Western economic and military alliances.

Meticulously planned by the Federal Press Office, the PR films drew upon the production capabilities and creative experience of the Deutsche Wochenschau newsreel company. Clandestinely founded in Hamburg in late 1949, with the Adenauer administration's direct involvement,[2] the company was a de facto government mouthpiece.[3] Relying heavily on the formal language, voice-over narration, and style of the *Neue Deutsche Wochenschau* newsreels, the PR films celebrated Adenauer and his politics by offering consistently favorable coverage of real-world events and political decisions.

The two films analyzed in this chapter were produced in 1961, a significant point in the FRG's development. Adenauer, who had led the nascent postwar state through a series of political milestones toward self-administration and integration in Western economic and defense networks, began his fourth and final term as federal chancellor that year. Under his leadership, the FRG had adopted a social market economy and a social welfare structure upon its founding in 1949, laying the groundwork for the country's widespread postwar socioeconomic recovery, culminating with an economic boom that became known as the *Wirtschaftswunder* (economic miracle) in the second half of the 1950s. By the early 1960s, as an affluent member of the Western Cold War alliance, the FRG was engaging in economic and development aid to the newly emerging postcolonial nations.

Both films discussed here use the FRG's status as a globally engaged, Western nation to illustrate the successes of Adenauer's policies to their domestic and foreign audiences. After several failed attempts to influence the FRG's independent broadcasting system more directly,[4] by the early 1960s the administration was increasingly using PR films to document and promote the "success story" of Adenauer's leadership.[5] Heavily centered on West German education, technology, economic strength, and expertise, the films examined in this chapter establish these traits as forming the "core brand" of the FRG, which, the films imply, thereby set it apart from the German Democratic Republic (GDR) and other Eastern Bloc opponents.

The following analysis of the narrative logic and visual tropes used in *Zu Gast in Bonn* and *Koblenzer Straße* shows that these films emblematize PR strategies of the period, by using West German exports to recently decolonized African and Asian countries as a yardstick for the successes of Adenauer's policies, portraying these exports as the fruits of West Germany's political and economic postwar ascendancy. This endeavor involves contextualizing these films as part of an ongoing Cold War filmic dialogue, in which the East and the West sought to highlight the superiority of the socialist and capitalist systems, respectively. In the process, this chapter illuminates the FRG's efforts to compete for geopolitical influence among decolonized Asian and African nations that were perceived to be under the threat of falling under the influence of the Soviet sphere.

A deeper examination of the aesthetics and ideology that underpin the films' portrayal of West German development aid and surrounding activities, however, shows that they belie underlying conceptions about postcolonial countries that run counter to the PR image of the FRG as a modern Western state that has left its roots in fascism and racism behind. In the films, for example, recently decolonized African and Asian countries are patronizingly referred to from a Eurocentric perspective as "young nations." One must also note that terms such as "developing country,"[6] "development aid," and "economic assistance" equally embody a hierarchical discourse on the world that prioritizes Western models of economic growth and development. Hence, these terms are used here with an awareness of their inherently problematic nature and for the lack of widely accepted, less problematic terms.

In fact, the PR films—particularly *Koblenzer Straße*—convey underlying assumptions about "young nations" and their inhabitants that are based on stereotypes, reductive opinions, and racialized perspectives that oscillate between the spectacle of the exotic and a worldview based on racial supremacy. Their perspective thus embodies what Fanon dubbed the colonialist "white gaze,"[7] as it denies recently decolonized countries a standing as equals due to the skin color of their citizens. Consequently, the films display no interest in lending the citizens or leaders of these "young nations" a voice or agency of their own. Instead, they remain voiceless "extras" in the films, reduced to projection screens for the Adenauer administration's PR ambitions. While this constellation is evident in both the films and the correspondence around their production, it ironically also shines through the lack of any archival records regarding reception in the countries being described.[8]

PR Films as Cold War Political Communication

That the two films were part of a Cold War dynamic—the competition between the GDR and the FRG for political influence in postcolonial nations—is central to the following analyses. The films deliberately sought to convey to their domestic West German, East German, and foreign audiences that the

FRG's socioeconomic development under Adenauer made the young state a role model for other nations, aligning it with the West as the self-proclaimed winners of the ongoing Cold War competition. The year in which the films were produced, 1961, was highly meaningful in Cold War history, as it saw a starkly increased East-West confrontation resulting in both the building of the Berlin Wall on 13 August and the Soviet-American standoff at Checkpoint Charlie on 27 October. The political climate deteriorated further with the Cuban missile crisis, from 16 to 28 October 1962. Apart from their commentary on the geopolitical constellation, the two PR films under discussion focus on the West's ostensible leadership in technology, economics, education, and science—to wit the space race between the United States and the Soviet Union—and the standard of living, as illustrated by the famous "kitchen debate" of 1959.[9] *Koblenzer Straße* also deploys a (neocolonialist) discourse of medicine and hygiene. By advertising the export of technology, medicine, and an education "Made in West Germany," the FRG laid claim to a leadership role in the competition between systems in recently decolonized parts of the world.

In contrast to the registers of 1950s anti-communism or the rabidly antisemitic propaganda of the Third Reich, West German PR films thus derived their persuasive momentum from the decidedly softer language of political PR.[10] As creations of the FRG's first post-fascist government, the PR films did not seek to vilify "enemies of the people" like the Nazis had done. Instead, their integrative narratives were rooted in socioeconomic fact and sought to unify all Germans (with East Germans often explicitly included) and rally them around Adenauer's vision of Western integration as the path to freedom, security, and prosperity. More restrained than the election rhetoric of Adenauer's CDU party, instead of openly critiquing the GDR, Soviet Union, or socialism, they focused on the FRG's burgeoning economic prosperity and the benefits of its social welfare state; but there was a clear expectation that the situation in the GDR was common knowledge and subtle allusions to it would suffice.

West German PR films indeed became an integral part of the propaganda dialogue across the Iron Curtain, and they were met by similar messaging in East German film productions. As Matthias Steinle points out, "In the first decade of the German division, the FRG and the GDR developed a media-based rhetoric in competing documentary film practices that followed the stereotyped patterns formed during the Cold War."[11] The GDR newsreel *Der Augenzeuge* (Eyewitness), produced at the state-run DEFA studios from 1946 to 1980, is in many ways a counterpart to the West German *Neue Deutsche Wochenschau* newsreel produced by the Deutsche Wochenschau, presenting a tightly controlled, state-sanctioned, and celebratory master narrative of the East German state. As in the West German newsreels, images of recently decolonized nations featured prominently in the GDR's filmic interventions into the Cold War's geopolitical competition. Segments like the 1956 *Blickpunkt Afrika: Mit der DDR-Delegation im Sudan* (Spotlight Africa: With the GDR's delegation in Sudan) were common in *Der Augenzeuge*.[12] By the end of the 1950s, the GDR had also initiated the production of longer films intended for foreign (as well as domestic) audiences, the so-called *Auslandsinformationsfilme* (foreign information films). By then, the GDR had become aware of a "deficit in propaganda directed abroad, including film propaganda," and formed a "working group specialized in propaganda directed abroad . . . at the DEFA studio for Newsreel and Documentary Films."[13] The GDR Liga für Völkerfreundschaft (League for Friendship among Peoples) was the primary buyer of such films and would distribute them through partner organizations abroad. The GDR's foreign information films can thus be compared to the PR films released by the Adenauer administration, with similarities extending to their focus on the successes of the GDR as the new, post- and anti-fascist German state, as well as their production techniques and distribution through predominantly noncommercial channels.

West German PR films were regularly distributed to schools and universities through film outlets at the state level, the *Landesfilmdienste* (state film services). When screened domestically, they were shown in *Aktualitätenkinos* (newsreel theaters), before full-length films in movie theaters, during political campaigns, and in noncommercial educational screenings. They were also often released in translation (with the commentary recorded in foreign languages) and screened at FRG embassies and cultural institutions abroad, however, and there is ample archival evidence that the Federal Press Office spent a considerable part of their production budgets on foreign-language versions. *Koblenzer Straße*, for example, was also released in British English, American English, and French.[14] In this regard, the PR films functioned as important communications to the FRG's international allies and its Cold War opponents.

Postcolonial Countries Come to Bonn

With a twenty-three-minute runtime, the black-and-white film *Zu Gast in Bonn* was completed in April 1961; by 1962, four hundred copies had been made for release to domestic, noncommercial audiences.[15] Interestingly and contrary to expectations, given the film's focus on internationalism and development aid, no foreign-language versions of this film are known to exist.[16] The production correspondence suggests that *Zu Gast in Bonn* was mainly intended for a West German audience, as its declared goal was to introduce West Germans to their new provisional capital city, Bonn. The film's synopsis at the production stage explicitly mentions that, in 1961, most West Germans had never been to Bonn personally and had no clear visual concept of the new capital.[17] *Zu Gast in Bonn* also conveyed a powerful message of success, however, regarding the FRG's Western model of parliamentary democracy and social market economy. As such, the film was designed to speak to both West and East Germans. Prior to the building of the Berlin Wall on 13 August 1961, East Germans could see Western films at Berlin's border cinemas, and West Berlin reception centers also frequently screened PR films for East German refugees.[18]

The film portrays Bonn as an intricately connected, modern capital that was oriented westward. Located on the Rhine's western bank, the city's unique mélange of quaint provincialism and technocratic modernism serves as a visual anchor for a peaceful, post-fascist version of Germany. The FRG's new provisional capital, the film suggests, serves as an international center for economic and military matters in the Western Cold War alliance. The film was a project between the Federal Press Office and scriptwriter A. W. Uhlig, who also directed it. The Federal Press Office considered the film an adequate vehicle for communicating two ideas to West Germans: the success of Adenauer's integration of the FRG into the Western Cold War alliance, and Bonn's reputation as a hardworking, austere capital that invested federal funds sparingly.[19] The latter message was necessary because the new capital was frequently criticized as a waste of money; especially before the construction of the Berlin Wall lent the division of Germany an air of indeterminate permanence, many voices in the West German media maintained that a future return to Berlin as the pan-German capital should negate any further spending on Bonn.[20]

One of the film's main strategies for legitimizing the new capital is to underline its function as an international hub of knowledge, education, and economy attracting students and diplomatic delegations from developing nations. The theme of development aid further aligns the FRG with other prosperous Western countries, such as the United States and France, thus asserting that the country is on a par with other, affluent Western democracies. The theme of prosperity reflects the popular saying of the economic miracle *wir sind wieder wer* (we're someone again). The film visually communicates Bonn's internationalism by offering the audience an exoticized visual spectacle that exploits cultural contrasts by juxtaposing shots of People of Color that emphasize their "otherness," with shots of German industrial sites and the quintessentially "German" Rhenish landscape.

The initial sequence, set at the University of Bonn, presents the city's stature as an international center of education by displaying (but not interviewing) students of color in lecture halls and corridors. This sequence visually reinforces the film's voice-over, which intones that, of the university's ten thousand students, "more than one thousand are foreigners from all parts of the world."[21] The sequence thus underlines Bonn's internationalism with a simple and passing reference to racial diversity, suggesting that—after World

War II and the Third Reich—an education in (West) Germany is once more in high demand internationally and particularly desirable among what the commentary generically refers to as "developing nations." The fact that the international students do not have a voice in the sequence illustrates how their objectified presence in the film is meant merely as an illustration; they are reduced to eager disciples of the FRG, the well-known and -established *Kulturnation* (culture nation). Like the diplomatic and economic delegations that also appear without voices in the film's later scenes, the silent presence of the students is exploited to fortify a sense of international and educational prowess that will make the new capital city attractive to its audience.

This strategy is confirmed in a sequence that presents images from the state visit of India's president Jawaharlal Nehru in 1956. In this sequence, a contrast is established between the delegation's display of racial diversity and Bonn's quaint, provincial architecture and landscape, thus framing Nehru's visit as an exotic, diplomatic spectacle within a familiar and knowable context. Nehru, his staff, and his family are portrayed with the German delegation during a scenic boat ride on the Rhine. The sequence juxtaposes shots of Indian women wearing prominently displayed saris, marking them as foreign, with quintessentially "German" shots of the Rhine.[22] With this juxtaposition, *Zu Gast in Bonn* invites the overwhelmingly white West German audience to dwell voyeuristically on the contrast of cultures it visually describes.

The spectacle of foreign visitors through such racialized visual contrasts and comparisons is repeated in the film's concluding sequence, which shows a reception during the visit of Thailand's royal couple in Bonn's Hotel Petersberg in 1960. German culture, represented here most prominently by a shot of the Siebengebirge mountain range through a large window,[23] once again encounters the Thai couple as an exotic other. As with the scene of President Nehru's visit, this scene focuses on objects, not people: clothing, fabrics, fashion. In a shot of Sirikit Kitiyakara, the Queen of Thailand, welcoming guests, the camera slowly pans twice over her elaborate dress, coming to rest on a close-up of her exquisite bracelet.[24] The filmmakers' fascination with intimate, feminine, exotic detail contrasts strongly with the medium long shots used to capture members of the male German delegation in tuxedos. Diplomatic attire references masculine, Christian, and Western cultural values; Adenauer, for example, is wearing his Order of the Golden Spur medal, awarded to him in 1955 by Pope Pius XII and underlining his status as a practicing Catholic.[25]

The film's exoticist narrative becomes more explicitly linked to the topic of economic and development aid once the commentary describes Bonn's proximity to the Ruhr region, the FRG's industrial heartland. Here, images and voice-over commentaries extoll the FRG's remarkable economic ascent to world leader status in industrial exports. One comment—"the young nations in Asia and Africa, who seek political friendship with the Federal Republic, believe in the possibilities that aid by the West German economy could yield"—casts the FRG as a model for recently independent countries.[26] It also reaffirms a Eurocentric hierarchy that positions a technologically advanced West, dominated by the United States and Western Europe, above of an amorphous mass of "African" and "Asian" countries, addressed only in the vaguest of terms and portrayed as languishing and in need of Western aid.

The film then presents leaders from Pakistan, Malaysia, and other postcolonial nations as they visit West German industrial sites and production facilities, indicating the interest of developing countries in the knowledge and technical capabilities behind the export label "Made in West Germany." A central scene in this context depicts Pakistan's prime minister Ayub Khan visiting an engine factory. The commentary explains that Khan "meets young interns from his homeland who are training in the Federal Republic."[27] In the scene, the prime minister stops at a bench to watch a young Pakistani man work on an engine under the supervision of a German worker, who finally takes the tool to demonstrate the work step. A medium shot of Khan shaking hands with the Pakistani intern and a brief exchange between them, with the intern smiling and nodding, suggests that this transfer of German technological expertise is highly valued by Pakistan.

In another sequence, a delegation described as "ministers and politicians from several African nations" visit a rolling mill and steel plant.[28] All differences between the African countries and their representatives are flattened by this description. The sequence opens with images of German teams operating a blast furnace in unison, with accentuated classical music accompanying their stoking of the fire. This brief interlude in the film's narrative, crosscut with a medium shot of the African delegation wearing safety goggles and helmets observing the scene, visually celebrates a critical element of West German industrial production and a foundation of the economic miracle. The sequence about the Ruhr region closes with images of negotiations between a delegation of People of Color and German officials in Bonn. The commentary explains: "After returning to Bonn, talks continue with German politicians and economists in a relaxed atmosphere. For years, the Federal Republic has paid a great deal of attention to economic aid to developing nations."[29]

Such scenes from *Zu Gast in Bonn* are clearly meant to underline the FRG's economic ascendancy as an emerging "global player." Pakistan was certainly of particular interest within Cold War dynamics, as it had received heavy investment and aid from the Soviet Union in the 1950s and 1960s. Similarly, several countries in Africa—such as Guinea, which will figure prominently in the next film discussed—became battlegrounds in the competition for influence between the West and the Eastern Bloc. For domestic audiences in the FRG, the film's narrative about development aid validated both Adenauer's economic and political policies and the emerging European Economic

Community. The narrative also reinforces old racist and colonial stereotypes, however, as it elevates West Germany from being a loser of World War II to being a superior and benevolent white nation in relation to economically less-developed countries inhabited by People of Color.

By means of such strategies, the film thus relies on a narrative of exoticism, racialization, and essentialism and a filmic staging of visitors to Bonn as visual spectacle. These establish that people from developing countries are "voiceless," "foreign," and "other" in terms of their racial and economic status. Much like the students of color at the University of Bonn, *Zu Gast in Bonn* portrays postcolonial nations as eager students of the FRG's socioeconomic prowess. The discussion of the next film will further concretize these findings.

Bonn Goes Out to Developing Nations

The forty-five-minute black-and-white film *Koblenzer Straße 99–103* follows the narrative premise of an average workday at the West German Foreign Office, whose street address in Bonn is cited in the title. Deutsche Wochenschau produced the film in 1961, with Heinrich von Tiedemann as director, and reworked it in 1963 to reflect changes in Foreign Office leadership and include recent political events in the film's opening sequence.[30] In addition to German, it was initially planned for release in French, English, Spanish, Arabic, and Portuguese, with a total budget of 65,000 Deutsche Marks for foreign-language versions; budget restrictions ultimately limited the foreign versions to English and French, however.[31] The Foreign Office and the Federal Press Office commissioned the film. The film narrates the Foreign Office's diplomatic daily routine, which includes everything from mundane events, such as employees parking in the morning, to actual diplomatic work. These reports form the narrative framework, which is interspersed with several inserts about diplomatic activities worldwide.

The film celebrates the FRG's increasing affluence by showcasing two types of exports: cultural and educational programming at German schools and Goethe Institutes abroad; and development aid, including technology, and medical and administrative expertise. To this end, the film advertises the FRG's successful model of education and technology, enabled by its social market economy. Intense international interest in these export goods, the film argues, serves as a benchmark for the success of the Adenauer administration's policies. In *Koblenzer Straße*'s celebratory narrative, the FRG's first postwar chancellor has been responsible for the country's progress, and its unique constellation of democracy, Western alignment, and market economy is respected worldwide. The strategy of highlighting this successful development is particularly significant when set against earlier films, such as *Ein*

Mann wirbt für sein Volk (A man advocates for his people, 1953), which had attempted to convince the United States that West Germany was worthy of being rehabilitated after World War II.[32]

Koblenzer Straße's longest story (thirteen minutes) focuses on development aid sent to Guinea. After the nascent West African republic gained independence from France in 1958, it had become socialist while still engaging in trade with the West. Against a backdrop of an FRG/GDR Cold War competition for political and economic influence in the emerging nation, *Koblenzer Straße* highlights the impact of West German aid in the form of cultural policy and of engineering, medicine, and hygiene. This segment asserts that the FRG is continuing Germany's long tradition of science and technology along with its humanistic ideal of *Bildung* (education, self-cultivation), which warrant admiration and should be exported to the "young nation." In this respect, *Koblenzer Straße*'s perspective on development aid neatly overlaps with *Zu Gast in Bonn*'s. As we shall see, however, by entering into the topics of health and education, the film strays deeper into problematic territory.

To assess the image of education portrayed in this key segment, it is helpful to examine how the discourse of *Bildung* is represented in a prior segment of the film. The topic first arises in a report on FRG educational programs in other Western European countries. Over images of the inauguration of a new German school in Milan, Italy, the voice-over emphasizes that the school is one of "about four hundred German schools abroad." Now showing images of schoolchildren receiving *Schultüten*—large cones of candy and gifts traditionally given to German children as they enter school for the first time—the voice-over elaborates: "In some areas, the ratio of German children is only 10 percent."[33] This scene suggests that non-German children receive the much-desired "gift" of German education and that other Western European nations value the German language and German culture highly. It is difficult to overstate to what extent, in 1961, such a depiction proposed a complete rewriting of the problematic notion of German culture after the Holocaust and World War II. With this and other scenes—featuring the Goethe Institute in Cairo and fellowship recipients at Villa Massimo in Rome[34]—the film manufactures an image of a *Kulturnation* self-assuredly showcasing a whitewashed version of its cultural heritage, alongside its industrial output, as a desirable export commodity. A mere sixteen years after the end of German fascism left a "rupture of civilization" in Europe,[35] the film carefully expunges all overt associations with Nazism. Instead, West Germany is shown in an international context of friendly, peaceful cooperation and economic development.

Contrary to this careful messaging, the following analysis will reveal that *Koblenzer Straße* still reflects a racist worldview based on the idea of European white supremacy. Its underlying racism directly contradicts its ostensibly enlightened tone and points back to the Nazi past and colonial times. This discrepancy, which extends from the level of the film's production to its visual grammar and commentary, ultimately undermines the filmmakers' ambition to render West Germany as a forward-oriented nation that has left its problematic roots behind.

In addition to the FRG's high level of economic development, standard of living, and status within the Western Cold War alliance, *Koblenzer Straße*'s problematic representation of German *Bildung* helps shape the sequence about Guinea discussed above. According to the film's logic, the FRG's success in these areas is responsible for the sustained interest of what it neutrally describes, at first glance, as "young nations." To West German audiences in 1961, this would have been a powerful validation of Konrad Adenauer's trademark policies. Sixteen years after the devastation that the end of World War II represented for Germans, *Koblenzer Straße* argues that the FRG is now in a strong position to bring the commodities of its unprecedented socioeconomic recovery to Guinea, a nation in need of economic and even—revealing the film's underlying racism—intellectual aid.

Apart from extolling the FRG's "developmental" success, compared to what it describes as an "underdeveloped" Guinea, *Koblenzer Straße* also references the diplomatic proxy warfare between the FRG and the GDR. Early on, the film comments on the competition of capitalist and socialist systems in Guinea by contrasting the US culture center in Conakry, which is "open all day," to its Soviet Union counterpart, which remains "closed due to biased propaganda."[36] After highlighting the "strong staffing" of diplomatic missions from the Soviet Union and China—and thereby underlining the threat posed to Guinea and West Germany by communism—the film presents images of the GDR's trade mission with this description: "massive propaganda against the FRG in the windows of the trade mission of the Soviet zone."[37] Referring to the GDR as the Soviet Zone, its name under Soviet occupation (1945–49), conveys the FRG's refusal to accept the GDR as a sovereign state, as expressed in the Hallstein Doctrine. The camera then presents images in the trade mission's windows—which depict Adolf Hitler and Wehrmacht generals, Stuka bombers, and Nazi officers executing prisoners—while the voice-over asserts that the FRG is "being depicted as warmongering and under the spell of Nazis."[38]

This sequence frames later segments about West German development aid as part of an ongoing Cold War struggle for dominance in West Africa. The film consistently suggests, however, that the FRG maintains the upper hand in its competition with the GDR due to its democratic system and superior economic and technological output. It was part of the film's overall design to use the FRG's development aid record as an indicator of western superiority. The correspondence during the film's production reveals that, during planning in 1960, the Foreign Office briefed director von Tiedemann about "atrocity propaganda against the Federal Republic" perpetrated by the GDR mission in Conakry and advised that von Tiedemann shoot in Guinea.[39] Correspondence between the Deutsche Wochenschau and the Press Office clearly states that the West German Foreign Office "would like this juxtaposition" between the FRG and the GDR "to be underlined" in the film.[40]

If the clash between East German communism and West German capitalism provides one framework for the Guinea segment, a close examination of the voice-over, editing choices, and production notes reveals another dichotomy—namely, between the film's underlying racism and its ostensibly enlightened, democratic, and humanitarian stance. The "young nations" terminology employed in the film inherently disregards Guinea's precolonial history and renders the country naïve, childlike, and in need of substantial guidance. Indeed, the discourse of guidance is the organizing principle of the entire story. The Guinea sequence opens with an exotic spectacle. Relying on a well-established colonial stereotype, a scene depicting young Guinean men and women in traditional costumes dancing to the sound of chanting and drumming, gives the audience exotic and essentializing images that establish an ostensible traditionalism and primitivism. Accompanying these colonialist tropes, the film's voice-over introduces the topic of decolonization by evoking the white supremacist discourse for colonialism itself, intoning: "In former times, the great white father wanted to lead his black children towards the sun. Now Africa is leading itself."[41] This commentary, while not openly evoking or disqualifying the "African" (Guinean) struggle for independence, revives the racist imagery of a "great white father" and his past attempt to "lead his Black children" to a better life. Of course, the concept of the "great white father" was used by colonial forces in India and Africa as well as by those who subjugated Native American tribes in North America.[42]

The uncritical, unambiguous use of this colonialist trope reveals the clash between the film's PR objective to showcase the FRG as humanitarian and post-racist, and its underlying racism. The white supremacist stance conspicuously refuses to regard Guineans as equals.[43] The voice-over's colonialist perspective is furthered by images of monuments, nostalgically introduced as vestiges of colonial times, "discarded like the stage set for a drama that has been taken off the theater program for good."[44] A close-up of a discarded memorial stone carrying the Latin words *lex* (law) and *justitia* (justice) suggests that Guineans abandoned these values, along with the monuments that symbolized them, in an irrational move ostensibly rejecting the values of the European Enlightenment.

The film's production notes make the white supremacist stance even more apparent. For example, in his correspondence with Deutsche Wochenschau, von Tiedemann alludes to the colonialist image of "uncivilized" local luggage haulers: "While in transit, during the loading in Bamako, the cheerful n— —

threw around our luggage so much that all the locks came off my private suitcases."[45] Concerning *Koblenzer Straße*'s preoccupation with the postcolonial state of affairs in Guinea, von Tiedemann also ridicules the country's capacity for self-administration, by hinting at widespread incompetency and corruption:

> The bank of Guinea is full of young boys with friendly smiles. They go about their work in a highly cheerful manner; from time to time, some money is missing from the treasury. Recently it was seven million. They had booked it wrong. *C'est la vie*! The polite black youngster who exchanges your travelers' checks at eleven may be already led away in handcuffs by twelve because they found him to have committed embezzlement. Who would hold such things against a political system that is so young?[46]

Von Tiedemann's dismissive irony and bias vis-à-vis independent Guinea, as well as his assumed cultural dichotomy of cultivated "whiteness" and uncultivated "Blackness," reveal as a façade the film's ostensibly neutral stance toward Guinean independence and supposedly enlightened and supportive tone as regards offering the country development aid.

The tension between the intended progressiveness of the film's rhetoric and the reality of its underlying racism continues in a short scene that depicts a high-level diplomatic visit to Guinea, which is received at Conakry Airport with military honors. The voice-over emphasizes that the military ceremony is a relic from colonial times: "Assimilated European ceremonial and innate African dignity unite."[47] The assumption is that the country has adopted colonial rituals emblematic of European civilization, like military honors. The commentary interweaves the concept of "European civilization" in counterpoint to the racist-essentialist idea of "natural African dignity," highlighting the "otherness" of Guineans, who are cast as eager students, or, less favorably put, witless copyists of what the film classifies as European standards of civilization. The ceremony thus assumes the character of a learned performance of "culture" adopted from the "great white father."

Another sequence, set at the FRG Embassy in the capital city of Conakry, emphasizes the value of an education in German culture and language. Young men are shown reading information boards that advertise university education and apprenticeship opportunities in West Germany. The commentary—that "German courses for interns augment the efforts for intellectual development aid"[48]—insinuates that the Guineans are intellectually inferior, and that the German language can enlighten them. A scene showing Guineans attentively reciting German words and phrases during embassy German lessons presents them as eager pupils of the teacher, seemingly a German national.[49] With its somber atmosphere, the seating arrangement, and a close-up of a textbook entitled *Deutschland*, the scene evokes a church service, as the "communion" of the German language is consecrated and shared with the students.

Following the racist logic of *Koblenzer Straße*, the commentary describes the role of the FRG ambassador to Guinea in terms of the "old Africa pioneers" of colonial times. Placed over a transitional scene depicting him prominently smoking a pipe (a symbol of Western civilization) while traversing the countryside in his Land Rover, the commentary frames the ambassador as a paternalistic teacher, a "great white father" carrying the light of civilization to what it indiscriminately describes as an uncivilized, Black continent. Over images of unpaved dirt roads lined with huts, the voice-over explains that the ambassador is traveling to "remote parts of the country that are waiting to be developed economically." The assertion that, "like him, the emissaries and experts of other rival nations also travel through the bush and the jungle" evokes the Cold War race for economic and political influence in decolonized Africa while strengthening the film's unironic observation that this race for political power represents another kind of colonialism.[50]

To a score featuring melancholic music on a West African stringed instrument, a river-crossing sequence continues the film's discourse on Guinea's developmental needs. It describes a traditional suspension bridge between two trees and a raft-like ferry, fashioned out of barrels and wood planks, as "romantic bridging and obsolete technology." Over images of an embassy Volkswagen van carefully navigating onto the raft, the commentary promises that "in a few months, a brand-new ferry from the FRG will connect here and there. If the German experts can get here quickly enough. Time is of the essence—everywhere in Africa."[51] While manually operated winches slowly haul the Volkswagen across the river, the audience is left to contemplate this ominous statement about the competition with rival nations. Here, the celebratory narrative of West German "civilization" conquering African backwardness through technology, expertise, and education becomes embedded in the view of the Cold War expressed by the "domino theory," making the FRG an integral part of the Western fight against the rapid spread of communism.

The scenes that follow further develop the film's discourse on progress and political competition by focusing on West German agricultural science, hygiene, and medicine. These were disciplines in need of a thorough realignment after the atrocities committed by Nazi medical doctors and researchers. The ambassador's first stop is an experimental plantation that produces cinchona bark, the raw material for quinine. Quinine, used to combat malaria, was central to a colonialist hygiene discourse, in which the white man comes to the aid of the inherently disease-ridden Africans.[52] The plantation itself is, of course, another colonialist trope. At the site, the ambassador encounters the "sole white man in the tropical mountain forest: a German agriculture expert." The voice-over presents the "young African district commander in the white suit, who took over the position of the former colonial official," as an "eager listener."[53] This comment establishes a clear hierarchy between the subservient Guinean official, in need of education and eager for initiation, and his German advisers.[54]

Images of the three officials touring the plantation, all clad in Western attire, are then contrasted with a scene that shows women in traditional garb using mallets to separate the cinchona bark from the tree branches. Strikingly, this scene replaces any diegetic sound with a nondiegetic recording of women chanting and drumming. The beat of this traditional music, entirely foreign to the scene, is likened to the women's striking of the tree branches. A scene depicting a manufacturing process is thus equated with drumming and chanting, a quintessential colonialist trope suggesting foreignness and primitivism that reinforces the clichés of dancing and singing at the beginning of the Guinea sequence. It also implies that Africans don't know the difference between work and play, and thus lack a work ethic that is necessary to succeed in the modern world.

Explicit references to tropical medicine abound in a scene that features a mobile X-ray truck donated by the FRG. The sequence shows images of locals waiting to be admitted for exams. The audience is shown a white, presumably German doctor handling X-ray documents marked in German, and his Black assistants. Images of the physician X-raying a Black patient transition into a close-up of a plaque in the truck that reads: "Presented to the Republic of Guinea in 1959 on the occasion of its declaration of independence." The statistic that more than 20 percent of the population are "severely affected by tuberculosis," placed over images of the white doctor examining a Black patient, introduces the declaration that "Africa needs young German doctors."[55] The possibility of Guinean doctors is not considered in the film, such that the audience comes away with the impression that the FRG is bringing its medical technology to bear on medical problems that Guinea is incapable of handling on its own; only West German doctors can step in to fight tuberculosis and "lead the black children to the sun."

The final sequence in the Guinea story reports on a Deutsche Wochenschau PR film about a Guinean state visit to Bonn being screened in a remote Guinean village, which sums up the colonialist imagery and reasoning of the preceding sequences. It also provides a visual climax and concluding allegory. The scene presents the "mobile Jungle Cinema," operated by the FRG Embassy and backed by Deutsche Wochenschau technical expertise. It directly contrasts West German high technology, embodied by the mobile film-screening truck, and the ostensible primitivism of the nameless Guinean village. The symbolism of the flickering cinema light literally illuminating the nocturnal darkness of the remote village is deeply allegorical and drives home the message that the FRG dispenses enlightenment, through such "intellectual development aid," to people in what the film portrays as an underdeveloped country, incapable of leading itself.

To introduce this sequence, the commentary mentions that the FRG ambassador will witness "one of these curiously impressive film showings in the African bush" (41:08). The subsequent elaboration that "it is often the first time that they see living images flickering on a screen" renders the Guinean audience as technologically uninitiated spectators who struggle to make sense of technological innovations like film. It echoes people's reactions to the earliest European film screenings of the late nineteenth century. The phrasing implies that the filmmakers have frequently witnessed reactions of shock and wonder in African audiences at such events.

The concluding sequence thus capitalizes on the juxtaposition between developed and undeveloped, between European and "African." Ultimately, the film allegorically uses images of villagers watching the film projected by the mobile screening truck in the village square at night to symbolize the cultural enlightenment of the "dark continent" by the West: "Here in the African bush, for a few hundred Africans, the world became wider than ever before."[56]

Conclusion

This chapter has focused on the ways that media and PR professionals working with and within the government of West Germany used FRG development aid to recently decolonized nations to celebrate the policy successes of the Adenauer administration, in particular for West German audiences, in governmental PR films. To do this, the filmmakers recycled well-worn filmic language and commentary, ranging from the spectacle of the exotic to racist and colonialist tropes of white supremacy. The resulting image of these "young nations" and their inhabitants is one determined by a lack of agency and voice. As eager recipients and, as portrayed, witless copyists of West German technology and culture, postcolonial countries and their people were reduced to a screen onto which the narrative of West German socioeconomic success under Adenauer, together with the assertion of a new German identity that had broken with its Nazi past and was an integral partner in the Western Cold War front, could be projected.

As an archive of West German imaginaries of both the FRG's rising status as a reformed postwar nation of technology and postcolonial countries and their people, the PR films are a cinematic treasure trove for Cold War discourses on progress as defined by prowess in science, technology, hygiene, and medical and economic export. The films' perspectives on development aid and the role of FRG science and technology exports are expressed using cinematic allegory. Ultimately, both films employ neocolonialist linguistic and visual tropes to portray the FRG as a vehicle bringing enlightenment to the "dark continent" of Africa. This "shedding of light," a classic colonial trope, is to take place in three sectors: engineering (to lay the foundation for Western standards of development and infrastructure); education (to provide "intellectual" information and training to what the films portray as inherently inferior societies); and medicine (to aid an ostensibly disease-ridden continent).

In the films, the FRG is framed as a respected member of the Western Cold War alliance and part of a Western vanguard in the race for influence in former colonies. Ironically, at the same time, the concepts and images enlisted to promote West Germany's role in the Western alliance were deeply problematic; they connected the films' visual language and voice-overs directly to both Nazi and racist white supremacist discourses of the past. By tapping into this problematic heritage, the FRG PR films demonstrate that, as late as the early 1960s, there had been no clean break with these concepts at either the Deutsche Wochenschau studios or the Foreign Office.

In particular, the Guinea segment in *Koblenzer Straße* significantly complicates our understanding of the FRG's fraught medial self-representation as a democratic nation that has transcended its roots in militarism, racism, and Nazism. Despite the film's carefully worded commentary, it becomes evident that neither the idea of German superiority over ostensibly primitive nations nor racialized discourses on health and hygiene had been discredited in 1960s West Germany. Clearly expressed in internal production notes, this perspective shines through at every turn of filmic language and cinematic symbolism and is used to convey the film's message of progress. With regard to the expectations of the production staff and government buyer of *Koblenzer Straße* in 1961, the film seems to have been perceived as effective, and its language adequate. It passed through close project supervision by the Film Division of the Federal Press Office and was reviewed in person by Foreign Minister Heinrich von Brentano; while Brentano requested a set of changes, none were concerned with the Guinea story.[57] Thus, apparently, the film received the blessing of the highest levels of government.

How are we to understand the continued use of racist and colonial perspectives in West Germany? Under the influence of the US Army in the late 1940s and 1950s, West Germans were urged to renounce antisemitism; but, as Heide Fehrenbach argues, they were simultaneously invited to rekindle their racism based on skin color. In her insightful study of the so-called "brown babies" of postwar West Germany, the binary between "Black" and "white" replaced Nazi notions of race, i.e., "Aryan" vs. "Jew"; crucially, this was reinforced by the racist and segregationist policies toward Black servicemembers of the US occupation forces.[58] A second explanation for the continued prevalence of racist thinking would be personnel continuities of pre-1945 career officials and former Nazis in West German governmental structures, famously including Hans Globke in the Federal Chancellery as well as others in the Foreign Office.[59] In fact, in a 1952 address to the Bundestag, Adenauer commented "that 66 percent of the diplomats in higher positions [were] former Nazis"; but, he added, he could "not build up a Foreign Office without relying on such skilled men."[60]

Not surprisingly, the continued presence of media professionals who had worked during the Nazi dictatorship extended to the Deutsche Wochenschau newsreel studios as well. A 1949 article in *Der Spiegel* magazine about the company's renewal as Neue Deutsche Wochenschau notes continuities from before 1945 in both personnel and the company name in a matter-of-fact tone. The article draws particular attention to the company's head cinematographer Erich Stoll,[61] who during the Third Reich had collaborated on the antisemitic propaganda film *Der ewige Jude* (*The Eternal Jew*, 1940) and had filmed the execution of the Wehrmacht officers involved in the 20 July plot to assassinate Hitler.[62] The neutral tone of the article captures the ubiquity and commonplace nature of such continuities in early West Germany. If we consider the proximity of the narrative premises of *Koblenzer Straße*'s Guinea story to Nazi cinematic discourses on exoticism, race, and hygiene—as seen, for example, in Douglas Sirk's *La Habanera* (1937)—we should not be surprised at how uncritically the FRG PR films accepted racist and white supremacist perspectives. Ultimately, both *Zu Gast in Bonn* and *Koblenzer Straße* emerge from these analyses as an important document of the problematic onset of the FRG's slow and ongoing process of emancipation from its racist and fascist roots.

Jan Uelzmann, associate professor of German at the University of Georgia, received his PhD in Germanic studies from the University of Texas at Austin. Merging approaches from cultural history, cultural studies, film studies, and literary studies, his book *Staging West German Democracy: Governmental PR Films and the Democratic Imaginary, 1953–1963* (Bloomsbury Academic, 2019) examines how "founding discourses" of the nascent FRG were reflected, reinforced, and actively manufactured by the government through PR films produced in conjunction with the West German newsreel company Deutsche Wochenschau. His research focuses on questions related to postwar democratization, governmental media policies of the Adenauer period, Cold War politics, and Weimar modernism. He has published widely on West German postwar cultural history, film, and literature, with articles that appeared in *German Studies Review*, *The German Quarterly*, *Monatshefte*, *Seminar*, *Colloquia Germanica*, and the *Journal of Cold War Studies*.

Notes

1. This chapter contains revised and extended sections from my book, *Staging West German Democracy*, and my article "Bonn, World City."
2. The company was founded under the name Neue Deutsche Wochenschau GmbH in 1949 and changed its name to Deutsche Wochenschau in December of 1955. Since the focus of this study is on the company's output in 1961, I will use Deutsche Wochenschau throughout. For more on the complexities of the company's founding, see Schwarz, *Wochenschau*, 94–133.
3. Newsreel historian Joachim Paschen ("Die Unabhängigste Wochenschau") uses the expression "governmental media outlet" (*publizistisches Regierungsorgan*). All translations mine.
4. One example would be the failed attempt to install a state-controlled television company— Deutschlandfernsehen GmbH—in an initiative that was ultimately shut down by the FRG's Federal Constitutional Court in 1961. For more details, see Von Hodenberg, *Konsens und Krise*, 150–51.
5. For more on the Adenauer administration's turn to PR films, see Uelzmann, *Staging West German Democracy*, 57–100.
6. While the term "developing country" seems to be falling out of use in the United States, it is still used by the German Federal Ministry for Economic Cooperation and Development: retrieved 5 September 2023 from https://www.bmz.de/en.
7. In his seminal work, the Afro-Caribbean psychiatrist and theorist Frantz Fanon develops this notion, which describes a white hegemonic perspective on the world, and which in turn essentializes, dominates, and subjugates Blackness (*Black Skin, White Masks*, 95–96).
8. The Deutsche Wochenschau production files for these films, which were reviewed for this study, did not contain any detailed records on either the circumstances of film screenings, or their reception in the countries in which they were shown. As the focus of this study is not on the films' reception but on the way they use postcolonial countries as passive entities upon which to project their FRG success story, the archival lacuna nevertheless serves my argument.
9. Regarding the space race, see Ettore Costa's chapter in this volume. The "kitchen debate" took place during the opening of the American National Exibition in Moscow on 24 July 1959. While touring the exhibit of a US model home outfitted with American consumer goods, US vice president Richard Nixon and Soviet leader Nikita Khrushchev got into an argument, during which they each extolled the benefits of their respective economic system.
10. For more on the Adenauer administration's conceptions of "PR" and "propaganda," see Rosumek, *Die Kanzler*, 32–47.
11. Steinle, "Visualizing the Enemy," 135.
12. "*Der Augenzeuge*, 1956/20," DEFA Foundation.
13. "Auslandspropaganda," DEFA Foundation.
14. Internal memo from Betz (Federal Press Office [FPO]), 27 September 1961, German Federal Archive B145/2816.
15. Letter from Thiel (Deutsche Reportagefilm [DR]) to FPO, 7 July 1961; internal memo (untitled) from Betz (FPO), 12 January 1962, German Federal Archive B145/1459.
16. Uelzmann, *Staging West German Democracy*, 231.

17. Film synopsis, "Dokumentarfilm 'Zu Gast in Bonn,'" n.d., Deutsche Wochenschau Archive, Hamburg.
18. This is documented in several instances, for example in the case of the 1953 PR film *Ein Mann wirbt für sein Volk* (*A Man Advocates for His People*), about Adenauer's first trip to the United States. Memo from Franck (FPO) 17 July 1953, German Federal Archive B145/1444.
19. Letter from Betz (FPO) to Uhlig, 8 December 1960, German Federal Archive B145/1459.
20. For a detailed account, see Krüger, *Die Finanzierung*, 43–79
21. *Zu Gast in Bonn: Ein Bericht aus dem politischen Leben der Bundeshauptstadt*, TC 2:44–2:50.
22. *Zu Gast in Bonn: Ein Bericht aus dem politischen Leben der Bundeshauptstadt*, TC 3:56–3:58.
23. *Zu Gast in Bonn: Ein Bericht aus dem politischen Leben der Bundeshauptstadt*, TC 23:23.
24. *Zu Gast in Bonn: Ein Bericht aus dem politischen Leben der Bundeshauptstadt*, TC 21:57.
25. "Exponat des Monats." Adenauerhaus.de.
26. *Zu Gast in Bonn: Ein Bericht aus dem politischen Leben der Bundeshauptstadt*, TC 17:23.
27. *Zu Gast in Bonn: Ein Bericht aus dem politischen Leben der Bundeshauptstadt*, TC 17:38–17:44.
28. *Zu Gast in Bonn: Ein Bericht aus dem politischen Leben der Bundeshauptstadt*, TC 17:51–17:52.
29. *Zu Gast in Bonn: Ein Bericht aus dem politischen Leben der Bundeshauptstadt*, TC 18:50.
30. Letter from Wiers and Purzer (Deutsche Wochenschau [DW]) to Betz (FPO), 29 January 1963, German Federal Archive B145/2816.
31. Internal memo from Betz (FPO), 27 September 1961, German Federal Archive B145/2816; Letter from Thiel (DR) to FPO, 9 September 1961, German Federal Archive B145/2816.
32. Uelzmann, *Staging West German Democracy*, 101–46.
33. *Koblenzer Straße 99–103: Der Alltag des Auswärtigen Amts*, TC 16:34–17:06.
34. *Koblenzer Straße 99–103: Der Alltag des Auswärtigen Amts*, TCs 19:21 and 20:08.
35. Jarausch, *After Hitler*, 3.
36. *Koblenzer Straße 99–103: Der Alltag des Auswärtigen Amts*, TCs 34:36 and 34:44.
37. *Koblenzer Straße 99–103: Der Alltag des Auswärtigen Amts*, TC 34:54–35:18.
38. The GDR did not hesitate to point out the actual presence of former Nazis in positions of West German leadership. In addition, however, the Eastern Bloc as a whole adopted the ideological strategy of equating the Western class enemy with Nazism, in order to suggest a continuity of militarist and fascist strains in Western society. See Ettore Costa's chapter in this volume for the Italian case.
39. "Aufnahmeobjekte in Westafrika," 26 September 1960, German Federal Archive B146/2816; letter from Purzer and Schwamkrug (DW) to Kurt Betz (FPO), 11 July 1960, Deutsche Wochenschau Archive.
40. Letter from Schwamkrug and Beyer (DW) to Betz (FPO), 29 September 1960, German Federal Archive B145/2816.
41. *Koblenzer Straße 99–103: Der Alltag des Auswärtigen Amts*, TC 32:33–33:20.

42. Se, *Colonial Childhoods*, 34; Jacobs, "Crossing Intimate Borders," 168.
43. The continuation of a colonialist stance in *Koblenzer Strasse* toward Guinea, a former French colony, has clear parallels to the stance of French postcolonialist filmmakers. See Sarah Stein's chapter in this volume.
44. *Koblenzer Straße 99–103: Der Alltag des Auswärtigen Amts*, TC 33:05–33:10.
45. Letter from von Tiedemann to Deutsche Wochenschau, 30 November 1960, Deutsche Wochenschau Archive.
46. Letter from von Tiedemann to Deutsche Wochenschau, 2 December 1960, Deutsche Wochenschau Archive.
47. *Koblenzer Straße 99–103: Der Alltag des Auswärtigen Amts*, TC 33:54.
48. *Koblenzer Straße 99–103: Der Alltag des Auswärtigen Amts*, TC 35:54.
49. *Koblenzer Straße 99–103: Der Alltag des Auswärtigen Amts*, TC 36:41.
50. *Koblenzer Straße 99–103: Der Alltag des Auswärtigen Amts*, TC 37:14–37:38.
51. *Koblenzer Straße 99–103: Der Alltag des Auswärtigen Amts*, TC 37:55–38:50.
52. As Megan Vaughan argues (*Curing Their Ills*, 2), "In the post-Enlightenment European mind, Africa, it seems, has been created as a unique space, as a repository of death, disease, and degeneration, inscribed through a set of recurring and simple dualisms—black and white, good and evil, light and dark."
53. *Koblenzer Straße 99–103: Der Alltag des Auswärtigen Amts*, TC 39:00–39:13
54. In this regard, the film shares underlying premises with French postcolonial educational films from the 1960s and 1970s. (See Sarah Stein's chapter in this volume.) Both France and the FRG, it seems, focused on their country's role as a culture nation, whose undisputed expertise is eagerly accepted by the less educated postcolonial nation.
55. *Koblenzer Straße 99–103: Der Alltag des Auswärtigen Amts*, TC 40:43–40:49.
56. *Koblenzer Straße 99–103: Der Alltag des Auswärtigen Amts*, TC 41:08–43:13.
57. Letter from von Hase (Foreign Office) to Betz (FPO), 24 May 1961, German Federal Archive B145/2816.
58. Fehrenbach, "Learning from America," 109.
59. Globke served as Adenauer's closest adviser and chief of staff in the Federal Chancellery from 1953 to 1963. Under the Nazis, Globke authored a "judicial commentary" on the antisemitic Nuremberg Laws of 1935 that ultimately helped pave the way for the Holocaust.
60. Tetens, *The New Germany*, 48.
61. "Neue Deutsche unverfänglich," *Der Spiegel*.
62. Schwarz, *Wochenschau, westdeutsche Identität*, 89–90.

Bibliography

"*Der Augenzeuge*, 1956/20." DEFA Foundation. Retrieved 4 May 2022 from https://www.defa-stiftung.de/filme/filme-suchen/der-augenzeuge-195620/.

"Auslandspropaganda." DEFA Foundation. Retrieved 4 May 2022 from https://www.defa-stiftung.de/defa/geschichte/daten-und-fakten/filmwesen-der-ddr/10-international/1070-auslandspropaganda/.

"Caravanserais: Cross-roads of Commerce and Culture along the Silk Roads." UNESCO. Retrieved 5 November 2021 from https://en.unesco.org/silkroad/content/caravanserais-cross-roads-commerce-and-culture-along-silk-roads.

"Deutsche Geschichte suchen und finden—Filme aus dem Bestand des Bundesarchivs." German Federal Archive. Retrieved 19 September 2023 from https://www.filmothek.bundesarchiv.de/?set_lang=de.

"Exponat des Monats." Adenauerhaus.de. Retrieved 13 April 2023 from https://www.adenauerhaus.de/downloads/Exponat_des_Monats/Jan17.pdf.

Fanon, Frantz. *Black Skin, White Masks*. New York: Grove Press, 2008.

Fehrenbach, Heide. "Learning from America: Reconstructing 'Race' in Postwar Germany." In *Americanization and Anti-Americanism: The German Encounter with American Culture after 1945*, edited by Alexander Stephan, 107–23. New York: Berghahn Books, 2004.

Jacobs, Margaret D. "Crossing Intimate Borders: Gender, Settler, Colonialism, and the Home." In *Gendering Border Studies*, edited by Jane Aaron, Henrice Altink, and Chris Weedon, 165–91. Cardiff: University of Wales Press, 2010.

Jarausch, Konrad Hugo. *After Hitler: Recivilizing Germans, 1945–1995*. Oxford: Oxford University Press, 2006.

Judt, Tony. *Postwar: A History of Europe Since 1945*. New York: Penguin, 2006.

Krüger, Jens. *Die Finanzierung der Bundeshauptstadt Bonn*. Berlin: Walter de Gruyter, 2008.

"Neue Deutsche unverfänglich." *Der Spiegel*, 1 December 1949. Retrieved 5 November 2021 from https://www.spiegel.de/politik/neue-deutsche-unverfaenglich-a-d5cf6b8e-0002-0001-0000-000044439172?context=issue.

Paschen, Joachim. "Die Unabhängigste Wochenschau der Welt? Aus den 1950er-Anfangsjahren eines publizistischen Regierungsorgans." Haus des Dokumentarfilms. Retrieved 11 November 2012 from http://dokumentarfilmforschung.de/dff/cms/?p=528.

Schröder, Hans-Jürgen. "Wahlkampfbilder: Die Visualisierung von Adenauers Amerikareisen 1953 und 1957 in Propagandafilmen der CDU." In *Wahlkämpfe in Deutschland: Fallstudien zur Wahlkampfkommunikation 1912–2005*, edited by Nikolaus Jakob, 137–50. Wiesbaden: Springer VS, 2007.

Schwarz, Uta. *Wochenschau, westdeutsche Identität und Geschlecht in den Fünfziger Jahren*. Frankfurt am Main: Campus Verlag, 2002.

Se, Satadru. *Colonial Childhoods: The Juveline Periphery of India, 1850–1945*. London: Anthem Press, 2005.

Steinle, Matthias. "Visualizing the Enemy: Representations of the 'Other Germany' in Documentaries Produced by the FRG and the GDR in the 1950s." In *Framing the Fifties: Cinema in a Divided Germany*, edited by John Davidson and Sabine Hake, 120–36. New York: Berghahn Books, 2007.

Rosumek, Lars. *Die Kanzler und die Medien: Acht Porträts von Adenauer bis Merkel*. Frankfurt am Main/New York: Campus Verlag, 2007.

Tetens, T. H. *The New Germany and the Old Nazis*. New York: Random House, 1961.

Uelzmann, Jan. "Bonn, World City: Explaining the FRG's Provisional Capital through Government Commissioned Documentaries during the Adenauer Years." *Monatshefte* 108 (2) (2016): 202–32.

———. *Staging West German Democracy: Governmental PR Films and the Democratic Imaginary, 1953–1963*. New York: Bloomsbury Academic, 2019.

Vaughan, Meghan. *Curing Their Ills: Colonial Power and African Illness*. Stanford, CA: Stanford University Press, 1991.

Von Hodenberg, Christina. *Konsens und Krise. Eine Geschichte der westdeutschen Medienöffentlichkeit, 1945–1973*. Göttingen: Wallstein Verlag, 2006.

CHAPTER 11

French Educational Film Production for African Countries, 1960s–70s

Sarah Stein

Introduction

France's film policy with respect to its former African colonies in the 1960s and early 1970s covered the production of educational films. After gaining formal independence from France in 1960, the governments of nine African countries entered into contractual agreements with a French mixed-economy company, called International Audiovisual Consortium (Consortium Audiovisuel International, CAI). The French government established CAI in 1961 with the objective of providing African countries with newsreels and educational films, a necessary step as none of the former French colonies had film production facilities at their disposal. These countries were thus reliant on films supplied from abroad.[1] Beginning in 1962, CAI also sent French staff to live in African countries and conduct filming on behalf of governments and development agencies alike.

This chapter examines the case of two films produced for screenings in Upper Volta (today Burkina Faso). The CAI employee stationed in Upper Volta was Serge Ricci. Like other CAI filmmakers working in contract countries, Ricci sent the footage he shot to a CAI laboratory located in the French capital city, Paris. There, the footage was developed, edited into a final product, and returned to the African country from which it had originated. The CAI newsreels and educational films shown in African countries were thus created in a complex production process that the former colonial power organized across long distances. CAI's film output consisted primarily of newsreels; but, according to a 1971 report by CAI director Gaston Heurley, CAI also produced a total of thirty-five educational films from 1964 to 1971.[2] The films examined in this chapter, two of the few extant CAI films, can be regarded as examples of films produced and delivered by CAI as part of postcolonial French film policy. This chapter examines the worldview as well as the expert knowledge conveyed in the films and shows that they were only apparently objective. In doing so, it ties in with deconstructivist and postcolonially informed development policy research.[3]

Media and French Postcolonial Policies

Efforts to establish a French film policy date back to the period immediately following World War II, when the French government began to exercise a stronger cultural and political presence in its African overseas territories.[4] For example, in his 1946 book *Service Africain*, high-ranking colonial administrator Robert Delavignette wrote:

> Let us become active through the radio and educational forms of film; through methods of teaching; through the creation of technical educational institutions to promote economic and agricultural activities overseas; through teaching and research at the Musée de l'Homme in Paris; through the development of the French Institute of East Asian Studies, the Institute of Islamic Studies, the Institute of African Studies. So let us act thus, so that a community of values and ideas may be formed in schools overseas, based on our duties toward humanity and inspired by the humanism and civil society of metropolitan France. Then, civilization will also be renewed in metropolitan France![5]

Delavignette describes nothing less than the vision of a Franco-African community maintained through cultural-political relations in general, and mass media in particular. It is also obvious that, in Delavignette's view, the initiative to create such a community should come from France. A connection to France's colonial tradition of its "civilizing mission" (*mission civilisatrice*) is distinctly present in the outline of this cultural policy. Furthermore, Delavignette's vision is marked by a sense of superiority, described by Gary Wilder as an attitude of "colonial humanism" typical of postwar colonial administration.[6] Delavignette, like his colleagues in the colonial administration, felt that the French government had an instructive, civilizing duty to perform in its overseas territories. As is discernible from the following examination of France's postcolonial film policy, this attitude did not change fundamentally in the wake of the formal independence of African countries in 1960, and postcolonial film policy was still meant to promote a distinct, hybrid Franco-

African culture whereby the African side should be fostered by French cultural programs.

In the first months of 1961, members of the French government and high-ranking administrative staff convened regularly in interministerial councils (so-called *conseils restreints*) to discuss how relations with the former colonies ought to be conducted in the future. The French government and administration excluded any outside involvement in these debates, meaning that no representatives were present from the countries that had emerged from the former colonies of French West Africa (Afrique Occidentale Française, AOF) and French Equatorial Africa (Afrique Equatoriale Française, AEF). There is also no indication on record that the French administration consulted with the governments of these newly established African states or inquired about their specific needs.

At the interministerial councils meeting on 12 April 1961, a number of ministers and high-ranking administrators discussed how to structure and contour French policy regarding the presence of cinema, television, and radio in the postcolonial African countries. As noted in the minutes, the meeting was attended by French minister of state Robert Lecourt, minister of culture André Malraux, minister of information Louis Terrenoire, minister of finance Valéry Giscard d'Estaing, secretary general of the Communauté Jacques Foccart, and Jean Foyer, the secretary of state responsible for relations with the countries of the Communauté. Also present were Michel Fourré-Comeray, director of the National Center of Cinematography (Centre National de la Cinématographie, CNC), and Raymond Janot, assistant director of the state-owned French Radio and Television Broadcasting (Radiodiffusion-télévision française, RTF).[7]

A concept paper, entitled "Cultural Policy in the Underdeveloped Francophone Countries," was circulated in preparation for the meeting. This paper, which unfortunately bears no indication of authorship, indicates that French administrative officials framed their cultural policy as a cultural development policy and an act deserving gratitude. Implicitly, of course, this framing attested to the colonialist belief that African states suffered from a cultural lag and needed support. In accordance with the viewpoint of colonial official Robert Delavignette described above, the paper asserts that "cultural relations should [now] take the place of the institutional ties [of the colonial era]." The document continues in this hegemonic tone, declaring: "This policy should aim to intensify cultural influence, but it should also develop local cultures and bind them more closely, through a syncretism constituting the common heritage of the Francophone states, to our [French] culture."[8] While French officials thus advocated for a cultural policy that would bring about a hybrid Franco-African culture, however, the project only demanded change from the African countries. The underlying assumption was that, by adopting French culture, Africans would attain a higher level of cultural development. Of course, this was a far cry from the understanding of cultural hybridity described, for example, by postcolonial theorist Homi K. Bhabha, whose conception of hybridity is based on a fluid, open process instead of a planned intervention.[9] In contrast, the interministerial councils apparently planned concrete measures for a film policy that was based on theoretically hegemonic assumptions of cultural hybridity.

By broadening the scope of research, however, I found that the consultations and decisions made by the interministerial councils were not solely intended to replace former colonial-institutional ties with cultural policy initiatives. Another factor in their deliberations was the role played by US film companies. French officials were encountering competition from Hollywood regarding film-related interventions on the African continent. Represented by the Motion Picture Association of America (MPAA) lobbying group, well-known US film companies including Columbia, Walt Disney, MGM, Universal, United Artists, and Warner Brothers were seeking active involvement in the newly decolonized regions of Africa.

Initially, the CNC and the French Ministry of Culture were open to cooperation with Hollywood, and they maintained a lively exchange with the MPAA from 1961 to 1963. The shared objective of this Franco-US exchange was to explore opportunities for developing a robust African film industry, thereby turning the African continent into a new, vital market for the French and US film industries. This was due to the fact that during colonial times, two companies—the African Industrial and Commercial Cinematography Company (Compagnie africaine de Cinématographie industrielle et commerciale, COMACICO) and African Film Company (Société d'exploitation cinématographique africaine, SECMA)—had invested in cinema construction and had established a film distribution system that continued to operate in a monopoly position even after decolonization. As Odile Goerg states, the monopoly position enabled COMACICO and SECMA to operate with a relatively limited and antiquated film stock.[10]

A paper prepared for the interministerial councils by CNC employee Michel Carrière indicates that consideration of such a Franco-US development initiative took place against the backdrop of profound changes in the United States and, especially, French film markets. "[The] development of the African film market," Carrière wrote, "would allow the producers of both countries to compensate to some extent for the decline in [their] income in the developed countries." The further comment that "competition from television is not expected to become noticeable [on the African continent] for several years" indicates that Carrière saw television as a relevant pressure in the French film market.[11] Indeed, as television technology spread in most industrialized countries, it exerted a profound change in the postwar world of media. Film historian Marcel Huret notes the massive increase in the number of television receivers in France in this period: "Whereas in 1950 there were just 3,794 television sets, by 1958 there were already over a million. This development posed great challenges for [cinematographic] production companies."[12] And so the French film industry, which from the beginning of the twentieth century had been one of the world's most vigorous, found itself confronting a significant decline in the postwar period, in the face of both increased television viewership and US competition.[13]

At the interministerial councils meeting, the CNC promoted its plans for a partnership with Hollywood on the African continent. In their attempts to convince political decision-makers, representatives of the French film industry also resorted to the rhetoric of the Cold War. By way of illustration, in a 1962 report the CNC wrote of the planned cooperation: "It is necessary to make sure that mainly French and foreign films that correspond to the values of our civilization are distributed [in African countries]. In contrast, films that contradict our worldview must be fought off with the greatest possible effort."[14] While the report does not explicitly state which films are to be considered "foreign," a confidential administrative note dated July 1963, which circulated among the interministerial councils, does offer greater specificity on this point, stating that the increased screenings of films from "Egypt, India, and countries of the East" must be opposed.[15] On one hand, this note demonstrates a worldview shaped by the competition of systems and values that marked the Cold War. On the other hand, it epitomizes a Eurocentric pattern of interpreting the world that was highly effective throughout colonial history, described by Stuart Hall as "the West and the rest."[16]

However, the plans for Franco-US cooperation had reached an impasse at the end of 1963. An administrative note submitted to the interministerial councils shows that Carrière tried to convince the French government to enter into joint negotiations with the US film industry for the purchase of COMACICO and SECMA. He complained about getting no response from the Ministry of Cooperation; though the general director of the CNC had approached the ministry to request assistance with a credit inquiry into the Central Fund for Cooperation (Caisse Centrale de la Coopération) in spring 1963, no response had been received by the end of November 1963. "Now," Carrière continued indignantly, "the Americans have negotiated with the owners of COMACICO and SECMA on their own." In his passionate appeal to French government officials, Carrière used anti-imperialist rhetoric directed at the United States to portend irreparable harm to the French economy: "Above all, American control over film distribution . . . in Francophone Black Africa means that movie theaters are flooded with productions from the United States, with newsreels, with propaganda and with advertising for products from across the Atlantic."[17] Despite Carrière's pleas, however, CNC plans for a joint venture project with the MPAA foundered, and by year's end, the exchange between the French and US film industries ground to a halt.[18]

It seems possible that the joint venture between the French and US film industries was primarily terminated due to the deterioration of the already fragile Franco-US relations after 1963. For one thing, the postwar rise of the United States as a new world power was accompanied by the US proclamation of values that helped delegitimize colonialism, which posed a challenge to France's self-image.[19] In addition, the plans of French president Charles de Gaulle to reform NATO's organizational structures, in order to position France as a world power of equal stature alongside the United States, were considered a failure by the end of 1963. Franco-American relations grew distinctly and observably more adversarial following that year.[20]

Though the CNC project failed, the interministerial councils of the French government had in the meantime developed a postcolonial film policy. The government had founded the CAI International Audiovisual Consortium, a mixed-economy enterprise consisting of France's three private newsreel companies—Gaumont, Pathé, and Eclaire—and the state-owned newsreel company, Les Actualités Françaises. CAI started establishing contractual relations with postcolonial African governments in 1962, with financing coming proportionally from the contracting countries and the French state development fund FIDES. CAI can thus be regarded as a typical example of the type of private-public partnership that replicated the business relations that had undergirded colonial infrastructure projects during the interwar period.[21]

Postcolonial Production of Educational Films

By the end of 1963, the French administration, which hitherto had exclusively produced newsreel films, was considering film production of educational material. In December 1963, the French Ministry of Cooperation promulgated a directive according to which the CAI was henceforth to show "fewer state visits and inaugurations" in its films and instead make "short informational sequences or films with instructive content on everyday issues (hygiene, cleanliness, nutrition, saving money, work methods, implementation of [development] programs, etc.)."[22]

The educational films that CAI produced in the following years were used in very different social contexts. Some of them were used in rural development projects. They often had an instructional character and aimed to prompt the audience to adopt working methods or behavior patterns described in the films. Those films were driven overland and screened on cinema buses as a way to reach more remote villages. Other CAI educational films—as we can see by analyzing the films—addressed urban audiences instead. They intended to provide the impetus for audiences to adopt certain behaviors and modes of living. Although the films were used in very different social contexts, some consistent common characteristics can be delineated. The first is the use of a dominant voice-over. In all the CAI educational films, a narrator describes the events shown in the film, provides information about the feelings of the film's protagonists, and summarizes their dialogue. There are no direct recordings of the filmed characters. The second characteristic is the docufictional nature of the films; while purportedly offering a transparent image of real life, all educational films contain fictional elements in addition to depicting elements in the world external to filming.

I will now examine two of these films, both of which were shot by CAI employee Serge Ricci in Upper Volta. In order to interpret the films, I will use a method formulated by media theorist Eef Masson, who proposes three main steps for media analysis. The first involves identifying the implicit viewer addressed by the film.[23] The second involves exploring the film's rhetorical means of attracting and motivating viewers. Lastly, Masson suggests observing references to the nonfilmic world.[24] Using this method of analysis, I will draw conclusions regarding both the context of use and the (ideal) audience intended by the filmmakers.

Building Wells: *Arms for Water*

The forty-minute color film *Des bras pour de l'eau* (*Arms for Water*), made by Ricci in 1964, reports on a well-construction project in a village in Upper Volta.[25] There are several indications that the well-construction project is part of a larger development campaign. For example, at one point in the film, the voice-over states that the French government's Aid and Cooperation Fund (Fond d'Aide et de Coopération, FAC) "offers a program for the construction of wells" and "grants social loans for this purpose." Additionally, the film depicts representatives of a development aid organization arriving in the village with materials needed for the construction of the well—casting molds, metal rods, and cement. Finally, the film's credits note that "François Etchegut, SATEC expert," participated in the film's development.[26]

Despite the inclusion of elements that actually existed in the world prior to filming, the realization of the well-construction project is presented in the film as a fictional example in an unnamed village in Upper Volta. This docufictional form combines a documentary presentational format with fictional and, as will be shown below, clearly idealized elements. Opening in this fictitious Upper Volta village, the film emphasizes deficiencies and deprivation in the lives of the local people. The cinematic images show people scavenging water from puddles and hauling it across the open landscape back to their villages.

Then viewers see women washing grain in the laboriously harvested puddle water. The voice-over explains:

> In the dry season, the water supply becomes a problem in Africa. There are waterholes several kilometers away from the villages, but wells are rare, and because they were built with traditional means, they often collapse. As a result, polluted water poses a permanent threat to the population and is responsible for a large number of diseases that affect entire villages. In most villages, neither the necessary knowledge nor the necessary materials are available to construct a well.

This description of a village environment characterized by danger and hardship evinces a need for intervention and improvement. Accordingly, the voice-over points out an opportunity to solve this precarious situation: "Upper Volta has found a good solution. In the villages, inhabitants are motivated to construct wells independently, with only a minimum of materials and technical equipment."

Following this opening sequence, the film presents the different steps involved in constructing the well: excavating a hole, extracting sand from the loamy soil, placing and fixing the casting molds, and mixing and pouring concrete into the well wall. The voice-over explains these stages of the construction slowly and in easily intelligible language. Images of villagers working together are presented in the film in what may be interpreted as an attempt to promote the spirit encoded in the community development (*développement communautaire*) policy. The construction project is thus staged as an endeavor carried out by a village collective in an exemplary manner. Everything proceeds in a very structured way, and the participants perform their labors in orderly harmony. For example, the (male) villagers are filmed forming a human chain to the remote waterhole in order to transport the water needed to loosen the dry earth. To the rhythm of drum music (nondiegetic), the villagers wield their hoes and till the soil, gradually excavating a deep hole.[27]

Toward the end of the film, the cinematic narrative shows village life in the period following the successful well construction. The newly built well has brought forth a state of ideal social progress. This is expressed through a change in the filmic use of color. Whereas previous scenes were dominated by brown and red tones of dusty earth, colorful images now appear. Flowers and verdant green lawns grow around the completed well. The earth itself seems to have changed completely, from a loamy to a nutrient-rich soil. Colorfully dressed women and their children are now shown fetching clean, clear water from the well. In this context, the narrator proclaims that the abundance of water enhances the "well-being of the people" and "opens up new possibilities." The supply of water "for consumption" as well as "for everyday domestic needs," accompanied by an "increase in buildings," will certainly improve "the existence of the inhabitants in different ways." To this end, the film shows how the newly available well water enables the production of clay bricks, which in turn are used to construct a school building. Finally, the closing sequence shows the completed school building and children, first watering a newly planted school garden and then joyously playing with clear well water that glistens brightly in the sun.

The overall impression conveyed by this filmic presentation of the well-construction project is of committed villagers whose joint dedication to the village community leads to far-reaching infrastructural and social progress. The clearly idealized presentation of the outcome, in a virtually paradisiacal setting, is starkly contrasted to the circumstances prevailing in the village at the beginning of the film: disease, danger, and an absence of knowledge and materials. Created for use in SATEC's development campaigns, the film possesses a clear docufictional character, presenting the technical knowledge of building a well within the framework of a salvation narrative. As such, it can be regarded as an attempt to motivate audiences living in rural village communities to emulate the film narrative by placing themselves at the service of the community and building a well within the framework of the community development program organized by French development agencies.

Science Festival: *The Third Scientific Fortnight of Ouagadougou*

The black-and-white film *La 3ème Quinzaine Scientifique de Ouagadougou* (*The Third Scientific Fortnight of Ouagadougou*) reports on a two-week science festival, organized through Franco–Upper Voltaic cooperation in 1972 and held in Ouagadougou, Upper Volta's capital.[28] The opening sequence of the film begins with outdoor shots in front of the Franco-Voltaic Center (Centre Franco-Voltaïque) in Ouagadougou, where the opening ceremony of the festival is captured. Built in 1963 as a French cultural institution in Upper Volta, the Franco-Voltaic Center continues to be part of France's diplomatic presence in Burkina Faso today.[29] These scenes are followed by images of men conversing in front of the building. The voice-over introduces them as the Upper Voltaic minister of national education and culture, the French ambassador, cultural attaché, and head of France's Permanent Mission for Aid and Cooperation (Chef de la Mission Permanente d'Aide et de Coopération). The entire opening sequence is accompanied by symphonic music, which can be understood as signifying an atmosphere of official festivity and harmony. Shots of luxury automobiles and a large number of motorcycles parked in front of the Franco-Voltaic Center indicate the presence of many festival guests and can be read as markers of a modern, prosperous urban lifestyle.

This initial sequence is followed by shots inside the Franco-Voltaic Center. The French diplomatic corps, teachers from Upper Volta in lab coats, and

French instructors active in development projects have all gathered there. They listen attentively to the speeches opening the science festival. Summarizing a speech by an Upper Voltaic teacher involved in organizing and hosting the festival, the voice-over emphasizes a self-critical assessment as well as a statement proving "progress": "In 1971, the organization of the science fair was still relatively modest and experimental; but the 1972 festival is already organized in a much more structured way." At the same time, the film shows the attending officials listening to his remarks with interest and applauding approvingly. This opening sequence presents the science festival as a cultural-diplomatic event organized in close collaboration between France and Upper Volta. By communicating this critical evaluation of the 1971 festival, the voice-over presents the Upper Voltaic organizers of the festival as France's docile partners.

This sequence is followed by a second, longer segment of the film, focusing on the lively atmosphere at one of the festival exhibitions. It is underscored by so-called "oriental" music, a typical sound cliché of the period created by European film composers to evoke the mystical and magical—hardly an appropriate atmospheric element for a science festival. This anomaly can be explained by the film's underlying composition, which mirrors a narrative pattern that Charles O'Brien analyzes in French colonial feature films. According to O'Brien, the opening sequences of such films always present a location conveying a "sphere of power and order" and inform the audience about the film's fundamental worldview; the cinematic narration then propels movement toward a space characterized as the "sphere of the anthropological," in which the protagonists—ensconced in a mystical environment—experience adventures.[30] Apparently, by deploying "oriental" music over images of the science festival, the filmmakers sought to present the exhibition hall as a space where it would be possible for visitors to experience adventure.

The film remains trapped in colonial patterns in its further presentation of events surrounding the festival. After showing images of various people visiting the exhibition and their social interactions, the voice-over explains that "one of the most interesting booths at the exhibition is that of Gérard Rumèbe, head of the physics department at the Palace of Discovery." In doing so, the narrator presumes that film viewers are familiar with the Palais de la Découverte, a science museum in Paris. The film then depicts a white Frenchman, Rumèbe, performing experiments with a Faraday cage, during which he uses gestures to present himself as a magician, wielding the rod he holds as a kind of magic wand to dispatch electricity. The film focuses on this "physicist-magician" Rumèbe with facial close-ups. By gazing into the camera, he addresses the film audience directly, which contributes to the striking effect of this sequence. Finally, the film shows the scene in a wide-angle shot. As the Faraday cage moves into the frame, it is revealed that a young black exhibition visitor is inside the cage; the "physicist-magician" discharges electric flashes to the cage from his magic wand. Such a film sequence effectively conveys the stereotypical colonialist depiction of a white man equipped with agency, on one hand, and, on the other, a Black person passively surrendering and trusting in the white man's abilities.

Another event highlighted by the film is the visit of Nobel Prize laureate Alfred Kastler. We observe the tall, gaunt man surrounded by an entourage as he enters the building and walks down a long corridor. Young people from Upper Volta explain some of the exhibition stands to him. Though the Upper Voltaic youths are depicted as active local agents for the first time in this sequence, the filmic images place Kastler at the center of attention, and the narration refers primarily to Kastler: "It was his first contact with Black Africa. Professor Kastler is an attentive and passionate visitor. His youthful spirit allows him to always feel comfortable among young people." The narrator establishes another link to the French scientific and institutional landscape by explaining that Kastler is a member of the Institut de France. Following these two longer sequences, presented as highlights, a shorter sequence shows a group of students from Upper Volta demonstrating what they have learned by presenting chemical experiments on the influence of tobacco consumption on the human body.

It can be argued that this film contains fictional elements as well, insofar as the science festival is presented as a mystical adventure. One can also reasonably assume that this mode of presentation was intended, at least in part, to motivate the country's youth to take an interest in science education. Furthermore, it is worthwhile to note that the film depicts the science festival in a manner that emphasizes French protagonists and institutions; by mentioning specific French names, the film refers explicitly to the French as primary agents. Conversely, the Upper Voltaic people depicted in the film are either students, demonstrating their recently acquired knowledge, or exhibition visitors shown as astonished audiences or anonymous crowds.[31] Even the Upper Voltaic officials shown at the beginning of the film are not mentioned by name. The selection of the Franco-Voltaic Center as the science festival's location, the presence of the French diplomatic corps, and references to French science institutions all underscore French authority in the domains of education and science. The cinematic presentation of the science festival highlights the close ties between France and Upper Volta, affording the former colonial power a special position.

The film implicitly addresses an audience of Upper Voltaic people willing to accept or even welcome postcolonial assistance (and entanglement) in the educational and scientific sectors. That such acceptance was, in reality, rather low became apparent a few months after the production of the film. In January 1973, teachers' unions in Upper Volta organized extensive strikes, demanding, among other things, the "Africanization" of the Ministry of Education. These strikes were of a magnitude great enough to trigger a political crisis in the West African country.[32]

Conclusion

World War II and, especially, the subsequent wave of decolonization, posed a fundamental challenge to the self-image of the French. French officials hoped to counter this crisis by increasing their cultural-political engagement with former colonies. To this end, they sought to address people in the countries of the former French Empire through a "cultural development policy." While documents distributed to the interministerial councils show that French film industry representatives used Cold War threat scenarios in trying to convince the councils to support their planned cooperation with the US film industry, the shaping of French film policy in fact remained largely untouched by the Cold War.

Establishing CAI in the immediate wake of political decolonization, the French government created cultural-political structures that allowed cooperation with African governments and supplied them with films. One of CAI's tasks was the production of educational films, which were part of broader French cultural policy and used in development campaigns as well as elsewhere. Both films examined in this chapter focus on technical and scientific, or expert, knowledge.[33] As is evident in the analysis of the two films, CAI educational films were meant to motivate their audiences to engage in specific activities and foster particular behaviors and modes of living. The 1964 film *Arms for Water*, for instance, imparted technical knowledge and addressed rural village communities. Its intention was to improve the infrastructure in rural areas. To this end, the film encouraged villagers to devote their labor to the service of the community and thus contribute to the success of the overarching development project. As a means of motivation, the film describes as a result a village in which people live in a paradise-like ideal state. In contrast, *The Third Scientific Fortnight of Ouagadougou* addressed an urban audience and intended to promote interest in higher education among young people. By creating an atmosphere of the mystical and magical, the film aims to interest young people in science. At the same time, the film promoted the French presence in the educational system and encouraged the population to take advantage of French services and knowledge.

Sarah Stein completed her PhD in history at the Department for Modern and Contemporary History at the University of Kassel, Germany. Her research focuses on film, the arts, and decolonization in the African-European context, particularly in relation to French politics. Her publications have addressed the history of cinema in the Congo, African filmmakers in the 1960s, French cultural policy, and the cooperative production of educational films in post-colonial Africa.

Notes

1. During the colonial period, the French administration had given little thought to film policy and had left film distribution largely to private companies. These mainly supplied French and US entertainment films, which were shown in commercial cinema houses in the colonies. For more on cinema culture in the French colonies, see Goerg, *Fantômas sous les tropiques*.
2. Rapport CAI 1971, Archives Nationales de France, Cote 19930381/16.
3. Critiques of the expert knowledge disseminated in the context of post–World War II development policy can be found in numerous relevant works by postcolonial authors. They argue that the theory and practice of development policy continued and reproduced the tradition of colonial knowledge and power structures; see, for example, Escobar, *Encountering Development*. For some years now, younger authors have begun to expose and deconstruct the links between western development policy and colonial knowledge and power structures; see, for example, Hongler, *Den Süden erzählen*.
4. János Riesz, for example, presents clear efforts in furtherance of a state-organized cultural policy after World War II. He points out that, from 1953 onward, a large number of cultural centers were founded in the territory of the French Union. Whereas there were twenty-eight centers in French West Africa in 1954, by 1955 this figure had risen to ninety-eight. According to Riesz, the goal of these organizations was to introduce broad segments of the local population to the educational level of "the evolved" (*les évolués*)—which meant educating people on the basis of the French school system. See Riesz, "Erfindung der 'Frankophonie,'" 233–34. Philippe Lane, who likewise notes increased French cultural policy efforts after World War II, regards dissemination and consolidation of the French language as one of the primary policy objectives. He also emphasizes the significance of cultural policy for the French government against the backdrop of decolonization efforts, and rightly describes technical assistance as an area that the French government understood as cultural policy as well. By the formal end of empire, technical assistance had become increasingly important and included training for fields like public administration as well as technical management in agriculture. See Lane, *Présence française*, 22–24.
5. Delavignette, *Service Africain*, 275–76.
6. See Wilder, *The French Imperial Nation-State*, 43–117.
7. *Compte-rendu du Conseil restreint du mercredi 12 avril 1961, 15h30*, Archives Nationales de France, Cote 20050584/41.
8. *Documents diffusés en vue du Conseil restreint du 12 avril 1961, Une politique culturelle dans les Etats sous-développés d'expression française*, Archives Nationales de France, Cote 20050584/41.

9. See Bhabha, *Location of Culture*.
10. See Goerg, *Fantômas sous les tropiques*, 32.
11. Note de M. Carrière relative à la vente des circuits COMACICO et SECMA, 30 décembre 1963, Archives Nationales de France, Cote 20050584/41.
12. Huret, *Ciné Actualités*, 141–42.
13. Jan Uelzmann's chapter in this volume also shows that, in the combined context of the Cold War and decolonization, films were seen as an important means of political influence in the competition for so-called developing countries.
14. Note sur les possibilités d'une entente franco-américaine en ce qui concerne la distribution et l'exploitation de films dans les Républiques africaines d'expression française et en République malgache, Archives Nationales de France, Cote 20050584/41.
15. Note sur la distribution et exploitation commerciales de films dans les Républiques Africaines d'expression française, 26 juillet 1963, Archives Nationales de France, Cote 20050584/41. By naming Egypt and India, the report cites countries whose heads of state at the time—Gamal Abdel Nassar and Jawaharlal Nehru—were among the leaders of the Non-Aligned Movement. Having emerged in the 1950s, the Non-Aligned Movement had been gathering support mainly in formerly colonized countries. It advocated an ongoing process of decolonization and attempted to develop ideas that could offer an alternative to the global politics of both the United States and the Soviet Union, as well as to the policies of the former colonial powers. See Dinkel, *Bewegung der Bündnisfreien Staaten*.
16. Stuart Hall's programmatic text, first published in 1992, does not describe the West as a geographically localizable entity but rather understands it as an effective idea and the result of historically explorable communication processes. See Hall, "The West and the Rest."
17. *Note de M. Carrière relative à la vente des circuits COMACICO et SECMA, 30 décembre 1963*, Archives Nationales de France, Cote 20050584/41.
18. The exchange of letters between the two associations representing the interests of the French and American film industries ceased entirely at the end of 1963. The correspondence in the relevant CNC archival holdings resumes only in 1969, the year Charles de Gaulle retired from the French presidency.
19. For a general overview of the US role in the decolonization process, see Cary Fraser as well as Raymond F. Betts. See Fraser, "Decolonization and the Cold War," 469–85; Betts, *Decolonization*. For more specifically on the Declaration of the Right of Self-Determination of Peoples by US president Woodrow Wilson and its impact on anti-colonial movements, see Erez Manela. See Manela, *The Wilsonian Moment*.
20. Vaïsse, *La grandeur*, 363–412; Woyke, *Die Außenpolitik Frankreichs*, 42–52.
21. Wilder, *The French Imperial Nation-State*, 49–50.
22. *Réunion des Reporteur-Opérateurs du CAI, Programme de travail, 12 décembre 1963*, Archives Nationales de France, Cote 19930381/16.
23. The assumption that implicit recipients can be identified in texts extends back to literary studies theories on the aesthetics of reception and grew from Wolfgang Iser's work on the implicit reader. By extension, the implicit viewer of a film can be understood as a characteristic of the filmic text's structure, for which an adequate or even ideal reception is assumed. Consequently, it is necessary to work out what ideal-typical viewers the films were intended to address. Of course, the thinking figure of the implicit viewer is a theoretical construct and can say little about actual individual readings of a text by recipients, and inquiries about implicit viewers therefore remain

abstract. However, analysis of CAI instructional films makes it possible to work out structures that may not have been intended but were unconsciously implied by CAI filmmakers.

24. See Masson, *Watch and Learn*, 127–43.
25. *Des bras pour de l'eau*, Upper Volta, CAI (1964), Archives françaises du film du CNC, Bois d'Arcy.
26. The Technical Assistance and Cooperation Society (Société d'aide technique et de cooperation, SATEC) was a French authority that provided assistance in the form of technical expertise for development projects.
27. The use of drum sounds as a marker for traditional, underdeveloped, and backward "African life" is also noted in Jan Uelzmann's chapter in this volume.
28. *La 3ème Quinzaine Scientifique de Ouagadougou*, CAI (1972), Cinémathèque Africaine de Ouagadougou.
29. Today, the institution is known as the French Cultural Center Georges Méliès (Centre Culturel Français Georges Méliès).
30. O'Brien, "The 'Cinéma colonial.'"
31. Jan Uelzmann describes a similar portrayal of African students in his chapter in this volume.
32. Englebert, *Burkina Faso*, 48.
33. In this depiction of technical knowledge and progress, one can once again draw a parallel to the findings of Jan Uelzmann with regard to West Germany.

Bibliography

Betts, Raymond F. *Decolonization*. London: Routledge, 2004.

Bhabha, Homi K. *The Location of Culture*. London: Routledge, 1994.

Chaubet, François, and Laurent Martin. *Histoire des relations culturelles dans le monde contemporain*. Paris: Armand Colin, 2011.

Connelly, Matthew. "Taking Off the Cold War Lens: Visions of North-South Conflict during the Algerian War for Independence." *American Historical Review* 105 (2000): 730–69.

Delavignette, Robert. *Service Africain*. Paris: Gallimard, 1946.

Diawara, Manthia. *African Cinema: Politics and Culture*. Bloomington: Indiana University Press, 1992.

Dinkel, Jürgen. *Die Bewegung der Bündnisfreien Staaten: Genese, Organisation und Politik (1927–1992)*. Berlin: DeGruyter Oldenbourg, 2015.

Durand, Pierre-Michel. *L'Afrique et les relations franco-américaines des années soixante: Aux origines de l'obsession américaine*. Paris: L'Harmattan, 2007.

Englebert, Pierre. *Burkina Faso: Unsteady Statehood in West Africa*. Boulder, CO: Westview Press, 1996.

Escobar, Arturo. *Encountering Development: The Making and Unmaking of the Third World*. Princeton, NJ: Princeton University Press, 2012.

Fraser, Cary. "Decolonization and the Cold War." In *The Oxford Handbook of the Cold War*, edited by Richard H. Immermann and Petra Goedde, 469–85. New York: Oxford University Press, 2013.

Goerg, Odile. *Fantômas sous les tropiques: Aller au cinéma en Afrique coloniale*. Paris: Vendémiaire, 2015.

Hall, Stuart. "The West and the Rest: Discourse and Power." In *Formations of Modernity*, edited by Stuart Hall and Bram Gieben, 275–331. Cambridge: Polity Press, 1992.
Huret, Marcel. *Ciné Actualités: Histoire de la presse filmée. 1895–1980*. Paris: Henri Veyrier, 1988.
Hongler, Patricia. *Den Süden erzählen: Berichte aus dem kolonialen Archiv der OECD (1948–1975)*. Zürich: Chronos 2019.
Jeancolas, Jean-Pierre. "The Reconstruction of French Cinema." In *France in Focus: Film and National Identity*, edited by Elizabeth Ezra and Sue Harris, 12–22. Oxford: Berg Publishers, 2000.
Lane, Philippe. *Présence française dans le monde: L'action culturelle et scientifique*. Paris: La documentation française, 2016.
Manela, Erez. *The Wilsonian Moment: Self-Determination and the International Origins of Anticolonial Nationalism*. New York: Oxford University Press, 2007.
Masson, Eef. *Watch and Learn: Rhetorical Devices in Classroom Films after 1940*. Amsterdam: Amsterdam University Press, 2012.
Mingant, Nolwenn. "La Motion Picture Export Association de Jack Valenti (1966–2004), corps diplomatique des majors hollywoodiennes à l'étranger." *Revue française d'études américaines* 121(3) (2009): 102–14.
O'Brien, Charles. "The 'Cinéma colonial' of 1930s France: Film Narration as Spatial Practice." *Visions of the East: Orientalism in Film*, edited by Matthew Bernstein and Gaylyn Studlar, 207–13. London: I. B. Tauris Publishers, 1997.
Riesz, János. "Die Erfindung der 'Frankophonie': Koloniales Erbe und globale Perspektiven im Widerstreit." In *Imperialismus in Geschichte und Gegenwart*, edited by Richard Faber, 223–40. Würzburg: Königshausen & Neumann, 2005.
Turpin, Frédéric. *De Gaulle, Pompidou et l'Afrique (1958–1974):* Décoloniser et coopérer. Paris: Indes savantes, 2010.
Vaïsse, Maurice. *La grandeur: Politique étrangère du général de Gaulle 1958–1969*. Paris: Fayard, 1998.
Wilder, Gary. *The French Imperial Nation-State: Négritude and Colonial Humanism between the Two World Wars*. Chicago: University of Chicago Press, 2005.
Woyke, Wichard. *Die Außenpolitik Frankreichs: Eine Einführung*. Wiesbaden: Verlag für Sozialwissenschaften, 2010.

Index

www.ingramcontent.com/pod-product-compliance
Ingram Content Group UK Ltd.
Pitfield, Milton Keynes, MK11 3LW, UK
UKHW021851210725
2444IPUK00004B/14

9 781805 396352